Principles and Practice of WASTE ENCAPSULATION

Jack A. Caldwell
Charles C. Reith

LEWIS PUBLISHERS
Boca Raton Ann Arbor London Tokyo

Library of Congress Cataloging-in-Publication Data

Caldwell, Jack.
Principles and practice of waste encapsulation / Jack A. Caldwell, Charles Reith.
p. cm.
Includes bibliographical references and index.
1. Waste disposal in the ground. I. Reith, Charles C.
II. Title.
TD795.7.C336 1993
628.4'456--dc20 93-9434
ISBN 0-87371-992-1

LEWIS PUBLISHERS
121 South Main Street, Chelsea, Michigan 48118

PRINTED IN THE UNITED STATES OF AMERICA
1 2 3 4 5 6 7 8 9 0

Printed on acid-free paper

To my wife, Diane, who has moved around the world as I practiced engineering, and to the three men who taught me to be an engineer -- Professor Jennings, Oskar Steffen and Syd Hillis.

Contents

SECTION II: ENCAPSULATION SYSTEM DESIGN

SECTION III: ENCAPSULATION SYSTEM PERFORMANCE MODELING AND ASSESSMENT

SECTION V: CASE HISTORY

Jack Caldwell has worked for over 25 years as an engineer and manager in the operation and remediation of contaminated sites. He has designed, constructed, operated, and reclaimed hazardous and radioactive waste disposal and cleanup facilities in Africa, Europe, and North America. Mr. Caldwell holds a baccalaureate degree in Civil Engineering, a Masters in Geotechnical Engineering, and a postgraduate law degree, all from the University of the Witwatersrand in Johannesburg, South Africa.

Currently with Jacobs Engineering, Mr. Caldwell is a Project Director at the Department of Energy's Fernald Environmental Management Project in Ohio. Prior to that he was the Manager of Operations for the Jacobs Engineering Kansas City office; in addition, he was the Program Manager of the ARCS Program that involved environmental support to the Environmental Protection Agency on tasks ranging from site characterization through contaminated site remedial design and cleanup. For five years, he was the Manager of Engineering for Jacobs Engineering on the Department of Energy's Uranium Mill Tailings Remedial Action Project, and in that capacity he oversaw the development of the design for the 24 cells in ten states that make up the program. The ideas that are the philosophical heart of this book were formulated in the years he spent on the Uranium Mill Tailings Remedial Action Project.

Charles Reith is the Director for Environmental, Safety, and Health on the Strategic Petroleum Reserve, a large-scale geologic storage facility for crude oil along the United States Gulf Coast. Dr. Reith has managed a wide variety of environmental activities, especially remedial actions, for Jacobs Engineering Group Inc. He has directed or contributed to the development of disposal strategies for hazardous, mixed, and radioactive waste. On the Uranium Mill Tailings Remedial Action Project, he spearheaded a program to integrate ecological considerations into the design of disposal cells, and he has lectured and published widely on the ecological perspective in cell design. Early in his career, Dr. Reith conducted research and developed plans in the reclamation of coal and uranium mines. He co-edited books entitled *Deserts as Dumps?* and *Principles and Methods of Reclamation Science*, the latter of which was a distillation of his doctoral research in ecology at the University of New Mexico.

Foreword

During the performance of a conceptual design supporting study for a disposal cell at the Department of Energy's Weldon Spring Site in Missouri, the concept of expanding the study into a book occurred simultaneously to several of the contributors. It soon became apparent that the procedures and design considerations for evaluating and selecting the most effective disposal cell cover for mixed waste materials had been assembled into a very useful document.

Jack Caldwell and I approached DOE's Project Manager, Steve McCracken, with the concept of using this method to assist the project in meeting its objective for dissemination of developed technology among DOE officers, as well as furnishing an excellent reference book for those involved in the disposal of radiologically contaminated waste. We are grateful for Mr. McCracken's support and approval to proceed with this effort.

This book builds upon the basic framework provided in the Weldon Spring Study to include chapters on radon control and case histories on disposal cells constructed at Durango, Colorado, in Illinois, for the UMTRA Project, in Spain, South Africa, Canada, and North America.

The contributions from diverse authors have been skillfully blended by Jack Caldwell to create a practical, intermediate, and usable compendium of philosophy, design and application to specific structures. This book will become a primary reference for the design of future mixed waste disposal cell covers.

Jack Cooney

Acknowledgments

Before the reader has the opportunity to continue with the remainder of the book, we would like to pause and record our profound thanks and deep gratitude to those who made this book possible: Jack Cooney for initially suggesting the idea; Jan Deckert who, more than any, kept it moving and, in truth, produced it, controlling and facilitating production, writing, editing, typing, and graphics; Donna Keele and Michele Doskey, who produced the superb figures; Susan Fox, who did the typing and layout; and our colleagues and the management in Jacobs Engineering who indulged our fancy and provided generous use of corporate resources to make this a reality.

We would like to credit the writing of Chapter 23 to Juan Santiago, with whom we have had the honor of working during the past five years. Recognition is given to Charles Burt and Steve Cox, who contributed extensively to Chapter 18.

Preface

Environmental protection and restoration are the compelling challenges of the late twentieth century. As populations grow and technologies advance, ever-more wastes are produced in ever-growing quantities. Of the many and varied environmental problems, the one that has captured our interest and our careers is the topic of our book: encapsulation of wastes in order to restore sites and protect the environment.

Encapsulation is not the solution to the waste problem. The solutions are source reduction, waste minimization, and resource recovery. Yet, for some wastes and for many existing sites, encapsulation is the only option: permanent isolation from the accessible environment. And while we wish to in every way encourage the nobler, more permanent disposition of waste, we believe there is, and always will be, a need for facilities to permanently entomb the residues of the past and the present.

The philosophy and the practice of design and construction of encapsulation systems that control dispersal of the waste is the topic of this book. The requirements for an encapsulation system are basically the same, whether the waste is municipal refuse in a landfill, hospital debris in a low-level waste dump, or mixed wastes at a remediated defense facility. A waste encapsulation system consists of four integrated and interacting parts: the foundation, the liners, the perimeter encapsulation system and the topslope cover (Figure P.1). The combination of these parts we call the cell, a holistic system for encapsulating the wastes within. The essential goal of a waste encapsulation system, and a goal that is pervasive in this book, is the isolation of the wastes from the accessible environment for a very long time, up to and beyond 1,000 years.

Encapsulation systems are as varied as the environments in which they are built, and the components of an encapsulation system are as multiple and complex as the wastes therein. Yet the common demands upon any encapsulation system for any waste are: to prevent inadvertent human or animal intrusion to the waste; preclude removal and dispersal of the waste; minimize the ingress of precipitation that could ultimately seep from the waste and degrade groundwater; and control the egress of gases or other products of the waste that could enter and impact the atmosphere.

This book describes the many types of encapsulation systems that have been designed and constructed, and that are now being formulated for disposal facilities ranging from simple dumps to sophisticated engineered disposal facilities for mixed (i.e., radioactive and hazardous) wastes. The book describes and integrates the diverse technologies and scientific disciplines that are involved in the design, construction, and maintenance of a durable encapsulation system. The book sets out the technical approaches that have been developed in the last decade to provide objective quantification of the long-term (up to 1,000 years)

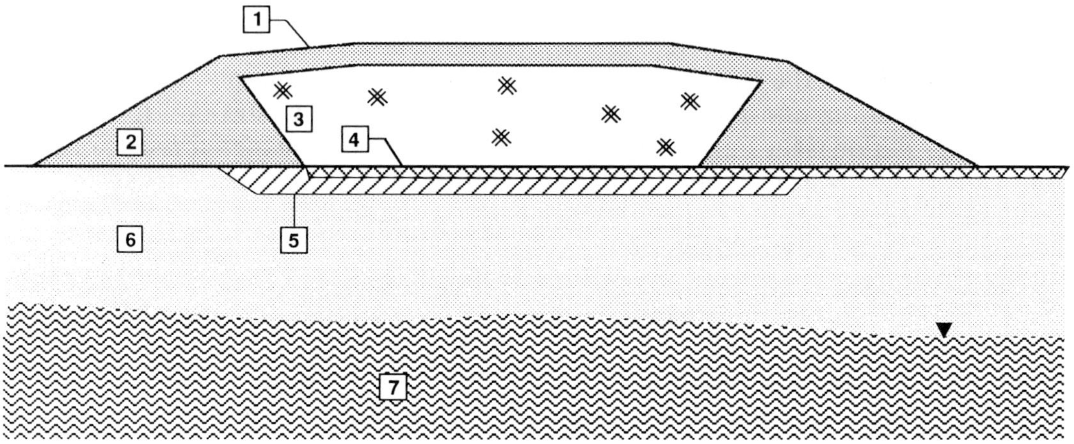

LEGEND: CELL COMPONENTS

1. COVER
2. DIKES
3. WASTE
4. BASAL LINERS AND DRAINS
5. FOUNDATION
6. VADOSE ZONE
7. GROUNDWATER

The essential parts of a waste encapsulation system which isolate waste and its emanation products from the environment are the cover, dikes, basal liners, drains, foundation soils and rocks.

FIGURE P.1

WASTE ENCAPSULATION SYSTEM COMPONENTS

performance of encapsulation systems. Finally, case histories are included so that the subtle adaptations needed to take theory through the art of design to the reality of soil and rock may best be described and appreciated.

This book is a synthesis of two perspectives on encapsulation. These perspectives largely complement each other and, when compatible, synergize into what we believe is the most enlightened and responsible way to permanently isolate noxious waste. Occasionally the views clash: we sometimes approach the contradictions with a sense of humor and professional parochialism; but we honestly believe that the truth in this matter is encompassing, and we think we are serving the reader well by revealing the full spectrum of perspectives. Where we agree, have the highest confidence. Where we disagree, choose your own course.

The two perspectives are that of the engineer, represented herein by Jack A. Caldwell, and that of the ecologist, represented by Charles C. Reith. The ENGINEERING PERSPECTIVE is one of conscientious planning, inspired design, and confident demonstration. These are the elements that brought us the pyramids of Cheops, the Panama Canal, and the Channel Tunnel, among the many marvels of human creation.

The ECOLOGICAL PERSPECTIVE is one that begins with the most thorough observation of nature and proceeds with a largely intuitive effort to imitate all that one has seen. This perspective has long underlain the fine arts, and has been exercised informally as part of agriculture, forestry, range management, and land reclamation. But never has the ecological perspective had so much to offer, especially when integrated meaningfully with the engineering perspective, as it does when applied to waste disposal. This is because the higher goal of encapsulation is perpetual isolation; the ideal cell will last as long as the local landscape, or as long as the wastes within remain hazardous. To achieve such longevity, we must look to nature and develop structures that are as durable as natural terrestrial features.

We believe that an integrated engineering/ecological perspective offers great promise as we attempt to design structures that will protect humans and the environment for generations from the hazards associated with noxious waste. Thus, a fundamental objective of our book is to elucidate and, where possible, integrate the two perspectives into a comprehensive approach toward the design of near-natural systems for perpetual encapsulation.

Many specialists have contributed to this book. Their efforts are a true reflection of the many sources of ideas, expertise, and energy that go to make the topic of waste disposal facility design and construction so dynamic, vital, and expanding a field of professional endeavor.

We who have contributed to this book envision our audience as being multidisciplinary, including the following:

- Engineers charged with designing, building, and maintaining new encapsulation systems and refurbishing existing waste disposal facilities.

- Scientists who provide the knowledge, expertise, and fundamental understanding of natural processes to support the planning, design, and performance assessment of long-lived disposal cells.

- Regulators and public authorities who write and enforce the laws and regulations that motivate encapsulation projects and dictate their design.

- The general public and all who in one way or another are involved with, interested in, or potentially affected by the wastes that must be encapsulated.

The authors wish, once again, to assert at the outset that we subscribe to the ideal of a society that minimizes, nay eliminates, wastes and, hence, the need for waste disposal and encapsulation systems. Nevertheless, as practical individuals, we recognize that there is, and always will be, some need for waste disposal. We are accordingly drawn to the challenges and the rewards of doing something, however little it may be, to fix things, and contribute to environmental improvement and protection of human health.

This book consists of five major sections. Each section consists of subsections or chapters on specific and interrelated topics. The sections are:

- The Philosophy of Encapsulation and Isolation

- Encapsulation System Design

- Encapsulation System Performance Modeling and Assessment

- Monitoring and Maintenance

- Case History.

Principles and Practice of WASTE ENCAPSULATION

SECTION I

THE PHILOSOPHY OF ENCAPSULATION & ISOLATION

Chapter 1

ENCAPSULATION: ENGINEERING PERSPECTIVES

WHAT IS DESIGN?

Philosophy is the art and science of thinking about things. Engineering is the art of applying scientific principles to benefit mankind. Because the design, construction, and operation of waste disposal facilities demand both good science and good engineering, applied with a certain artistic flair, we start this book with a detailed examination of the philosophy of the encapsulation and perpetual isolation of wastes. This section describes the engineer's perspectives; the next gives the perspective of the ecologist.

We believe that we are striving for the same thing even though the words are different and the perspectives and professional prejudices differ. Philosophically we aim in the same direction and seek the same truth. Our differences are simply a manifestation of different roads to the same goal; maybe they are the same road driven in different vehicles, and maybe it is the same car on the same road, and the only difference is the personalities of the passengers. We offer these two perspectives in the hope that they frame the scope of the debate and provide multiple insight about the topics which constitute the greater part of this book.

A considerable part of this book is about the design of encapsulation systems. Because the idea of design plays so large a role in the development of encapsulation systems, I consider it prudent to devote a few pages to thinking about the meaning, activity, and objectives of design. I will explore a little the philosophies of design that guide me, and, because philosophy so often shades into ethics, will move to the moral issues that confront those of us who work in the disciplines that are concerned with wastes. Hopefully this somewhat obtuse, abstract, and probably wordy digression will make it possible to set the stage for action and cleanup that are the gut stuff of a consultant's life.

The very word "design" is an opportunity to argue and disagree, for the word itself has many subtle shades of meaning. As a verb, "to design" implies the conscious act of conceiving of the layout of a cover or any other part of a complex waste encapsulation facility and translating that concept to reality via words and drawings. To design a cover means to do something, to produce a tangible reality, a concrete expression of an idea. The design engineer reads, reflects, visits the site, defines what is needed, conceives of and formulates the response, and hence prepares drawings, specifications, calculations, and reports. In short, the engineer prepares the design. But pause and note that the use of the word "design" as a noun exemplifies another completely different meaning of the word, which needs to be clarified to pursue a discussion on the philosophy of design.

The first obvious meaning of the word "design," when used as a noun in the phrase "prepare the design," is that the design is a collective phrase that describes the drawings, specifications, calculations, and reports that describe what is to be built. But we must acknowledge that the word "design" encompasses something more than a set of papers. The concept of the (noun) design is that it covers the intangible, the aesthetic, the form, the functional response of the physical to the needs of the surroundings. Thus a beautiful design implies more than the wood, or the concrete and metal, or the soils and gravels of the cover. The design beauty lies in the grace, the appropriateness, the fit of the physical entity. It lies in the skill of using materials for purposes not their common end. It lies in the skill of the intellect that identifies a need and provides that which is needed using what is available.

The design of a disposal cell can be thought of as the combination of elements, the soil, the rock, the vegetation; the design of the cell is the integration of those disparate materials into a unity that has form and function. The design of the cell is the honesty and strength of the intent and performance of the whole in protecting the wastes. The design of the cell is the appropriateness of the physical disposition of the various materials, alone and in combination, to the site, to the wind and the rain, to the temperatures and the animals and plants that will be a part of the encapsulation forever.

The Sources of Design

We may ask where the inspiration for design arises, where the ideas come from. We may get an answer faster if we first ask, why do we design (verb) or create designs (noun). The simplest answer, to my mind, is that we evolved as, are now, and will continue to be *homo faber*, man the maker. Our long distant ancestors whether from Africa and the Rift Valley, two or one million years ago, or only 200,000 years ago (when we are reported to have a common mother), or our ancestors from the time some 35,000 years ago as Cro-Magnon Man displaced the Neanderthal during the waning of the ice age, or 10,000 years ago with the coming of settlement and agriculture, the spread of the wheel and Indo-European languages 4,000 years ago, have been makers. The ability to make something is what put us on, and kept us on the course of change that brings us to where we are now. We are bred of those who survived because they designed and made something. Whether it be those who fashioned a cutting or killing tool from rock, or those who selected the stones to build a wall to defend the citadel, or those who raised a tower in faith, they are all our ancestors, and we follow our instinct from them to design and build when we select rock to encapsulate our wastes. We follow them when we use soil to encapsulate our residue just as they moved earth in pursuit of glory and gratification.

Designing is possibly the purest of the existential acts of self-expression. I do not wish to get into the arguments that swirl around the multiple layers and ethics of existentialism, but for me quite simply, living is being, and designing is an act that defines the reality of that existence. The indescribable act of self-satisfaction that accompanies the self-expression embodied in a design is part of a greater whole. Of course it may be no more than a chemical reaction akin to the high we experience after a long run or, (as I am told) more ephemerally and dangerously, in the use of drugs. Regardless of whether the self-expression-existentialist pleasure comes from body-chemistry or some genetic wiring or some higher spiritual norm, it exists, of that there is no doubt. And so we design, and we build, and we create.

My old professor told us as students that engineering should be fun to be worthwhile. An examination of the philosophy and physiology of fun is beyond the scope of this book. But we all know fun when we see or enjoy it. Some prefer words such as joy, stimulating, exhilarating, and fulfilling. Others talk of a work spirit, work ethic, or simply morale. All are valued. All must be goals of designers. All are reasonable motivators for the designer, but to me fun is what it is all about. So let no more be said than that designing should be fun, and a good design should have in it an element of fun.

On a recent trip to Dresden in the old East Germany, I took particular note of the art. The state spent great sums on "art" in that failed social experiment. The art was without a single trace of fun, humor, or happiness. It was unmitigatedly ugly. The design of the sculptures and the public plazas may have been in tune with the former times but, I believe, their ponderous crassness stems from the simple absence of a smile.

To be good designers and to produce good designs, we must approach our tasks with humility, and with enthusiasm, and with awe, and above all with a touch of humor. For only by synthesizing all human emotions will we produce something that is truly human and, hence, good.

Judgment in Design

The art of design may best be conceived of as synthesizing facts, scientific theory, and judgment to produce a plan of action. But what, you may well ask, is judgment.

To the engineering scientists, whose lives are numbers and computers, judgment is a crutch, a poor substitute for sophisticated analytical procedures. To the junior engineer and to many an unimaginative practitioner, judgment is no more than an impressive name for guessing. The cynics amongst my colleagues brand judgment as a poor substitute for collecting hard facts and for rational thinking.

As any practicing engineer and designer of action knows, however, judgment is an indispensable attribute when problems cannot be solved by numerical analysis. In fact, judgment is probably even more crucial when there are numerical analyses aplenty. Judgment is required to inject reasonableness into numerical analysis; for numerical analysis without judgment leads to very poor decisions and bad design. We know there is an attribute called judgment. We recognize its presence, we are pained by its absence, and we grieve over the record of our past lapses of judgment.

All attempts to define the essence of judgment come down eventually to this simple definition: judgment is a good sense of proportion.

Engineering judgment can be a gift; something you have or do not have. I prefer to think that judgment, particularly as it applies to encapsulation, is based on:

- Observation of natural phenomena
- Familiarity with the truths of science
- Thorough analysis of available information, including collecting and examining all the data, doing all the calculations, and evaluating all reasonable options
- Comprehensiveness, breadth of knowledge, and distinguishing and extracting important information from a plethora of irrelevance

- Balanced consideration of multiple concerns and thoughtful reconciliation thereof

- Examination of past mistakes

- A "gut feeling" brought about by age, panic, a schedule deadline, or a drink.

I have always subscribed to the definition of engineering as the art and science of harnessing the forces of nature for the benefit of mankind. I know that this is an old-fashioned definition rooted in Victorian arrogance, but to me it has a ring of truth and purpose, and it is general enough to cover the more modern ethics of protecting the environment; for I know that a degraded environment debases the human spirit and does not benefit mankind.

To cultivate engineering judgment, within the meaning of the term as I define it above, we can do no better than to draw from Peck (1969) who recommends the following:

- Make every assignment count: observe attitudes and techniques, facts, sizes, and methods.

- Teach your brain to register what your eyes see: keep a notebook, write reports on observations, and compile inspection reports.

- Consciously evaluate the size of things: ask how big is the column, how deep can frost penetrate, how many drums would be filled per day at a seepage flux of 10^{-7} cm/sec.

- Read: read technical journals, books, papers; but don't only read the formal contents, also study the advertisements for they are what is happening in practice.

- Study case histories for they are the record of past successes and failure.
- Consult and communicate with people from all disciplines.
- Recapitulate: frequently examine one's past contributions with a critical eye.
- Strive for continuous improvement and innovation.

Factors That Affect Design

I have talked long enough about the origin of and impulse that we as humans have to design. Let me now turn to the factors that affect the designer in his or her compilation of a design.

A significant factor that affects a design, that molds what the designer does is precedent. The good designer is always aware of what was done in the past, but they are also aware of that which other designers are doing at the same time. This is often called fashion or style. There were fashions or styles of pyramids in Egyptian and Greek temples, much as there are Paris or New York fashions as expressed by the length and breadth of a hemline. There are styles and fashions of cover design; (the rock covered pile is as much a fashion founded on precedent as any plunging neckline).

We apply the title genius to those who can break out of the grip of fashion, who can design something that derives from but is not restricted by precedent.

A repeat of the past done without understanding, or grace, or humor is scored as poor design. Add judgment, applicability, beauty, and new insight to precedent and we have good design. Hence precedent is essential, can be a force for good or bad, and is the anchor of design as much as precedent is the stability in a just legal system.

The law, and the regulations, and the standards that flow from the law also set the design and guide or constrain the designer. Good laws can lead to good designs; bad laws most often produce bad designs. A whole essay could be written on the impact of law on design. Suffice it here to say that waste encapsulation systems are inherently a product of the laws they follow. Throughout this book we refer to the RCRA cover, which is a basic stratigraphy of vegetation, soil, and plastic. These outer layers of a cell are based on the Resource Conservation and Recovery Act requirements. Even the name of the cover comes from the governing law.

Lawyers are seldom engineers, and those who write the laws almost never have extensive technical understanding. Laws perforce must encapsulate a spirit, a desire, an intention. They cannot, and should not, be technical directions or engineering codes. Hence laws may be vague and open to interpretation.

A good law strengthens the designer's hand in that it enables him to do what he knows is right but which politics or economics discourages. A bad law restricts the designer for it, hobbles him, and reduces his ability to be creative.

Most of the technical advances discussed in this book have been made because some law mandated that an agency, institution, or engineer do something to improve or protect the environment. I do not decry the modern laws that have driven the changes in waste disposal and remediation in the last few decades. They reflect the desire of society, and they have created the attitudes that have forced and enabled engineers and scientists to do new and necessary things. We are creatures of society, and society is to a large extent what our laws direct. We design to the laws and their spirit, and we design in accordance with the legal dictates. Our designs are the laws incarnate.

A third factor with a fascinating impact on the design is the sum of personal experiences and preferences of the designer. I am convinced that I can predict the political outlook of an engineer on the basis of his preference for a rock or a vegetated cover: the Republican inevitably prefers rock; the Democrat prefers the soil and grass. The Democrat views rock as a brute-force approach, resisting erosion at the expense of ecological integrity with the surrounding landscape. The Republican views the vegetated cover as severely compromised, accommodating lesser concerns such as biointrusion and water balance at the expense of the required guarantee of durability, of the status quo. (What that says about national politics, I leave to others to decide.) At one extreme we may say that each designer has his or her own preferences and style or fashion, and that is the basis of the impact of personality on design.

But I suspect that it goes deeper than that. We all have unique ways of conceiving of reality. Whatever the origins of our specific perspectives on reality, they do influence how we view the issue of waste encapsulation: what we consider adequate, what we find satisfactory. Some of my work on covers was once reviewed by an engineer versed in agriculture. His comments all related to the fact that I was not using the U.S. Department of Agriculture erosion design standards on a disposal cell. To him a farm, nay the greater whole of agriculture, is forever, and so that which is done on a farm must be good for 1,000 years, at least on a disposal cell.

Conversely the same material was reviewed by one who had recently designed a RCRA cover and who had no doubt spent considerable time selecting the right geomembrane. All that reviewers' comments related to geomembranes in covers, even though we specifically excluded geomembranes from the cell cover because of their unproven long-term performance.

Cover design and encapsulation of wastes is no doubt an art, and like all art the signature of the artist's personality comes through the painted canvas, the rock, and soil. This is good if the artist's spirit is beneficent; it is bad if the intent is lowly or poorly conceived. This book will, hopefully, expand the scope and understanding of those who will be tasked with encapsulating wastes, thereby contributing to and kindling the expansion of the spirit of the beneficent artist resident in every designer.

The Ethics of Encapsulation

A consultant or contractor who is called on to design and build a waste encapsulation facility must, if they are ethical and imbued with a moral sense, do at least the following:

- Honestly represent the best interests of the client
- Comply with all pertinent laws and regulations
- Accept that in waste disposal society has established a system of checks and balances that may provide results that differ from one's subjective evaluation of what is proper and prudent
- Encourage Peer Review of all design as a way to establish current best practice and standards.

The best interests of the client may differ from those of the reviewer, of the regulators, and even of society. I talk of subjectively evaluated best interests when I say that we must first represent the client's best interests. That is why the second principle is that we must break no laws or regulations. I talk only of representing a client's best legally sanctioned interests. That is why we are retained and that is why we must do those things which our client seeks (legally) to do.

If we believe in our hearts that we cannot normally represent a client's best (legal) interest, because, for example they conflict with some personal moral perspective or life objective, then, I believe we are professionally bound to stop working for that client. I reiterate that

I am talking only of doing for a client what is legal; if it is illegal, then we have no business doing that work. But I recognize that there are many situations in waste encapsulation when the subjective preference may conflict with a client's objective. In such a case, we may choose to persuade our client to our perspective. If we succeed, it is good. If we fail, there is, in my opinion, no particular opprobrium associated with continuing to work for that client. We may, of course, simply prefer to remove ourselves from the project, if that is what our conscience dictates.

I take it as a given that the laws and regulations that govern waste encapsulation are the best current expression of a society's will in waste control. That does not mean that I believe all current laws and regulations are good or prudent or needed. I recognize that change is often needed and that we should fight for the change we deem necessary. But in waste encapsulation we cannot bow and bend to every whim and fancy. We cannot impose our judgment on society's laws. Until the law and the regulation changes, we must follow it.

I know that laws and regulations are often open to interpretation. I have already said that our clients' best interest should be our guide when we interpret an ambiguous law or regulation. A fundamental reason for believing in this dictum is the elaborate system of review and checks and balances that society (at least in the U.S.A.) has established for reviewing and discussing and critiquing a waste encapsulation project. If we perchance present a biased perspective, then I am comfortable in the knowledge that other engineers will review what I do or propose and will in the fullness of their own ability concur with me or rapidly stop it by their review comments or by withholding the necessary concurrence to proceed.

One of the most difficult ethical traps that faces the designer is the positive feedback that may develop with a regulator. They get to know each other and reinforce each other's views, meritorious or otherwise, while working through difficult problems and obscure regulations.

A course of action valid for one situation is adopted, and expedience precludes a change in a different situation that demands a different solution. The designer seeks the regulator's approval and so avoids the harsh task of upsetting a working procedure.

Even more discomforting is the regulator and designer who, consciously or subconsciously, colludes to squelch minority opinions that conflict with their past decisions and agreement. I have seen this happen and have wondered at the obduracy of otherwise decent men who have had to face an about-face of their long held positions. All I can say is that the designer should avoid such situations, and if you get into one, call for peer assistance to negate the potentially unethical consequences.

Longevity, Design Life, and Performance Period

I include a discussion of the longevity of encapsulation systems in this section on The Philosophy of Design; I do this because longevity in the context of a disposal cell is more philosophy than law or the use of a design equation.

In this book we use five periods, defined as follows, when thinking and talking about the longevity and design life and performance period of encapsulation systems (see Figure 1.1):

- The Construction Period: This is the period during which the cell is constructed and the wastes are placed.

- The Short-Term: This is the period during which we can rely on active oversight, maintenance, and repair. I believe that this period is no longer than 50 years.

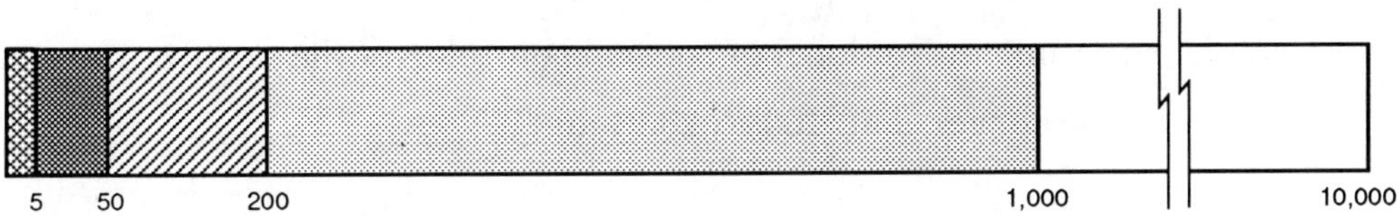

YEARS	PERIOD	PERFORMANCE
0 - 5	CONSTRUCTION	
5 - 50	SHORT-TERM	OPERATION AND ACTIVE MAINTENANCE
50 - 200	MEDIUM-TERM	MONITORING AND PASSIVE MAINTENANCE
200 - 1,000	LONG-TERM	POSSIBLE MAINTENANCE
1,000 - 10,000	VERY LONG-TERM	IN EQUILIBRIUM WITH ENVIRONMENT

The performance periods of a disposal cell range from the first 50 years for active control to the legally mandated 1,000 and the desired much longer time period.

FIGURE 1.1

DISPOSAL FACILITY PERFORMANCE PERIOD

- The Medium-Term: This is the period for which it seems reasonable to anticipate some institutional or societal control over the facility. I suspect the period is no longer than 100 to 200 years.

- The Long-Term: This is the period to which we should turn our attention and think how the encapsulation systems will perform without human intervention, as it responds quietly to the forces of nature. Surely this period extends from 200 to 1,000 years.

- The Very Long-Term: This is the period from 1,000 to 10,000 years which is the period for which humans have built dwellings, cultivated the land, and had diverse languages. In considering this degree of longevity, we must contemplate the possibilities of climate change, loss of institutional control, and intrusion by humans who have adopted languages and cultural perspectives far different from those of today.

We need to distinguish between the design life of an encapsulation system, and the performance period of a cell component and the recurrence interval of a design event. Confusing the three leads to misconceptions and generally muddied discussions.

The design life may be defined as the period for which the disposal cell will encapsulate the wastes with a desired degree of certainty. The design life may be conceived of as a reasonable estimate of the median period for which the majority of the components that make up the encapsulation system will function as intended. The component performance period need not be the same as the encapsulation system design life.

The performance period of a component may be very short or very long, and it need not have a one-to-one correlation with the cell's design life. For example, a geomembrane in a cell may have a performance period of only fifty years. The design life of the cell may still be more than 1,000 years if the design provides for a clay layer to control seepage when the geomembrane fails.

Neither the design life nor the performance period should be confused with a particular design event recurrence interval. For example, the recurrence interval of the site design earthquake is not the same as the design life of the disposal cell. The design earthquake may have a recurrence interval of one million years; that does not mean that the design life of the cell is one million years, or that the sideslopes will perform satisfactorily for so long a period. All that it means is that there is a one in a million chance of the design earthquake occurring in a given year. In brief, there is a one in a million chance the sideslopes will be subjected to the forces imposed by the design earthquake each and every year of the 1,000 or more years of the design life.

The design life for disposal facilities in the U.S. may range from as short as 30 years to as long as 10,000 years (see Figure 1.2). The required design life may be established by law, regulation, technical desirability, or custom. Design lives as short as 30 years have been specified for the cover of municipal landfills. Design lives of up to 500 years are required for low-level radioactive waste disposal facilities. Design lives of up to 1,000 years, to the extent reasonably achievable, are mandated for the covers of uranium mill tailings piles. A design life of 10,000 years is applied to deep-geologic disposal facilities for certain high-level or transuranic radioactive waste. Designers of near surface disposal facilities for transuranic contaminated soil have adopted the standard as a general design objective.

In some European countries where the works of man constructed 2,000 and more years ago still stand stable, there is considerable sentiment to adopt even longer design lives. Desirable as this may be, it is simply not technically possible or feasible to make accurate or substantive conclusions about the design life of disposal cell covers. Precedent and technical experience to make provable claims about the life of alternative covers does not

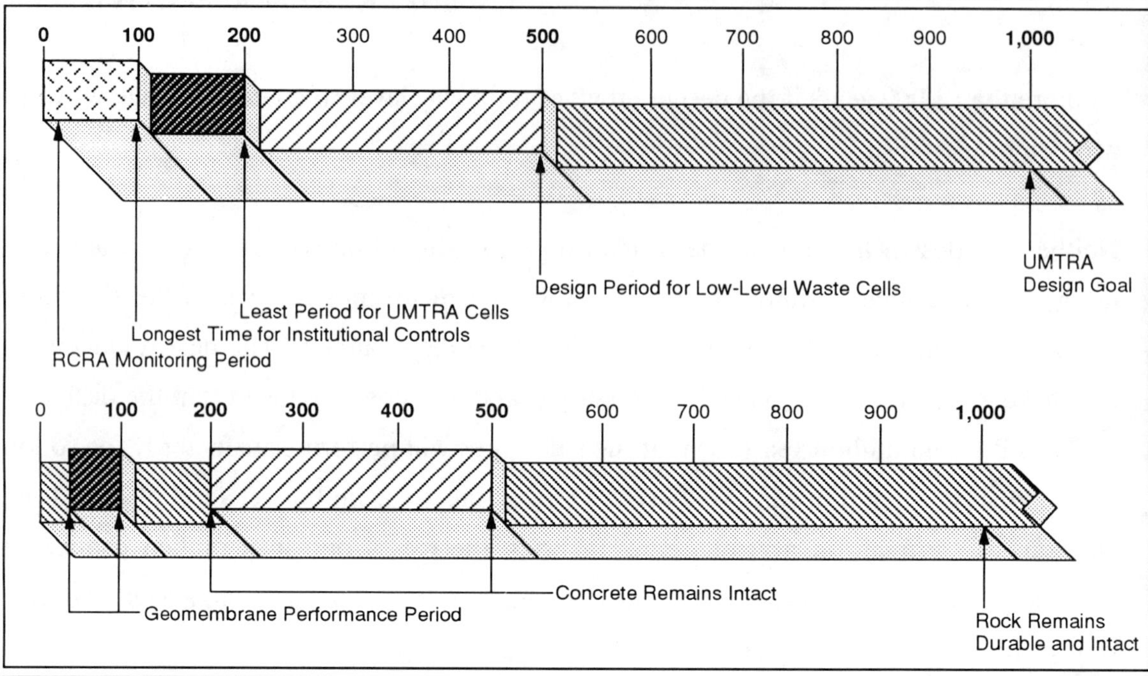

Different laws call for differing lives, which are compared here to the assured performance period of some cell construction materials.

FIGURE 1.2
CELL DESIGN LIVES AND COMPONENT PERFORMANCE PERIODS

exist. Natural analogs provide ample evidence for optimism and confidence that engineered covers will endure for long periods, but there are no direct, long-lasting examples or replicas of the covers and liners that are proposed and now being built to encapsulate wastes. Therefore, conservative designs should be provided which are likely to resist both extreme and cumulative destructive events throughout their design lives.

The following design life is a reasonable objective for most waste encapsulation facilities: "1,000 years to the extent reasonably achievable, and at any rate for 200 years." This design life is consistent with the philosophy that very little, if any, maintenance should be needed to achieve such a design life.

A design life of 1,000 years to the extent reasonably achievable, means that within the bounds of reasonableness and technical practicability, the great majority of the disposal cell components will function as intended for a median period of 1,000 years. In practice most of the cell components will function satisfactorily for considerably longer than 1,000 years. For example none of the construction materials, the sands and clays, will "deteriorate" in that period. The rocks will still be rock and the gravels will still be gravel. The slopes will not fail and unacceptable deformation will not occur. Run-off will still occur as planned and vegetation will continue to grow.

Some of the disposal cell components may change in 1,000 years. Hence, we may talk of the performance period of individual components or elements of the disposal cell. Foundation soils may consolidate, the cell contents may rust, and the geomembranes at the base of the cell will disintegrate. For that reason, there is a second part to the design life requirement: to the extent that 1,000 years assured performance is not reasonably achievable or demonstrable, the components (construction materials) should last, and the cell should perform satisfactorily, at any rate, for 200 years. Design details should be chosen so that the disruptive forces reasonably foreseeable in a 200-year period will not disrupt the integrity of the cell even though certain components may not perform for a period equal to the longest design life. For example, if the precipitation run-off from the design storm, which has a reasonable likelihood of occurring in the first 200 years of operation of the cell, may just initiate erosion of or gullies in the soils and gravels of the top part of the cover, then the planned maintenance of the cover should provide for repair.

To establish that it is not reasonable to provide a performance period of 1,000 years, and to justify a design life of more than 200 years, the approach set out in the U.S. Department of Energy (DOE) Technical Approach Document (DOE, 1989) may be used. Briefly, this involves establishing one or more of the following:

- The design details needed to achieve 1,000 years of assured performance are not technically practicable (e.g., durable materials are not available).
- The 1,000-year design necessitates features that result in a greater negative impact than the benefit associated with the longer design life.
- The cost of providing for 1,000 years is excessive by comparison with the benefits.

An indefinite design life is theoretically possible if the facility operator undertakes an active maintenance philosophy, and can establish with assured credibility that it will continue active maintenance indefinitely.

The U.S. Nuclear Regulatory Commission (NRC) considers 100 years the maximum period that a promise of active institutional control may be the basis for complying with the U.S. Environmental Protection Agency (EPA) regulations governing waste disposal. It is likely that society, in general, will also accept a promise of active maintenance to be reasonable for no more than 100 years. In practice, therefore, 100 years may be the maximum period to which a philosophy of indefinite active maintenance may apply. Thus the minimum active maintenance design life may be said to be 100 years.

Technical Factors Affecting Design Life

The three technical factors that most significantly affect design life are: design storm; material performance; and surveillance and maintenance plans. Briefly: the design storm establishes erosion control features; material performance primarily influences the use or avoidance of geomembranes; and surveillance and maintenance plans affect cover vegetation choice.

The design storm is usually defined as the recurrence interval of precipitation that causes run-off which will not initiate erosion of the uppermost component of the cover. For example, a 200-year design storm means that the run-off from precipitation that has a one in 200 chance of occurring in a given year (i.e., a 200-year recurrence interval) will just start to erode the cover soils. Note that in a 200-year period, the probability that a 200-year recurrence interval precipitation will occur is about 67 percent.

Design storms with recurrence intervals of up to a maximum of 500 years are established by statistical examination of the precipitation record in the area of interest. Design storms of such a recurrence interval may be adopted for cover design because it is practical to provide vegetation-supporting soil that will resist erosion by run-off from such precipitation.

The probable maximum precipitation (PMP) in an area may be established by considering the most extreme meteorological and climatic conditions (atmospheric moisture) that could conceivably occur and lead to intense precipitation. The statistical probability of such an event occurring in a given year is thought to be less than one in a million. Such extreme precipitation may be adopted as the design storm because the resulting cover (usually with a durable rock upper layer) has a very high factor of safety or degree of assured stability against erosion.

At a project for which there are no laws or regulations that mandate adoption of a particular design storm for waste encapsulation facilities, the technical approach generally adopted is to calculate the effect of storms of varying recurrence intervals on the disposal cell, assess the consequences, evaluate reasonable response actions, and implement feasible surveillance and maintenance programs. If it can be demonstrated that it is reasonably practicable to provide for the probable maximum precipitation, for example by limiting erosion with very flat slopes, durable rock, or maintaining suitable vegetation, then the probable maximum precipitation may be designated the design storm. When, however, the design details required for aesthetics or vegetation control are incompatible with preventing erosion by the probable maximum precipitation, then the 200-year to 500-year recurrence interval event is adopted as the design storm.

Aesthetics

A disposal cell becomes a permanent part of the landscape. Ideally, it should blend into the area; visually, it should be aesthetic. As beauty is, to quote the cliche, in the eye of the beholder, it is difficult to establish unequivocal criteria for an aesthetic or beautiful cell that complements and even enhances the visual environment. Nevertheless, the following guidelines may provide some assistance:

- Generally, the more "natural" looking the cover is, the more it is perceived as aesthetic. In an arid environment, a rock cover may appear more "natural," particularly if the rock color is similar to or complements the color of the surrounding soils and rock. In a more humid environment, a vegetated cover may be more visually pleasing.

- A perimeter screen of trees may enhance the overall perspective of the site, obscuring the overall mass of the cell.

- Varied geometrical shapes may lead the eye around the cell or promote a feeling of geometric conformity to surrounding topographic forms.

- Siting of the cell away from easily accessible vistas could minimize the need for aesthetic considerations.

The design engineer's job with regard to the aesthetics of a waste encapsulation system is to act as the integrator of the various, often irreconcilable, aesthetic perspectives of the design team members. The designer must solicit the points of view and desires of those charged with establishing and maintaining the cell. He must hear the intuitive and qualitative desires of the owner and those who will have to look at the cell. He should deal sensitively but firmly with an irascible regulator who puts blind compliance with an insensitive or stupid law ahead of aesthetics; such a regulator should be reminded that his is but one of a matrix of perspectives that are integrated to design a facility that must perform properly well beyond the lifetime of those associated with its establishment.

Ultimately, the designer takes the credit or blame for the look and feel, the quality if you will, of the disposal facility. Thus he may be permitted some leeway in imposing his own perspective. Those who succeed have a worthy monument. Those who fail join the pantheon of desecrators of human sensitivity.

Design Standards

Once all factors have been considered, the team responsible for the design, construction and maintenance of the encapsulation system establishes a set of design standards. The set for a particular facility is an intelligent blend of the obvious local and national laws and regulations, prudent engineering practices, and the reasonable requirements of the community affected by the facility. Table 1.1 is an example of a set of standards adopted for the design of a system to encapsulate overburden soils from a uranium mine in Oregon. Although there are many numbers in this table, we must recognize that it incorporates a very specific philosophy and set of ethical demands on the designers, and understanding from regulators. In particular, we note the statement that all standards relate to design and not performance. This is a key point in design and construction of facilities for the long-term. Simply stated, this means that the success or failure of the design is not predicated on the performance of the constructed works. The designer and his company will not be around to monitor performance in 1,000 years. The requirement is that the design be compiled and its performance documented in calculations, with models, analysis and judgment, and by analogy with natural processes. We can do no more as engineers. We cannot design a facility for low maintenance and then rely on performance observation and response.

From a philosophical or professional ethics perspective, I believe we have to accept the following as fundamental propositions when working with, designing, and building waste encapsulation systems:

- Some release of contaminants in the long-term is inevitable.
- An acceptable release rate is one that will not adversely impact human health and the environment.
- All waste encapsulation schemes will ultimately fail (although the failure is not really failure if we can limit the release rate to below the hazard level).
- As engineers, we can hold back the forces of nature for a limited period only.

TABLE 1.1

PROPOSED ENCAPSULATION SYSTEM DESIGN STANDARDS

PARAMETER	STANDARD
DESIGN LIFE	Remain stable for 1,000 years to the extent reasonably achievable. Remain stable for at least 200 years, in any event.
DISPERSION CONTROL	Prevent dispersion of contaminated materials by wind and water erosion. Prevent inadvertent intrusion and relocation by humans and animals.
RADON CONTROL	Reduce surface radon flux to less than 20 pCi/m^2s. Reduce radon concentrations to less than 0.5 pCi/L at the site boundary.
LEACHATE	Minimize leachate generation from infiltration flux through the cover.
LONG-TERM MAINTENANCE	To the extent practical, reduce the need for long-term maintenance. Long-term surveillance and maintenance will be undertaken.
DESIGN STANDARDS	All standards are design standards, not performance standards.
REGULATIONS	Comply with ALARA principles, as applicable. Comply with applicable and relevant federal, state, and local laws and regulations.

- There is probably not enough money available to permanently secure all existing waste facilities.

To design to a given design life, all the engineer can do is honestly represent the best interest of his client, comply with all pertinent laws and regulations, accept that there is a societal system of checks and balances, and regulatory agency concurrence is the ultimate arbiter of compliance with longevity requirements. I recommend that the engineer encourage peer review of design to establish and document current practice and standards. That way future generations will respect the designer's decision as rights even if in 1,000 years they are shown to be wrong.

I have no more to say on the philosophy of the design of disposal cells and waste encapsulation systems. Suffice it that each designer, each professional, and every person involved in waste disposal think these issues through, act legally and morally, and try to leave the world a cleaner place when they leave than when they entered.

Chapter 2

PERPETUAL ISOLATION: ECOLOGICAL PERSPECTIVES

PERSONAL PERSPECTIVES

As an ecologist, I am a watcher of nature. I observe the landscape and the land forms that comprise it. I take special interest in the organisms that reside in the thin veneer of habitat that envelops our globe. From my perspective, the environmental crisis is a failure to maintain balance in the flow of materials throughout that zone of hospitability. The supplies of required resources such as fresh air and water now lag below the biological demand for them. The environment's capacity to absorb our noxious byproducts is overwhelmed. Hazardous substances such as oil and radionuclides have been drawn from their relatively remote and inaccessible deposits into our domain, where they have been processed, enriched, and stored behind precariously thin barriers. The spent residues of these toxic extracts must be returned to whence they came. If they were created of a human process, perhaps then they can be destroyed. But if they were simply concentrated from natural deposits, and they are as indestructible as for instance radioactive materials, then perpetual isolation is probably our only recourse.

As an ecologist, I can also observe many examples of nature's effectiveness in perpetually isolating materials from the accessible environment. If we consider that encompassing volume of all matter derived from the cataclysmic origin of our universe, all but the most infinitesimally small fraction is permanently beyond the reach of humans and our earthly co-habitants. Indeed our envelope of life is minuscule on such a scale, and our human endeavors to extend our reach - outward to the moon and beyond - inward to the depths of the deepest borehole - are barely mentionable. Unfortunately, our ability to export the unwanted residues of our consumptive lifestyle is also highly limited, resulting in the all-too-frequent revisitation of toxic end products to the accessible environment. The dilution-solution to pollution fails as well because the absorptive capacity of our atmosphere and hydrosphere simply falls short of the demands placed upon them by our burgeoning population.

Natural Analogs

Given the shortness of our reach when it comes to putting away unwanted substances, we must look close at hand for opportunities to permanently isolate waste. These opportunities, viewed from the ecological perspective, are Natural Analogs - natural phenomena or systems that can be emulated to achieve with confidence the desired degree of isolation. Consider for instance the uranium deposits that are mined for feedstock to the nuclear power and weapons industries that have contributed so much of today's most hazardous waste. Some of these deposits lie on or just below the surface in erosive environments, where uranium is readily and perpetually migrating from the site, providing highly detectable beacons that once attracted the early prospectors touting Geiger-counters. The uranium in such a readily detectable and minable situation may be in a porous media, surrounded by still more porous media, allowing the nuclide and its daughters to leach easily to groundwater and travel widely throughout the region. Such a deposit, though natural, is not a promising analog for waste isolation.

Now consider an equally fertile deposit where uranium is tightly held in an absorptive matrix such as humate. The immobility of the nuclide is enhanced by the fact that the deposit is bounded by layers of impermeable or highly absorptive clay-rich sediments. Though not deeply buried, the deposit lies beneath a depositional environment, where colluvial and aeolian sediments ever thicken the barrier between the ore and the accessible environment. Because of its natural isolation, such a deposit may be detected only by computer modeling, predictions of accumulation, and exploratory verification with instrumented boreholes. In fact, many such deposits likely exist, but remain undiscovered due to their remoteness and the effectiveness with which the material is naturally encapsulated. These are excellent analogs for waste isolation.

One of the mostly widely touted analogs, acclaimed for its attestation that radionuclides can indeed be immobilized for millennia, is the Oklo natural reactor in Gabon. Here, due to rather extraordinary circumstances, uranium was naturally concentrated by the percolation of uriniferous water through a highly absorptive medium. The result was a critical mass that sustained a low-grade fission reaction for half a million years, generating fission products and transuranic radionuclides that represented a natural source term. This so far unmatched deposit of concentrated nuclides remained largely intact, migrating no more than a few meters since their inception some ~2 billion years ago. Those who argue that nuclear waste can be safely interred, despite its noxiousness and longevity, can be safely disposed often point to the Oklo phenomenon as evidence of their viewpoint. Sadly, this extraordinary opportunity to study nuclides in natural media has been lost under the shovel. Shortly after its discovery, the deposit was mined for its remarkable concentrated uranium even as scientists worked feverishly to learn what they could.

Geomorphic Stability

The Oklo deposit was a highly unusual phenomenon of the like that may or may not be encountered again. Nature is full of analog systems that are anything but rare, yet are laden with information of value to the seeker, designer, and demonstrator of waste-isolation approaches. Any landscape anywhere that has been stable for a million years can provide us with a lesson on how to design for perpetuity. But first we must determine that the landscape is indeed as ancient and stable as we suspect, and that the long-lived stability is a result of the landscape having endured a full range of potentially disruptive events: earthquakes, floods, probable maximum precipitation, and so forth.

How do we certify a landscape's antiquity? The answer resides in a careful compilation of evidence from multiple disciplines such as geomorphology, paleopedology, dendrochronology, palynology, and other tools of science used to explore the past. Rather than dwelling myopically on the necessarily conservative models that describe a soil's response to the probable maximum precipitation, cell designers should mobilize multi-disciplinary teams to scour the landscape for clues of what has happened and how natural features have responded. For instance, buried soil horizons on a desert bajada can be dated to the pliocene, indicating that deposition has occurred perhaps not continuously, but with no regressive erosion that has exposed or removed the paleosoil. Buried pollen, desert varnishes, and even packrat middens can be convincing clues to the antiquity of a landscape.

Just as a stable landscape can guide the design of an impoundment to resist breach, so can other natural analogs guide us in designing to resist other forces that tend to degrade the isolation system or insidiously mobilize the contaminants within. For instance, if we intend to exclude organisms - plant roots, burrowing animals, and microbes - from the deep layers of an encapsulation system, then we must look to nature for examples of analogs that have thwarted biointrusion through the ages. Later in this book, we will discuss how favorable soil can be layered over inhospitable rock or coarse sand to create a stratigraphic sequence that, if properly proportioned and constructed, will discourage downward root advancement. Experimental plots have demonstrated that an appropriately placed loose-cobble biobarrier does indeed retard the advancement of both roots and burrowing animals, as well as invasion by small creatures such as ants and termites, for the months and even years of experimental observation. Test plots, as such, are useful to help engineers refine and demonstrate their construction techniques; however, the value of a few year's observation is quite limited relative to the virtually indefinite longevity requirement of a cell for long-lived wastes. Once again, the cell designer must turn to nature to find formations or soil stratigraphy that resist biointrusion. Researchers in the Hanford Barriers Program, where the challenge is to construct covers with 10,000+ year design lives, have discovered deposits of aeolian silt loam layered one or two meters deep over relatively coarse glacial sand and rock. Beginning many tens of thousands of years ago, the post-glacial deposition process has gradually created an abrupt transition in the soil between fine silt loam with favorable water and nutrient conditions and coarse media with conditions that were relatively hostile to biological activity. Compared to the aeolian silt loam, the underlying glacial fill lacked nutrients and capillary surfaces for water storage. Further, the fine soil/coarse media transition acted as a capillary break that, judging from calcium carbonate deposits in the soil, had long impeded the downward movement of water in the soil (except during unusually wet circumstances). The relative absence of roots and animal burrows in the deep layer, or of residual traces of past biological activity, attests to the effectiveness and longevity of such a stratigraphic impediment to biointrusion.

The careful study of biointrusion into natural analogs such as described above, as well as into ancient Indian Mounds with similar configurations, indicates that layers of soil and rock can be sequenced to provide a system that will resist biointrusion for many thousands of years. It is only by looking far into the past and documenting the long-term behavior of a natural system that we can be confident that an analogous man-made system will provide the required performance well into the future.

Longevity

The ecological perspective is unwaveringly centered on the concept that we must always look to nature and emulate natural phenomena in our design of encapsulating systems. Within this perspective, we can define specific elements or tenets that are in some cases compatible with the engineering perspectives expressed earlier in this chapter, and that are in other cases contradictory to them. Where conflict exists, we sometimes attempt a reconciliation and other times leave it to the reader to choose his or her own perspective from among or between the alternatives. It is certainly not our intent to confuse the student or practitioner of encapsulation; we rather seek to express the spectrum of viewpoints that must be considered in the truly conscientious design of an encapsulating facility.

The ecologist on a cell design team looks beyond the numbers that frame a disposal facility's design life. If a cell for long-lived waste is built to last for 30 years or even 1,000 years, then it is built to fail its fundamental mission of encapsulating waste long enough to protect future generations. I serve on a review panel for a program to build barriers that will protect from exposure or mobilization stabilized interred wastes that include transuranic radionuclides with million-year half-lives. Several of the engineers on the program are intent upon satisfying the regulatory requirement that generally applies to transuranic waste for a disposal strategy that will isolate the waste from the accessible environment for 10,000 years (10 CFR 61). Fine. Ten thousand years is a worthy and ambitious objective, greatly transcending the design lives for bridges, dams, and even baseball stadiums. But a 10,000-year design life isolates a million-year half-life waste for a mere 0.1 percent of the 10 half-lives that must transpire before that material is generally regarded as nonhazardous; what about the other 99.9% of the wastes' hazardous duration? And what about hazardous chemical wastes that are deemed permanent and immutable in their toxicity?

The answer that engineers most often give is that a finite design life must be specified to provide a target for the durability of materials and the longevity of the overall structure until the inevitable failure occurs. After all, every structure must someday fail, right?

This, too, represents a point of dichotomous (though perhaps not irreconcilable) thinking between the ecologist and engineer. The ecologist agrees with the engineer that all structures made by humans invariably change, but this change need not lead to failure - to the premature or excessive release of waste. The objective is to design so that change will enhance and perpetuate the encapsulating system. The engineer's perspective perhaps is biased by the fact that nearly all human constructs are erected with a certain neatness of appearance, perhaps even beauty, in mind. Change is often seen as deterioration or failure: the crumbling and burial of a temple in desert sand; the cracking and overgrowth of a Mayan pyramid by subtropical jungle. Ecologists view the deposition of materials and growth of vegetation as potential allies in the effort to encapsulate waste. This difference between the engineering and ecological perspectives is beautifully represented by a series of spontaneous debates that revolved around the issue of whether to use vegetated soil or rock on the outer layer of covers for stabilized uranium mill tailings piles around the U.S. Advocates of rock covers pointed to the durability of the rock against erosion that would otherwise breach the cell and expose the encapsulated waste, resulting in failure of the cell. I, as the primary advocate of vegetated covers, argued that vegetation would surely grow between the rocks and, because of the hostile biological conditions in the cover's upper layers, would extend their roots ever-deeper into the underlying layers, eventually damaging the encapsulating layers and initiating the transport of the waste. At one point during the debate, an engineer whom I greatly admire explained that his advocacy of rock covers was based on the fact that once built, the cover would remain static, exactly as it appeared on the blueprint and exactly as it checked out during the post-construction inspection. Change, he explained, is too hard to manage. Dynamic systems in nature have unpredictability that cannot be accommodated; that must instead be resisted with massive, immutable systems. Now, this engineer's assertion of a rock cover's immutability was by his own admission an overstated perspective, expressed to fully frame his position in the debate. But I think his remarks do indeed reveal the ingrained engineering instinct to design for long-term stability, in the sense of unchangeability, whereas ecologists, farmers, and range managers design their projects and facilities with change in mind. Biologists are trained from Natural Science 101 to respect the pervasiveness of dynamic process such as ecological succession and evolution, and perhaps are more inclined to recognize the inevitability of these processes and seek to design facilities that will not just accommodate them, but will benefit from them.

The goal, therefore, is to develop disposal systems that will change in harmony with the landscape in which it is sited. The cell must be located where a reasonably predictable and favorable change can be expected and taken advantage of. This might, for instance, be a zone of deposition on a montane pediment, where colluvial materials are reliably imported from upslope, or an area where a long-term, more or less continuous deposition of aeolian particles has occurred. The siting process for a disposal cell already includes the participation of a geomorphologist who performs analyses to ascertain that the landscape is fundamentally stable and unlikely to be affected either by a relentless erosion or catastrophic flooding. Similar analyses are performed of seismicity, climate, and so forth, to identify possible processes or events that could degrade the long-term stability and performance of a cell. The ecological perspective is that these analyses are highly worthwhile, but they should concentrate not just on the fatalistic objective of minimizing the likelihood and extent of failure, but rather should seek to identify locations where the most likely long-term scenarios are those that will enhance cell performance.

It is important to emphasize that the designer's vision for the future of a cell should recognize that changes will occur in many ways, with multiple, interrelated consequences. The cell must not be sited or designed to encourage change that will improve one performance characteristic, but degrade another. For instance, if a disposal cell is buried by a sand dune, this may well provide enhanced protection against exposure of the waste by erosion, but the deep sand over the cover may unfavorably alter the cell's water balance, encouraging infiltration by destroying the mechanism that evapotranspires meteoric water from the cell. This will happen if plants are buried and the overlying sand is devoid of the capillary pores that enable moisture to be evaporated from depth. Therefore, the cell's burial by a sand dune, while reducing the likelihood of breach, is an overall negative because of the increased production of contaminated leachate and consequential release of waste into the groundwater. A more desirable long-term change would be the gradual deposition of colluvial or aeolian sediments at a rate that would not reduce the vegetal productivity that may be key to the cell's long-term balance. The gradual rate of sediment deposition is the key, because the plant community on the cell should remain undisturbed. In fact, any vegetated cover will gradually increase its own thickness simply by capturing atmospheric carbon and fixing that material into the plants themselves, whose roots and shoots will eventually decay and assimilate into the cover's upper layer of soil.

All this leads us to the ecological perspective on longevity, which is that a well-conceived cell should be as long-lived as the landscape in which it is sited. If the landscape is stable, then so should be the cell that is integral to that landscape. If the landscape is a growing one, accumulating materials by sediment deposition and biological production, then the cell should grow with it. Only events that are so catastrophic as to reshape the landscape such as cataclysmic earthquakes or continental glaciation, should disrupt the cell that has been designed to be fully integral to the landscape.

Before proceeding to the other specifics of the ecological perspective, it must now be conceded that these are idealistic objectives that should supplement, but not supplant, the pragmatic tools of the engineer, including the traditional engineering approach of specifying a finite design-life and designing toward its achievement. The extraordinary and unprecedented challenge of designing for perpetual isolation compels us to transcend the traditional, if pragmatic, bounds of the engineering profession in favor of a more holistic, intuitive approach in which the enlightened observation and emulation of nature is integral to the entire design process. Although this transcendental approach to design may not lend itself to calculation sets and design documents, it should be a heartfelt goal of each member of the design team, engineer and ecologist alike.

Maintenance

The ecologist viewpoint is that anything short of a maintenance-free cell is nothing more than a "Superfund site waiting to happen." The object of our profession is to destroy hazardous waste, whenever possible, and to permanently dispose that which cannot be destroyed. Our obligation is to free subsequent generations of the responsibility for caretaking our hazardous residues, not to saddle them with housekeeping chores which, if neglected, will result in the re-pollution of the environment that we worked so hard to clean. Planners of long-term disposal systems have long recognized the difficulty of maintaining institutional control over the property where a cell is located, attempting to discourage drillers, developers, and other users of the land from impinging on the integrity of the cell. And yet, some of the same planners seem willing to accept that a perpetual commitment for gully repair, plant removal, or lawn manicure must be performed for the wastes to remain secure.

Another element of the rock-versus-soil debate on the Uranium Mill Tailings Remedial Action Project that elucidates the differences between the engineering and ecological perspectives involves the issue of maintenance. A number of engineers and regulators in the debate uncompromisingly support rock covers over vegetated covers even though they concede that the rock covers will require frequent and perpetual maintenance to remove unwanted deep-rooted vegetation. In their minds, the requirement for unassailable resistance to erosion is so compelling as to override all other considerations. As a consequence, several cells have been designed, approved by regulators, and constructed with full knowledge that deep-rooted plants will invade and that perpetual plant removal will be required. It is the view of this ecologist that, except in extraordinary circumstances, no cell should be built with a design that will assuredly require maintenance to remain intact. If the objective of perpetually isolating waste is to be achieved, then the design should be concurred upon by every member of a multi-disciplinary team. In other words, a design should never be sanctioned in which one member of the team has identified an inevitable failure scenario that can only be addressed by maintenance.

But is a maintenance-free cell even possible? Once again, we need only look to nature, where every example represents a maintenance-free system. Stable surfaces have remained so since the last glacial retreat. Meadows and prairies have remained treeless since the last change in climate or fire cycle. Our cell designs should emulate these systems, applying our observations of nature to develop conceptual approaches that will either resist or accommodate change in order to remain intact. There will always be some measure of uncertainty regarding the direction and extent of changes that must be anticipated, and therefore it is unrealistic to expect to design and build a cell that will perform perfectly in all possible circumstances. This should be our objective: a cell for which any reasonably foreseeable failure has been anticipated and designed for.

Some degree of maintenance or post-construction refinement may be necessary until the cell reaches a state of equilibrium in its integrity with the landscape. For instance, as will be discussed further in the chapter on vegetated covers, the plant community on the cell should be a climax or close analog thereto. A "climax" community, in ecological terms, is the type of plant community one finds in an area that has long been undisturbed. It is the terminal state in the stepwise invasion and replacement of plants that occurs during recovery from a perturbation to a condition of long-term stability. It is the goal of reclamationists, and should be the goal of cover designers, to stabilize surfaces with plant communities that closely emulate a selected local climax. Although it is not possible to actually install a climax community, it is possible to introduce a close analog and to hasten the successional process (stepwise evolution of a surface) so as to achieve the equilibrium conditions of an advanced-successional or climax community more quickly than might otherwise occur in nature. This is done by providing well-developed soil with the appropriate, rationed additions of fertilizer and irrigation that will accelerate the growth, reproduction, and establishment of plants. The point of all this is that some initial maintenance on a vegetated cell cover may be needed to facilitate the progression of the plant community to a condition analogous to climax. Follow-up fertilization, irrigation, and interseeding (of bare patches or areas with only one species) may be necessary to provide the desired degree of diversity, establishment, and durability of the plant community that stabilizes the cell. Other repairs to the cover's surface may also be necessary during this period of establishment, although these should be minor if the cell has been designed to be inherently stable and fully integral to the landscape.

Law and Ethics

At the risk of sounding cliched, I'd like to assert that the foremost law to be honored by the truly ethical cell designer is the law of nature. By closely observing the many facets of nature, the multi-disciplinary design team will discern what features should be emulated in the cell design and how cell failure might occur. All failure scenarios should be addressed regardless of legal and budgetary constraints, because there is no reason to build a cell that will no-doubt fail and therefore require perpetual maintenance. The goal is perpetual isolation; if current budgets and regulation prohibit the designer from pursuing this objective in earnest, then he should persuade his client to accept a temporary solution and wait until the law authorizes a permanent solution. This again may be idealistic, in which case the practical course of action is to explicitly accommodate the realities of budgets and regulators while never abandoning the objective of developing a maintenance-free, natural cell. The commercial or bureaucratic circumstances in which we build cells may sometimes appear at cross-purposes with our objectives of perpetual isolation. In most cases, however, the persistence, conviction, and persuasiveness of the engineer and his design team can be brought to bear on the situation, eventually convincing both clients and regulators of the worthiness of developing a truly permanent, unassailable system for isolating waste.

Chapter 3

INTERNATIONAL PRACTICE AND NATIONAL LAWS

GENERAL

Standard practice in waste disposal and encapsulation design is first to ask, "What laws and regulations govern our situation?" The practice of seeking out and analyzing potentially applicable, relevant, and appropriate regulations (ARARs) consumes a great deal of time and energy.

This must be done for individual projects on a site-specific, state-specific basis. Reasons why this is necessary are:

- Failure by the facility's proponent to fully identify all potentially applicable, relevant, or appropriate laws diminishes the confidence level of all who must approve the encapsulation facility in the design set out by the facility's proponent.

- The designer must know what specific scientific and engineering details and components are absolutely required because of laws and regulations.

- The public must be assured that their valid concerns and demands, as crystallized in the law, are being met and complied with.

- The client who pays for the encapsulation facility must be able to proceed secure in the knowledge that they are obeying the law and will not at a later date be called on to retroactively do at more expense that which could be done efficiently now.

We acknowledge and recognize the significant role laws and regulations have had and now play in setting the design of waste encapsulation systems designed for perpetual solution of noxious wastes. This is as it should be; without the laws we would not have achieved what we have, and we would not be able to achieve what we still have to.

That said, the temptation is to attempt in this section to set out and describe all the laws and regulations that govern the design, performance assessment, construction, and surveillance and monitoring of the encapsulation systems that are the topic of this book.

Reasons we do not do this include:

- Laws differ from country to country, from local jurisdiction to local jurisdiction
- Laws change as the political whims of the times change
- Laws are open to interpretation and manipulation by interest groups who espouse one perspective or another
- Laws applicable to a single situation may conflict or even be mutually exclusive.

Basically, we believe that in an enterprise that involves long-term encapsulation of noxious substances, we should be guided more by ethics, goals, and objectives that are in accord with the spirit and intent of the laws rather than by the laws themselves.

For these reasons, we do not attempt a full survey of the laws and regulations that may be uncovered in an ARARs analysis. Rather, we set out briefly the most significant codes, guidelines, laws, and regulations that have variously:

- Forced technology by making new demands on engineers and scientists
- Set design details by specific stipulations
- Established design philosophy or technical approach by the very boldness of their scope and requirements.

Our purpose is to describe the framework within which we must design and build as engineers and technical specialists. We do not try in what follows to write a legal tome for lawyers.

INTERNATIONAL REQUIREMENTS

The International Atomic Energy Agency (IAEA) (1990), establishes these as the main objectives of decommissioning and rehabilitating uranium mine and mill waste disposal facilities:

- Protect the health and safety of persons and the environment
- Contain and isolate wastes in a decommissioned area
- Minimize reliance on active institutional controls
- Establish a cover that controls contaminant releases (radon, metals, dust, seepage, and gamma radiation).

The IAEA notes that most jurisdictions in which encapsulation of radioactive waste are undertaken have regulations that cover the following:

- Longevity: The U.S.A. and Spain, for example, adopt 1,000 years to the extent reasonably achievable. The Canadians recognize only that structures degrade with time and that preference should be given to systems that degrade gradually.

- Verification: Both predictive and performance verification may be allowed. Prediction involves the use of generally accepted engineering principles and practices. Performance based verification involves post-construction monitoring for a defined period during which it must be established that the encapsulation is performing as predicted.

- Air Emissions: Most countries call for minimum radon release loads with 20 picocuries per square inch per second of radon - 222 being a representative figure.

- Misuse Restrictions: These generally take the form of site access control (fences and warning signs), and institutional controls (government ownership of the land of deed restrictions). The Canadians call for justification and demonstration of the need for institutional controls through cost-benefit analyses.

- Maintenance Requirements: The Spanish regulations call for the need to minimize the need for long-term maintenance. The Canadians recognize that it may not be economically practical to implement a decommissioning plan that permits an encapsulation facility to be abandoned. They acknowledge and accept applications for perpetual care options, although this does not extend to "controls which require ongoing human intervention (as would be required for effective treatment)."

- Water Protection: All countries party to the IAEA document have regulations that call for surface and groundwater protection. These generally take the form of specific numerical water quality standards, a defined point of compliance, and provision for alternative, less restrictive standards, if it can be demonstrated that there are no adverse environmental or human health effects.

- ALARA - The principle of As Low As Reasonably Achievable is called for in most jurisdictions. This implies an encapsulation system that incorporates all feasible and practical components and design details that reduce contaminant emission, to and below regulatory limits. The encapsulation system should be constructed to provide the greatest possible containment as is reasonably achievable given the availability of funding and current technological capability.

The IAEA notes that the design of uranium mine and mill decommissioning and rehabilitation works is an art that combines scientific and technical knowledge, engineering judgment, examination, and evaluation of facts and alternative layouts, and the search for a balance between (potentially incompatible) demands and constraints of cost, regulations compliance, social acceptability, protection of human health and the environment, and technical and engineering practicability.

The IAEA states that a successful rehabilitation or closure works design relies as much on the skill, judgment, and genius of the designer as on the use and implementation of:

- Standard operating procedures

- Adherence to design codes and standard technical approaches

- Implementation of quality practice procedures such as peer review, value engineering, total quality management, and Quality Assurance/Quality Control (QA/QC) procedures.

With regard to longevity, the IAEA notes that there is no basis or body of knowledge to predict the performance of design details of materials for design lives of 1,000 years or longer. To achieve socially and morally defensible results, the IAEA recommends:

- Use only natural materials

- Back up with natural materials any man-made components which are not otherwise avoidable (in that they may be required by law or regulation)

- Emulate natural topographic and geologic forms

- Design for the most extreme event (i.e., maximum probable precipitation of flood or maximum credible earthquake)

- Provide redundant features to augment the failure of a parallel but separate system

- Seek opportunities for synergistic design in which components work together to a mutually desired performance objective

- Provide for inevitable and continuous natural processes (e.g., the growth of vegetation).

The IAEA design guidelines may be considered a type of supra-national applicable, relevant, and appropriate regulation. At the least, these guidelines constitute a basis from which to assess the absolute adequacy of the covers discussed and recommended in this volume.

LOW-LEVEL RADIOACTIVE WASTES REGULATIONS

In accordance with DOE Order ORN 5820.2A, the disposal cell for low-level radioactive wastes will conform to the requirements of 40 CFR 192 (Health and Environmental Protective Standards for Uranium and Thorium Disposal). The EPA administers these regulations under the authority of Section 275 of the IAEA and the Uranium Mill Tailings Radiation Control Act (UMTRCA). These standards include the requirements that disposal cell control shall be designed to:

- Be effective for up to 1,000 years, to the extent reasonably achievable, and in any case, for 200 years

- Provide reasonable assurance that releases of radon-222 from residual radioactive material to the atmosphere will not exceed an average release rate of 20 picocuries per square meter per second

- Control seepage from the disposal cell so that specified groundwater protection standards are complied with.

In order to achieve the design life requirement, only natural materials (soil and rock) are used, and the cover is designed to resist extreme events such as:

- The probable maximum precipitation

- The maximum credible earthquake

- Extreme freezing and thawing (up to 200-year recurrence interval cold).

In addition, the design should avoid the need for active maintenance.

The NRC may, under the terms of 10 CFR 40, issue a license for the long-term surveillance and custodial care of remediated facilities. The conditions and procedures for long-term surveillance are to be described in a technical report prepared by the licensee.

The disposal cells designed and constructed in accordance with these regulations generally involve placement over the waste of a multicomponent cover of natural materials. Liners and basal drains have not been used. Groundwater protection is assured by selecting a site (relocating the tailings if necessary) at which seepage from the cell of contaminants does not result in an exceedance of groundwater quality standards. The long design life and the ALARA principle preclude collection and treatment of the seepage in leachate collection systems.

MIXED LOW-LEVEL WASTE REGULATIONS

Mixed waste is material that is considered radioactive waste as regulated under the International Atomic Energy Agency (IAEA) which also contains hazardous substances that are either listed as hazardous waste (40 CFR 261) or are regulated under the Resource Conservation and Recovery Act (RCRA).

In regard to the design and evaluation performance of a disposal cell or encapsulation system, the following are applicable or relevant and appropriate:

- Waste should be isolated to comply with the long-term performance objectives of 10 CFR 61
- The disposal site must be well-drained with respect to surface water, i.e., free of flooding or frequent ponding
- The site must be located so that upstream drainage areas are minimized to decrease the amount of run-off that could erode or inundate the waste disposal facility.

THE EPA REGULATIONS

The EPA (1987) provides design guidance for covers for landfills and surface impoundments. They recommend a multi-layer cover that includes the following elements:

- A layer consisting of a vegetated erosion control component and 600 mm minimum thick topsoil, or a nonvegetated surface that minimizes erosion and promotes drainage off the cover.
- A drainage layer of 300 mm thickness with a minimum hydraulic conductivity of 10^{-2} cm/sec. that will effectively minimize infiltration into the low-permeability layer, and have a final bottom slope of two percent after settlement or subsidence.
- A low-permeability layer that effectively minimizes water infiltration: this consists of a 20-mil-thick flexible membrane liner, and 600-mm-thick compacted soil with a maximum hydraulic conductivity of 10^{-7} cm/sec.

- Additional optional layers that may be used on a site-specific basis (e.g., gas vent layer; biotic barrier layer; and exclusion of the drain layer in an arid environment).

The design guidance for the soil/vegetated top component includes these requirements:

- Maximum erosion rate of 2.0 tons/acre/year as calculated with the U.S. Department of Agriculture Universal Soil Loss Equation (USLE).

- A surface drainage system capable of routing run-off without causing erosion rills or gullies.

The EPA recommends further that a geotextile or graded natural filter be used between the soil and drain to prevent migration of fines from the soil to the voids of the drain.

Finally, the EPA calls for placement of the low-permeability infiltration barrier layer beneath the depth of freeze-thaw.

EPA regulations call for the provision of basal liners, drains, leachate collection, and leak detection systems beneath the wastes in disposal cells. The requirement that the base of the encapsulation system preclude passage of waste leachate to the groundwater is based on a noble desire to protect groundwater resources. This technical requirement contradicts and is naturally exclusive of the international practice for uranium mine and milling wastes of providing long-term, ALARA encapsulation that protects groundwater by site selection and performance, rather than perpetual care and maintenance.

When the designer is faced with a situation in which simultaneous compliance with EPA's call for liners and short-term materials and the international ethic of long-term, zero maintenance is necessary, the only practical solution is to provide a facility that, in the short-term will function as envisaged by EPA regulations, but which in the long-term will function as a uranium mill tailings pile. Hence, we must provide basal drains and liners, and provide for collection and treatment of leachate. But, in addition, we must recognize and demonstrate that in the long-term when synthetic materials degrade, when drains block, and humans cease to collect and treat the leachate, that there will be no adverse environmental or human health impact.

This imperative, this reconciliation of contradictory philosophies and regulations is the topic of this book, so let us proceed.

AN AFTERTHOUGHT

Common practice in design of an encapsulation and in the reports that justify new systems is first to list ARARs. Next, the designer describes the encapsulation facility. All too often, the designer is thrown into confusion and despair because the laws are unclear, contradictory or of disputable application. The laws may fail to provide any rational guidance to the scientist or engineer, and so in due obedience to convention ARARs are listed and subsequently ignored or used as an excuse for inaction.

I believe this is unfortunate and the wrong approach. I propose the following process that avoids or mitigates the negatives of present practice:

- First, establish scientific and engineering goals and performance requirements that are ethical, prudent, practical, and in accordance with international and national norms and expectations.

- Compile a design and prepare the performance assessment to demonstrate how the encapsulation facility will behave both in the short- and very long-term (i.e., past 1,000 years).

- Demonstrate (by discussion and regulatory analysis) that the design of the encapsulation system and its reasonably anticipated performance will meet or exceed the spirit and intent of reasonably applicable, appropriate, and relevant laws and regulations.

SECTION II

ENCAPSULATION SYSTEM DESIGN

Chapter 4

DISPOSAL CELL COMPONENT GEOMETRY

GENERAL

The prime determinants of cell geometry are the size and layout of the proposed site and the volume of the wastes to be encapsulated. Generally these are fixed by factors that fall outside of the scope of this book.

A waste disposal facility consists of a disposal cell and the man-made features and natural topography that surrounds and affects the cell performance. The disposal cell consists of the wastes and the various features that encapsulate the wastes, such as the foundation, the liners, the cover, and the perimeter dikes, if used (see Figure 4.1).

This chapter discusses factors that may be varied to optimize the cell geometry and performance. In particular, we concentrate on the layout of the cell, the details of the foundation excavation, and alternative perimeter encapsulation systems.

The space within which waste is contained is comprised of volume above the natural ground topography and volume beneath the natural ground topography. This is often referred to as the "below-grade" part of the cell. The space occupied by the waste may further be subdivided into that volume beneath the topslope cover, and the volume beneath the sideslope cover. This latter part of the cell is often referred to as the cell embankment. In optimizing the layout of a disposal cell, and enhancing the performance of the encapsulating system, one usually concentrates on optimizing the cover geometry and design. This is because the cover plays so critical a role in the performance of the disposal cell. If a good

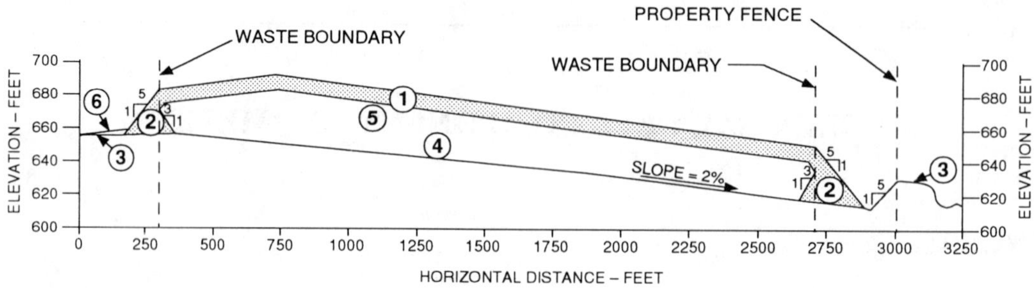

LEGEND

(1) TOPSLOPE COVER

(2) CLEAN-FILL DIKE

(3) NATURAL GROUND LEVEL

(4) EXCAVATION ZONE

(5) ENCAPSULATED WASTE

(6) FINISH GRADE OUTSIDE OF CELL

The cell section is based on the design of a facility for mixed waste.

FIGURE 4.1

SECTION OF A POSSIBLE CLEAN-FILL DIKE, PARTIALLY BELOW-GRADE CELL

cover layout is achieved, satisfactory disposal cell performance is achievable. For this reason a considerable part of this book discusses disposal cell covers, and the design and performance of covers are not considered further in this chapter. Only factors affecting cover geometry as controlled by the cell layout are discussed here.

CELL EXCAVATION GEOMETRY AND COVER DESIGN

Figure 4.2 shows that the disposal cell geometry may result in complete above-grade placement, below-grade placement, or partial below-grade placement.

At the most extreme, if the whole cell volume were below-grade, the cell cover would be coincident with or deviate little from the pre-existing topography. The cover would, in a sense, be no more than top cover and a simple restoration of the site. Figure 4.3 shows a layout of a disposal cell for which both at-grade and partial below-grade placement were considered. Figure 4.4 shows how the height of the cell shown in Figure 4.1 decreased as the depth of excavation increased. The main benefits of a lower cell height include:

- A greater part of the cover may be topslope as compared to sideslope; as will become apparent through this book, topslope performance is generally superior to sideslope performance.

- Lower height sideslopes enhance embankment slope stability, and reduce the erosion potential of upper component materials. Figure 4.5 shows the effect of cell height on the factor of safety of the cell shown in Figure 4.1 for increasing excavation depths.

- Excavation may produce materials that could be used in the construction of the cover. Less construction material reduces both the total effort and the environmental impact of construction. The more material that has to be

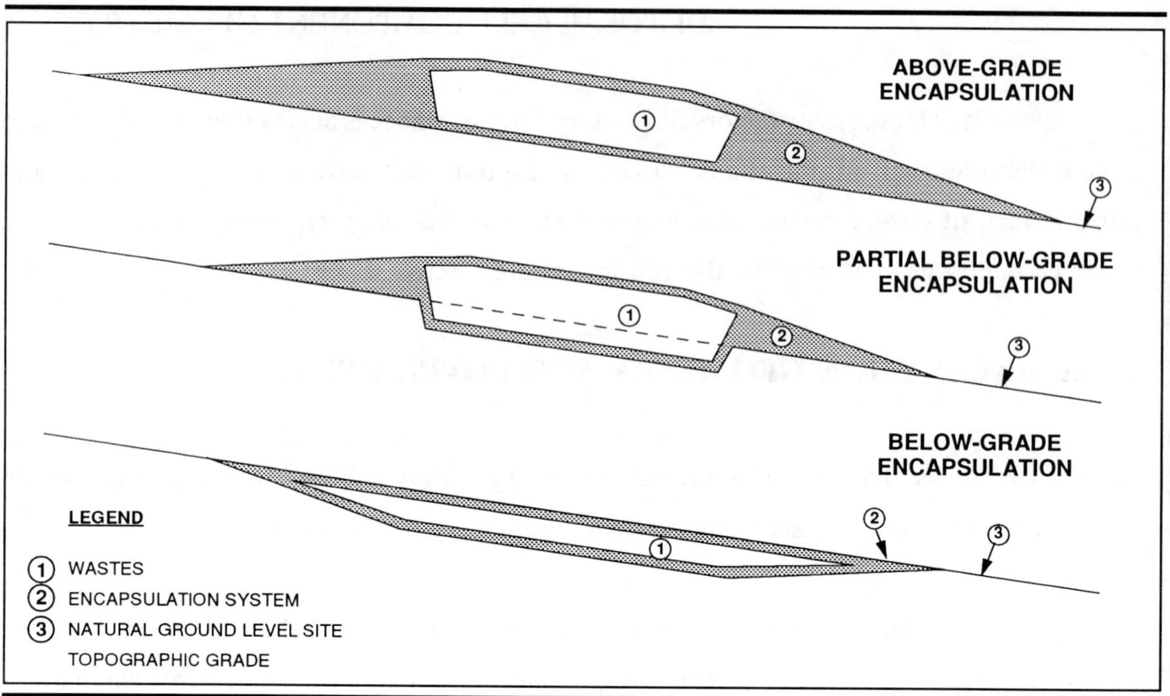

Waste may be encapsulated above, partially, or totally below natural ground level.

FIGURE 4.2
WASTE GEOMETRY

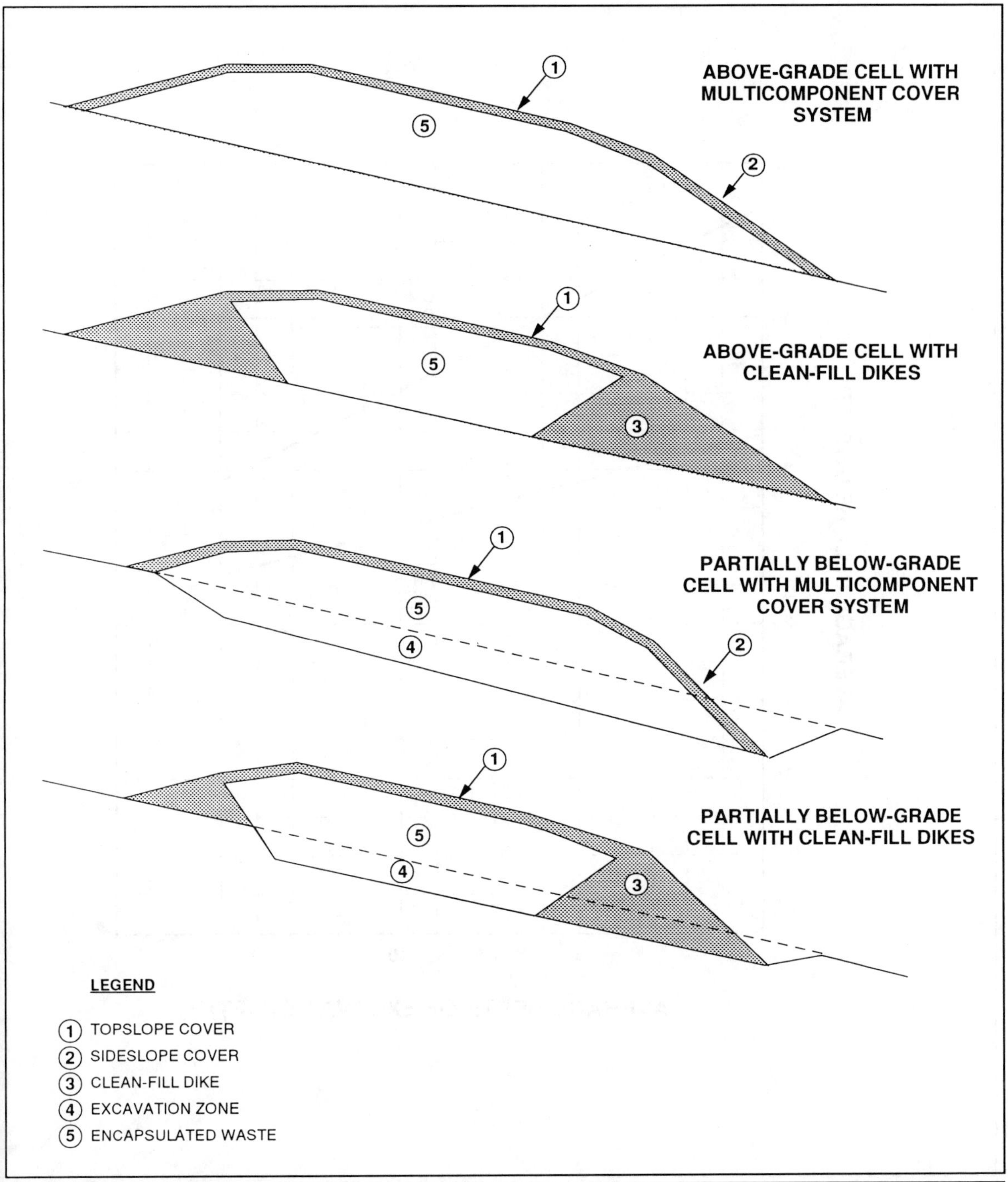

These are the four basic cell types that may be defined and evaluated when designing an encapsulation system.

FIGURE 4.3

BASIC CELL TYPES

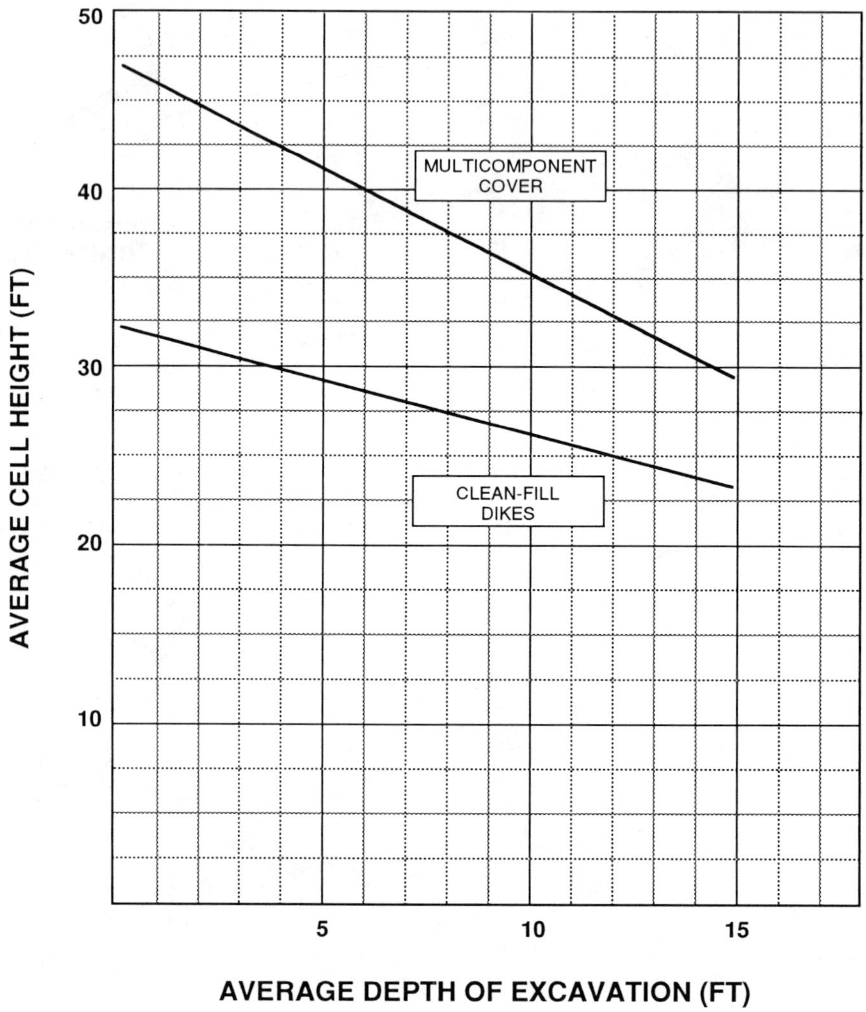

As the depth of the excavation increases, the cell height decreases.

FIGURE 4.4

CELL HEIGHT VS EXCAVATION DEPTH

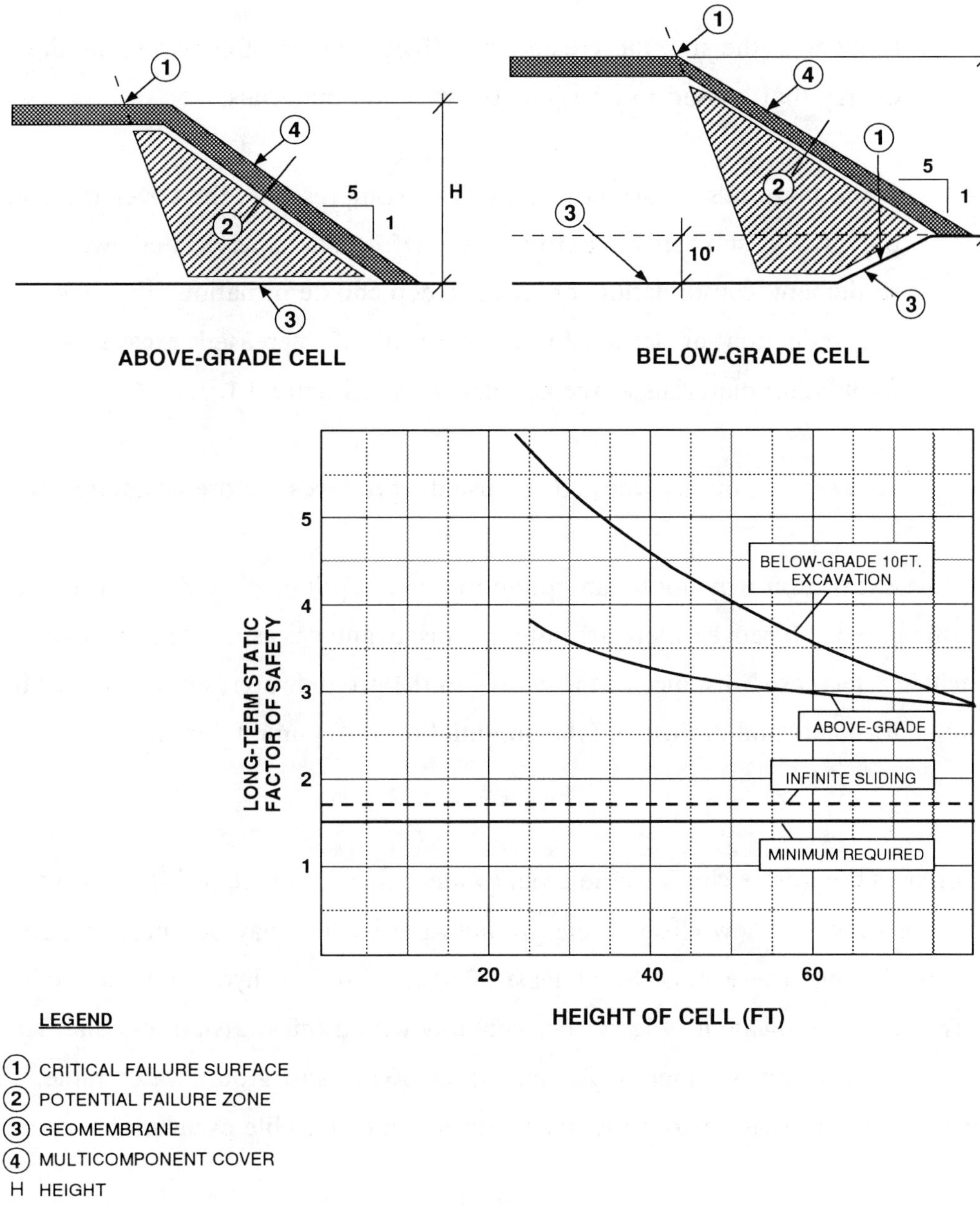

Reduced cell heights and below-grade liners increase the slope stability of the perimeter encapsulation system.

FIGURE 4.5

FACTOR OF SAFETY OF SIDE SLOPES

brought to the site, the greater the off-site impact of construction, the more energy that is used to transport construction materials.

- If the materials excavated are weak or compressible, the cover is no longer potentially subject to detrimental deformation associated with subsoil settlement, consolidation, or lateral creep and deformation. Figure 4.6 shows the reduction in settlement as a result of increased excavation of the foundation materials at the cell depicted in Figure 4.1.

- A lower height of cover and cell usually produces a more aesthetic structure.

We note that excavation may not be an option for cover optimization if an existing facility is to be remediated. Moreover, there are potential disadvantages associated with excavation. These include: decreased distance of the wastes to the groundwater; more complex design and detailing for liners and drains; and the potential need to pump to extract leachate from the cell.

Excavation puts the wastes closer to the groundwater table. In some jurisdictions there are statutory constraints on how close to the groundwater wastes may be encapsulated. For example, in Missouri there must be at least 30 feet of soil of hydraulic conductivity of 10^{-7} cm/sec. (or an equivalent of lesser permeability with a thickness no less than 20 feet) between the wastes and the highest feasible level of the site groundwater table. Such restrictions, while prudent, restrict the maximum depth of feasible excavation.

A further factor to consider with partial below-grade placement is the design of the leachate control system if one is used in the cell. Provision should be made for daylighting of the outlet pipes. Figure 4.7 shows how this may be done. The excavation is extended beyond the edge of the encapsulation system, and a large sump or swale that grades to drain may be formed. The leachate collection pipes go through the perimeter encapsulation system

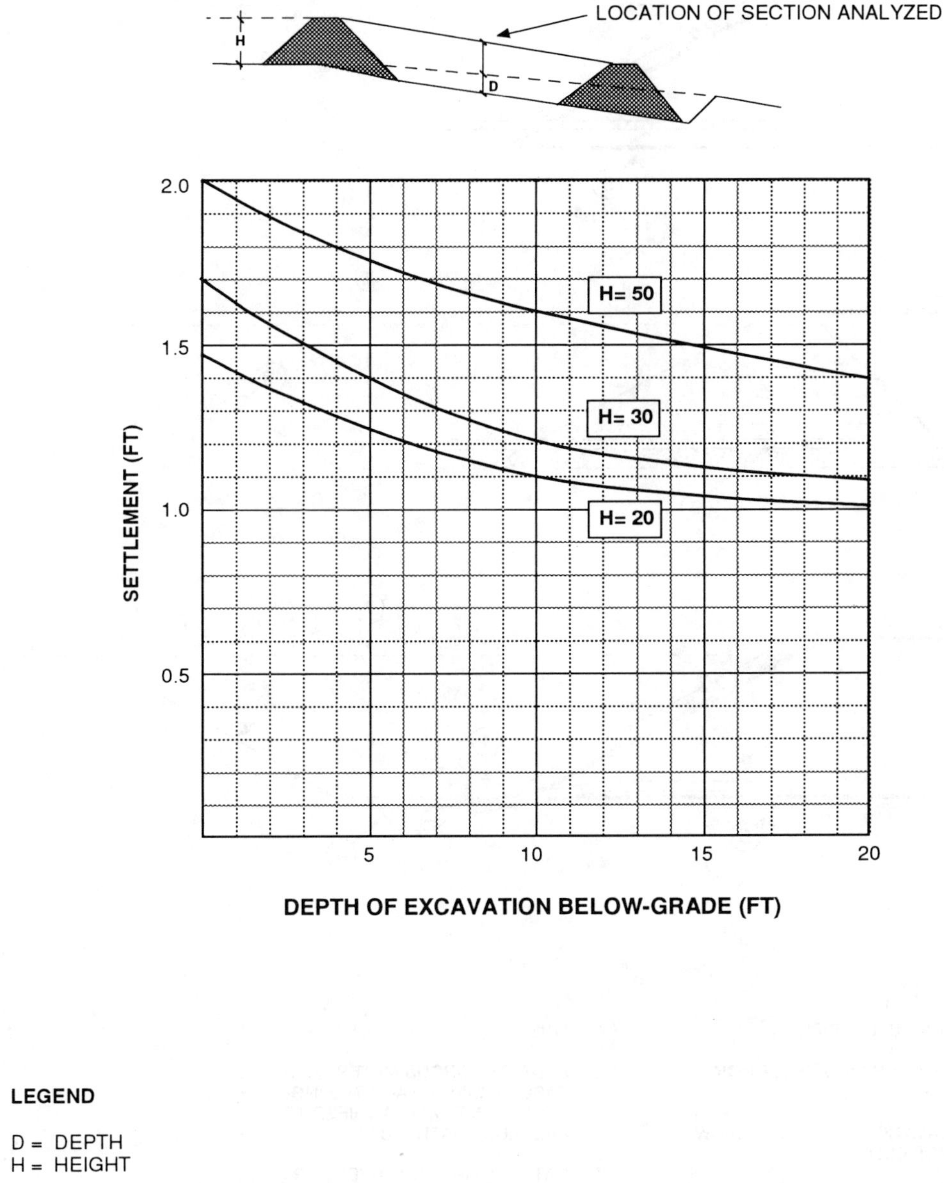

The greater the depth of excavation, the less the cover settlement.

FIGURE 4.6

SETTLEMENT VS EXCAVATION DEPTH

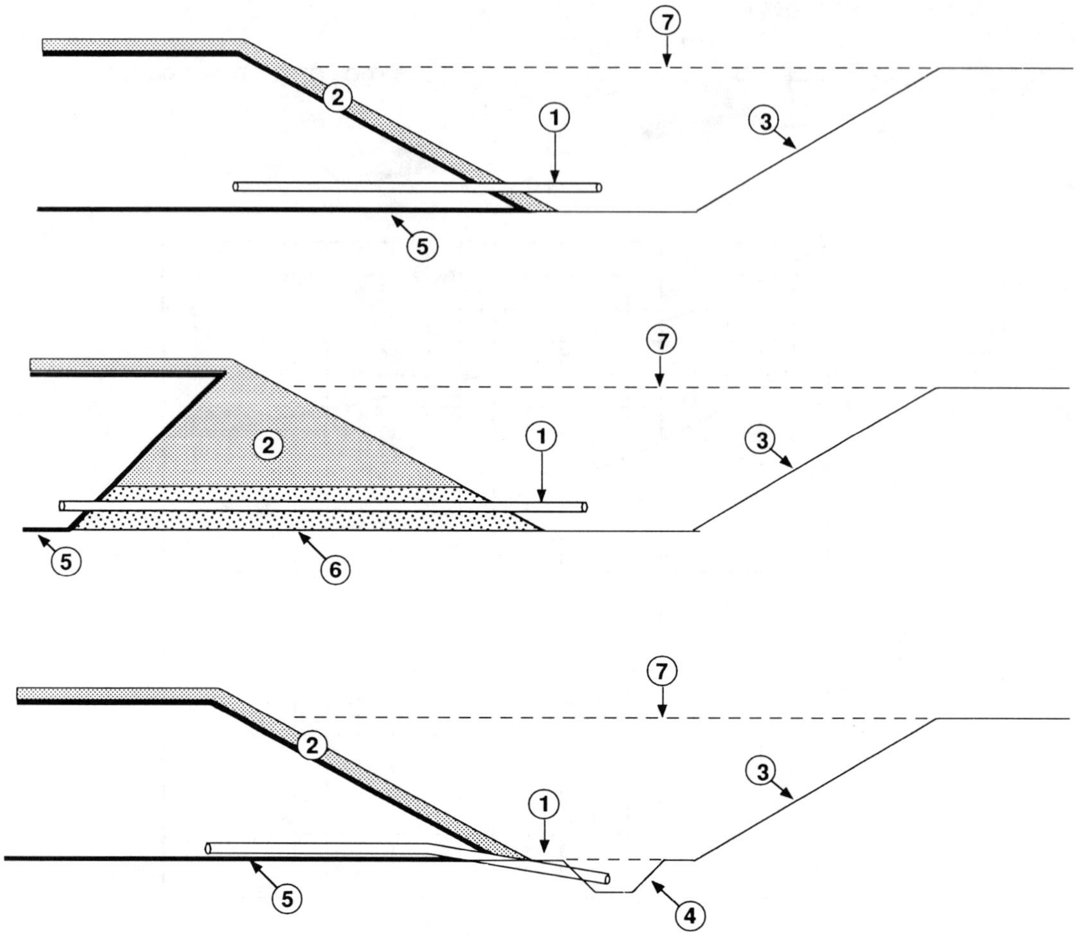

LEGEND

1. DRAIN OUTLET PIPE
2. PERIMETER ENCAPSULATION SYSTEM
3. EXCAVATION FACE FOR BELOW-GRADE CELL
4. COLLECTION SUMP OR TRENCH
5. LINER
6. ZONE OF POROUS MATERIAL AT BASE OF DIKE TO ACT IN LONG-TERM AS NATURAL AQUIFER TO PRECLUDE BATHTUB EFFECT
7. NATURAL GROUND LEVEL FOR BELOW-GRADE CELL

Different ways to daylight the leachate collection system outlet pipes above the topography.

FIGURE 4.7
PARTIAL BELOW-GRADE DRAIN OUTLETS

into the extended excavation where they daylight. Here the leachate may readily be observed, collected, and treated (if necessary) before discharge.

Figure 4.8 shows the results of a Value Engineering Workshop in which we undertook to compare above- and partial below-grade waste placement. Value Engineering is a formal procedure for identifying and evaluating alternative approaches, procedures, and designs. The main steps in a Value Engineering session, at least as is relevant to the results discussed here are: identify alternatives to be compared; define comparison criteria; assign weightings to the comparison criteria; compare alternatives by assigning scores based on the extent to which they achieve defined criteria; rank alternatives on the basis of the sum of the weighted scores; list criteria advantages and disadvantages; and finally provide decision-makers with recommendations for action. The results of the Value Engineering session confirmed the advantages listed above for partial below-grade waste placement.

COVER GEOMETRIC COMPONENTS

Once the above-grade area and volume of the cell are fixed, the designer may vary, adopt, or avoid one or more of the following to optimize the performance of the disposal cell, its topslope cover, and the perimeter encapsulation system:

- Sideslope inclination
- Height
- Slope length
- Concave surfaces
- Sharp divergent corners
- Converging planes
- Cell area.

ALTERNATIVES TO BE EVALUATED—WASTE PERIMETER ENCAPSULATION SYSTEMS

ABOVE-GRADE ENCAPSULATION

PARTIAL BELOW-GRADE ENCAPSULATION

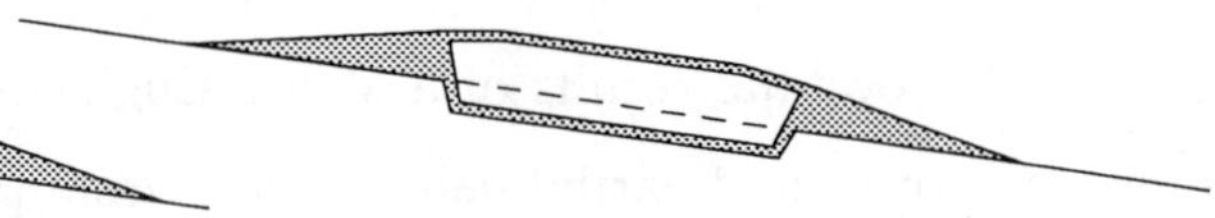

EVALUATION CRITERIA & WEIGHTINGS

SIMPLIFY CONSTRUCTION	**6**
MINIMIZE IMPORT OF CONSTRUCTION MATERIAL	**5**
LEACHATE CONTROL	**22**
REDUCE SETTLEMENT	**16**
REDUCE FOOTPRINT	**14**
ENHANCE CELL PERFORMANCE	**37**

ALTERNATIVES COMPARISON

ALTERNATIVE	SCORE (% OF MAX)
ABOVE-GRADE	**57**
PARTIALLY BELOW-GRADE	**76**

CONCLUSIONS

- **PARTIAL BELOW-GRADE PREFERRED:**
  - Enhances cell performance (Erosion, stability, settlement)
  - Reduces cell height (Visual impact)
  - Minimizes material import
- **ABOVE-GRADE ADVANTAGES:**
  - Leachate control
  - Construction

Value engineering used to compare alternative waste disposal cell options.

FIGURE 4.8
VALUE ENGINEERING SUMMARY AND RESULTS

Many of these are interrelated; there are probably an infinite number of combinations for a single site and a specified volume. For the purposes of clarity and simplicity of discussion, we discuss each separately below.

Area. Increasing the cover area reduces the height of the cell. If the increase of area does not lead to an unacceptable increase of cover materials (hence cost) and infiltration (which as a first approximation is proportional to cover area), then the enhanced cover performance associated with lesser heights is desirable and should be adopted (see discussion on height below).

Layout. In theory, a circular cell on a flat plane involves the most effective use of cover materials; i.e., the greatest volume of enclosed waste for the least volume of cover material. Layout factors that significantly affect the efficiency (vis-a-vis material use) of covers are the use of sharp divergent corners and too small a cell height relative to cover thickness. With respect to the first, place your hands together as though for prayer. The enclosed volume is essentially zero. Slowly open your hands by pivoting them about an imaginary hinge formed by your pointing finger. When your hands are at 90 degrees to each other, they are as the conventional geometry of two sides of a cell with a square or rectangular base. You will readily note that the volume thus enclosed by your hands is reasonable. Conversely, if you close your hands once again, you will notice that the less the angle of separation, the less efficient the hands-cover is. For this reason, the prudent designer avoids sharp corners. The designer cuts them off even if it means increasing the number of corners and decreasing the cell footprint area.

With respect to the inefficiency of too small a cell height (particularly in the absence of below-grade excavation), the general rule is that the depth of covered waste should exceed twice the cover thickness. Lesser waste depths involve inordinate cover cost per volume of cell, and it is better to reconfigure the cell layout to avoid such poor cost-geometry. For specific sites, wastes, and cover designs, comparative cost evaluations may readily be compiled to establish the relationship between the cost of the cover (and the unit volume thus created) versus the cost of decreasing the cell footprint, increasing below-grade excavation, relocating in-place wastes, or whatever other parameter is relevant.

Sideslope Inclination. General practice is to construct disposal cell sideslope covers at an inclination of 20 percent or 5 horizontal to 1 vertical. This practice results from numerous site-specific studies which establish that such an inclination is a reasonable compromise between the following often competing factors:

- Volume enclosed
- Enhanced run-off
- Decreased slope length
- Slope stability
- Erosion control
- Practical construction.

The first three are best achieved by steeper slopes; the latter three are best achieved by flatter slopes. Slopes steeper than about 2.5 horizontal to 1 vertical should be avoided in the interest of long-term performance. While such steep slopes may be efficient in increasing cell volume, covers are difficult to construct, and we believe they have a low probability, in the longer term, of avoiding erosion and mass movement due either to soil creep or static and dynamic instability.

Slopes of 7 horizontal to 1 vertical are so flat that it is difficult to create a meaningful cell volume. I once designed and constructed a cell with slopes of 6 to 1. The cell footprint is about 1 kilometer by 1 kilometer. At an ultimate height of 15 m, the volume per meter of additional height is significantly reduced by the rapidly dwindling top area. This geometry is to be used only when, as was my case, the foundation soils are weak and slope stability concerns forced so flat an entity.

Site-specific comparisons may be undertaken to compare the effect of varying sideslope inclinations on critical design and performance parameters such as stability, volume, erosion resistance, and height (aesthetics).

Topslope Inclination. There is considerable debate, and certainly no agreement as to the proper inclination for the cover topslope. I have read comments on my design that the cover should be inclined at less than 1 percent. In reply, the practical contractor simply laughs and points to the real difficulties of placing soil and rock over great lengths at such slopes. To my mind, the ultimate argument against such flat slopes is the simple fact that even minor amounts of cell content settlement or deformation destroy and negate such a flat slope; the result is a reverse slope or depression that ponds water and increases infiltration.

Thus we have the first rule for establishing the inclination of the topslope cover. It should be sufficiently steep that deformation of cell contents will not affect or negate cover performance, such as run-off or control of infiltration.

If the cover is to be designed to resist erosion by run-off from the probable maximum precipitation, and the cover must support vegetation as a design element, the topslope cover inclination should be as flat as possible. In practice this is usually between 2 and 4 percent. Such an inclination, while difficult to control during construction, provides for some cell content deformation, sheds run-off without excessive erosion potential, and may consist of materials that balance the competing demands of supporting vegetation and erosion control.

Slope Length. The major determinants of the slope length are the area and height of the disposal cell. A greater slope length is undesirable as this increases the run-off and hence the erosion potential.

Concave Surfaces. These should be avoided if at all possible. The major reason for this is that where two plains come together, a channel or concentrated flow area is created. This is a zone of increased erosion, hence breach potential. The general rule should be: arrange cover planes to produce sheet flow run-off for optimum performance.

Case History 1

Optimization of cover geometry for a given disposal cell was the topic of a Value Engineering or Alternatives Assessments Study I attended. We sought to establish the best geometry for a disposal cell to contain treated mixed wastes. Table 4.1 lists the potential cell and cover geometric alternatives. In order to define the optimum cell layout, we defined a base case cell; a square footprint on level ground at the south end of the site and no below-grade excavation. Next we defined, as listed in Table 4.2, a series of changes that could be made to the base case cell. To evaluate which changes were preferable to optimize the cell, and in effect define the preferred and actual cell layout, we established and defined the evaluation criteria given in Table 4.3. Preference rating as shown in Table 4.4 produced a weighting for each evaluation criterion. This ruling was carried out by a group of six experts who discussed each point until we reached consensus. Finally, by examining each cover and cell geometric change for each evaluation criterion as shown in Table 4.5, we ranked actions in order of preference.

This shows that, at least for this specific case, cell and cover performance are best enhanced by increasing the depth of below-grade excavation. Next, it is preferable to increase cell area (for a given volume), hence decrease height, as a way to improve overall cell performance. The results on Table 4.5 show that changing slope inclination and length have relatively small effect on improving cell performance.

TABLE 4.1

POTENTIAL CELL AND COVER GEOMETRIC ALTERNATIVES

	ALTERNATIVES
1.	Allow concave surfaces
2.	Allow sharp corners
3.	Vary depth of excavation
4.	Allow converging planes
5.	Minimize corners
6.	Vary slope
7.	Vary height
8.	Vary width
9.	Vary length
10.	Vary area
11.	Base plane orientation relative to topography
12.	Topslope plane orientation relative to topography
13.	Clean-fill dikes
14.	Internal material geometry
15.	Stages of construction.

TABLE 4.2

POSSIBLE CHANGES TO THE BASE CASE CELL

	POSSIBLE CHANGES
1.	Move location north - from southernmost end of siting study area
2.	Add corners - from base case of four
3.	Increase slope - from base case of 5H:1V
4.	Increase height - grade brake elevation from 5 feet high
5.	Increase area - from base case
6.	Increase depth - excavate below existing ground
7.	Allow converging planes
8.	Allow base plane to deviate from topography - currently follows topography
9.	Allow grade brake to deviate from horizontal plane - change from sideslope to topslope to occur on other than horizontal plane
10.	Expand siting area (optional).

TABLE 4.3

EVALUATION CRITERIA

	CRITERIA
A.	Volume Flexibility - flexibility and opportunity for changing design and construction to accommodate changes in volumes, forms and characteristics of waste materials
B.	Aesthetics - public perception, overall height and appearance
C.	Land used - minimize cell area
D.	Simplicity of Design and Construction - minimize design and construction complexity
E.	Minimize construction materials - balance cut and fill and minimize construction material quantities
F.	Leachate Collection - facilitates leachate collection
G.	Settlement - facilitates minimizing settlement
H.	Slope stability - facilitates meeting slope stability requirements
I.	Infiltration/Flux - facilitates minimizing infiltration for groundwater protection purposes.

TABLE 4.4

CRITERIA RATING

FUNCTION BEING ANALYZED: GEOMETRY CRITERIA

CRITERIA	RAW SCORE	RANKING OF CRITERIA
A. Volume Flexibility	14.0	4
B. Aesthetics	5.0	7
C. Land Use	0.0	9
D. Simple Design & Construction	8.0	6
E. Minimum Material	8.5	5
F. Easy Leachate Collection	2.6	8
G. Reduce Settlement	22.7	2
H. Enhance Slope Stability	23.0	1
I. Reduce Infiltration	14.1	3

PREFERENCE RATING TABLE

A	A 3.5	A 3.3	A 2.2	A 3.0	A 2.0	G 2.1	H 2.6	I 3.0
	B	B 2.7	D 1.7	E 2.0	B 2.1	G 3.6	H 3.4	I 1.8
		C	D 3.0	E 3.1	F 2.6	G 3.7	H 3.7	I 2.6
			D	E1/D1	D 2.2	G 2.9	H 3.0	I 1.8
				E	E 2.4	G 3.4	H 3.7	I 2.2
					F	G 3.1	H 3.6	I 2.7
						G	G 1.0	G 2.9
							H	H 3.0

IMPORTANCE RATING SCALE

4-Major preference
3-Medium preference
2-Minor preference
1-No preference, score one point for each letter

NOTE:
1) DROP CRITERIA WITH A RAW SCORE OF 2 OR LESS
2) IN PREFERENCE RATING TABLE, LETTER DENOTES PREFERENCE CRITERION; NUMBER IS IMPORTANCE RATING
3) RAW SCORE IS THE SUM OF CRITERIA IMPORTANCE RATINGS FROM PREFERENCE RATING TABLE

TABLE 4.5

CELL OPTIMIZATION ACTION RATING

ALTERNATIVES Criteria	Rel. Wt.	Move North 1	Add Corners 2	Increase Slope 3	Increase Height 4	Increase Area 5	Increase Depth 6	Allow Converging Planes 7	Allow Base to Deviate from Topo 8	Allow Grade Brake to Incline from Horizontal 9	Increase Site Area 10
A. Volume Flexibility	14.0	1/14	3/42	2/28	2/28	5/70	4/56	1/14	3/42	2/28	5/70
B. Aesthetics	5.0	1/5	2/0	1/5	1/5	1/5	3/15	2/10	3/15	1/5	1/5
D. Land Use	8.0	1/8	1/8	2/16	1/8	4/32	2/16	1/8	4/32	1/8	4/32
E. Minimum Material	8.5	2/17	1/8.5	2/17	2/17	1/8.5	4/34	1/8.5	4/34	1/8.5	1/8.5
G. Reduce Settlement	22.7	1/23	1/23	1/23	1/23	4/91	4/91	1/23	4/91	2/45	2/45
H. Enhance Slope Stability	23.0	1/23	1/23	1/23	1/23	4/92	5/115	1/23	3/69	2/46	2/46
I. Reduce Infiltration	14.1	1/14	1/14	4/56	3/42	1/14	1/14	1/14	2/28	2/28	2/28
Total		104	129	168	146	313	341	101	311	169	235
Rank		9	8	6	7	2	1	10	3	5	4

1) NOTES RATING Excellent - 5 Very Good - 4 Good - 3 Fair - 2 Poor - 1

2) 1/14 - THE 1 DENOTES THE RATING. THE 14 IS THE PRODUCT OF THE RATING AND THE RELATIVE WEIGHT

Chapter 5

COVERS

GENERAL

A cover is the most significant part of a waste encapsulation system. The cover may be either a topslope or a sideslope cover. Because of the significantly different inclinations of the two, different designs for each on the same cell may be appropriate. Many alternative cover designs have been proposed and used. Many more cover designs are feasible and could be used at selected sites to meet specific needs.

This chapter first describes some alternative covers that have been used in the past or are currently on the drawing board. Next, we proceed to establish a rational procedure for formulating a specific cover for a specific site. Finally, the chapter addresses the issue of comparing alternative covers and ranking them when more than one cover variant is feasible at a site.

PRECEDENCE

Numerous alternative covers for a great variety of waste disposal facilities have been designed and constructed. Each represents a unique design response to a unique regulation or site. Figures 5.1 to 5.5 show five alternative covers that have been proposed and used in the past, particularly for disposal cells for radioactive wastes. The following is a brief description of each.

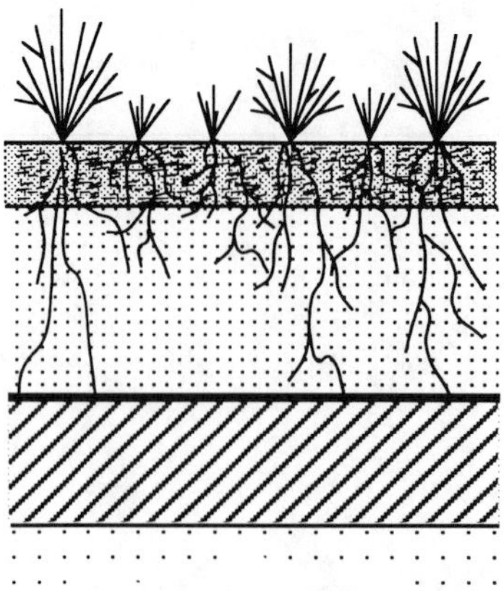

	VEGETATION
1.0'	**SOIL***
3.0'	**FROST PROTECTION AND ROOTING ZONE: Random Soil (Clay, Silt, and Sand)**
	GEOMEMBRANE
1.5 - 2.0'	**INFILTRATION AND RADON BARRIER: Silt and Clay**

***NOTE: ALTERNATIVE TOP COMPONENTS**

*0.5' EROSION RESISTANT ROCK
*1.0' EROSION ROCK WITH SOIL IN VOIDS
*1.0' GRAVEL MULCH (SOIL/ROCK MIX)

This cover is cheap to construct and does isolate the wastes from human contact. The geomembrane may be omitted as in the long term it is not effective.

FIGURE 5.1

ALTERNATIVE COVER 1: THE SIMPLE VEGETATED/ROCK COVER

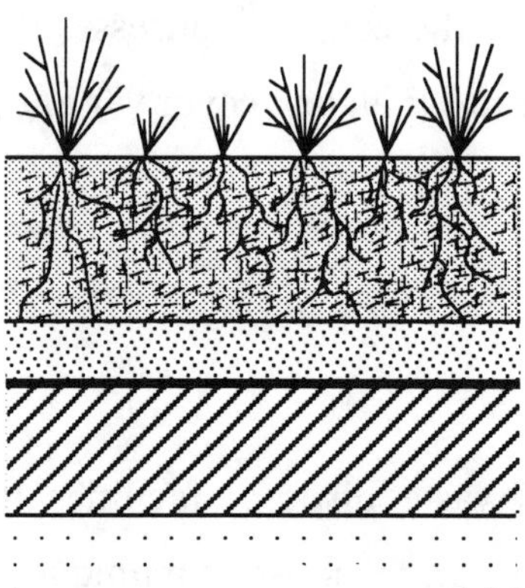

	VEGETATION
2.5'	**SOIL: Random or selected***
1.0'	**FILTER/DRAIN/BEDDING: Sand** **INFILTRATION BARRIER: Geomembrane**
2.0'	**RADON/INFILTRATION BARRIER: Silt & clay**
	WASTE

***THESE MAY BE COVERED WITH ROCK MULCH.
ALTERNATIVELY, THE UPPER 0.5' MAY BE COARSER GRAINED GRAVEL.**

This cover complies with the minimum requirements for a cover for a hazardous waste disposal facility.

FIGURE 5.2

ALTERNATIVE COVER 2: THE RCRA COVER

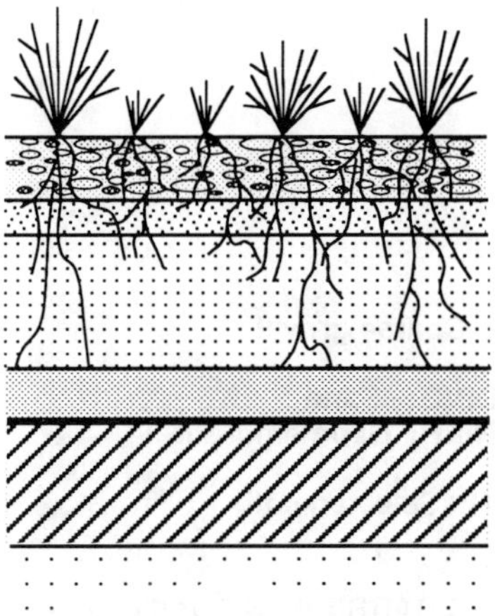

This cover may be used on the sideslopes of a disposal cell where significant resistance is required.

FIGURE 5.3

ALTERNATIVE COVER 3: THE DOUBLE DRAIN COVER

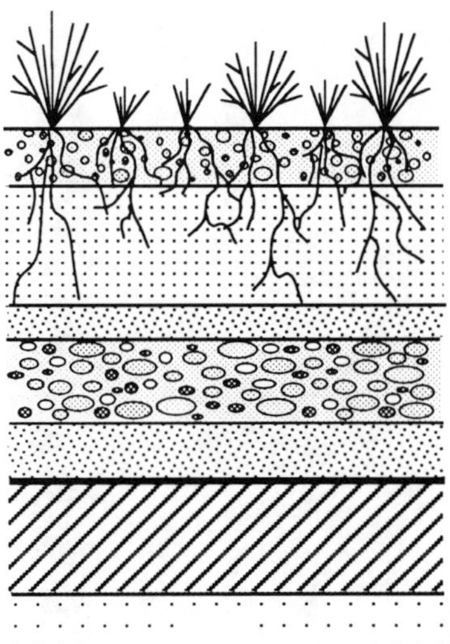

This complex cover provides significant resistance to animal and plant intrusion; it reduces erosion and infiltration; it is, however, expensive.

FIGURE 5.4

ALTERNATIVE COVER 4: THE MULTICOMPONENT COVER

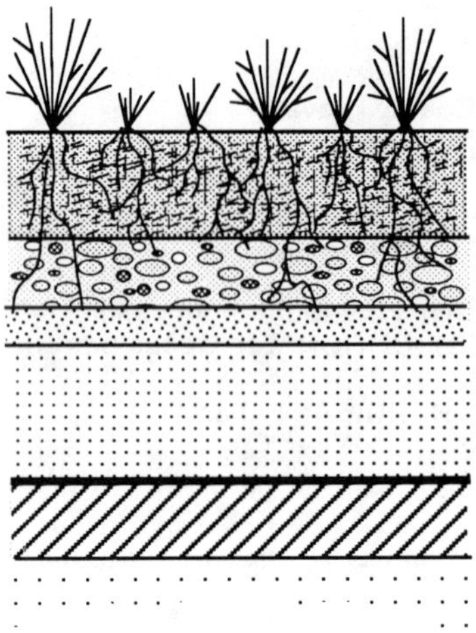

VEGETATION

0.5' - 1.5 SOIL

1.0' EROSION BARRIER: Coarse gravel

0.5' FILTER/DRAIN/BEDDING: Sand
INFILTRATION BARRIER: Geomembrane

2.0' FROST PROTECTION: Selected soil

0.5 - 1.0' RADON/INFILTRATION BARRIER: Silt and clay

WASTE

This cover provides the best opportunity for a lush vegetation growth. It also provides erosion protection against gullies that may develop in the soil.

FIGURE 5.5

ALTERNATIVE COVER 5.
THE EROSION RESISTER

Cover 1: The Simple Cover

Figure 5.1 shows this cover. This combination of cover components, or essentially similar components, has been used on many landfills. As its name implies, it is the simplest of the five covers evaluated; it provides a lower cost, least function, least complex, and shortest design life cover. The lowermost layer, directly above the encapsulated materials, is the infiltration and radon barrier. For this and other covers, a minimum thickness of 450 mm is adopted. This is the least practical thickness for reliable functioning: placed in three, 150 mm lifts, it provides reasonable assurance that there is an insignificant probability of overlapping deficient zones. Overlaying the radon/infiltration barrier is a geomembrane or a geomat which may have to be included because of EPA regulations or to function as a low permeability infiltration barrier. The frost layer is placed three feet thick. This, plus the overlying one foot of vegetation-supporting soil, provides an adequate frost protection cover over the radon infiltration barrier.

The upper foot is a soil intended to support vegetation. This is a nominal thickness for reasonable vegetation, but roots (even of shallow-rooted species) may penetrate to the frost protection layer. Alternative erosion protection systems are noted.

Cover 2: The RCRA Cover

Figure 5.2 shows the components and their arrangement for this alternative. It is called the RCRA cover because it complies with the requirements established by the EPA for the cover of a disposal cell containing RCRA wastes. If the RCRA regulations are applicable or relevant and appropriate for a given site, this cover may be considered the minimum regulatory compliance cover.

From the encapsulated materials up, the first component is the radon/infiltration barrier. This is covered by a geomembrane, as is called for in the EPA guidance documents.

Over the geomembrane is a drain bedding layer of sand. This layer serves first to cover and protect the geomembrane from puncturing or tearing by the sharper particles in the overlying soil. Second, the sand, which must have a hydraulic conductivity greater than 10^{-2} cm/sec., acts as a drain, shedding water that percolates through the overlying material. Finally, the sand may act, in appropriate circumstances, as a capillary break to preclude water from the overlying soil percolating down into the drain. The capillary break will generally deter root advancement because roots do not readily grow from relatively moist media into relatively dry media.

Above the drain is a 600 mm thick layer of soil for vegetation. If required to augment the erosion resistance of the uppermost surface, rock mulch, gravel, or coarse-graveled soils may be added to the top of this layer.

The EPA regulations also call for frost protection, but do not specify a frost protection layer. The soil layer for vegetation can be thickened to provide frost protection and provide improved conditions for plant growth by providing greater nutrient and water retention capacity.

Cover 3: The Double Drain Cover

Figure 5.3 shows the details of this cover. Its name derives from the presence of two drains. The first, or lowermost layer, is the radon barrier. Over this, as for the RCRA cover, are a geomembrane, a drain, and random soil. The geomembrane is optional and may be omitted if the hydraulic conductivity of the radon barrier is low enough for it to double as an infiltration barrier.

Thus far, this cover looks very much like the RCRA cover. The significant difference is in the upper two layers and the effect they have on the functioning of the cover. The first of the upper two components is the second drain. This also functions as a filter to preclude erosion of the underlying soil by lateral flow through the drain. As a bedding layer, this layer protects the underlying soil from being punctured by rock particles and, as discussed later, augments the erosion resistance of the overlying rock.

The uppermost layer of durable rock is sized to resist erosion by flow from the design precipitation. This layer provides maximum achievable erosion protection. The drawback or potential downside consequence is that the rock reduces evaporation from the cover, therefore potentially increasing infiltration. Also, a layer of rocks favors the invasion of deep-rooted, woody species that will have to be removed perpetually to avoid root damage to the underlying layers. To counter this, the voids may be filled with soil, but the resultant vegetation is likely to be sparse and/or require maintenance (mowing).

One final advantage of the uppermost rock layer is its potential to deter burrowing animals and inadvertent human intrusion. Conversely, a good quality rock may be an economic resource that future generations might choose to remove and use elsewhere, thus rendering the cover vulnerable to erosion.

This particular cover has been extensively used in the sideslopes of UMTRA Project disposal cells. On that project, it has been selected as an economical way to achieve maximum sideslope erosion resistance, frost protection, slope stability, and infiltration and radon flux control. At some sites the growth of volunteer plants is such that frequent maintenance (at least annually) will be necessary (see Chapter 18).

Cover 4: The Multicomponent Cover

Figure 5.4 shows the configuration of the multicomponent cover. As for the two previous covers, from the waste up, the first three components are the radon barrier, the infiltration barrier, and a drain. At this point, a new component is encountered: the biointrusion barrier, a layer of large gravel and cobbles whose particles are sized to impede burrowing animals. The thickness is selected to deter root penetration and possibly discourage inadvertent human intrusion. Being covered, it is unlikely to be removed for alternate use.

Over the biointrusion barrier is a filter to prevent migration of fine particles of soil from the overlying soil into the voids of the biointrusion barrier. An alternative to this filter is to "choke" the uppermost surface of the biointrusion barrier. This is done by running a heavy vibratory compactor over the rock to crack the uppermost particles and produce fines to fill, block, or choke the layer voids. In effect, the filter is created on-site by in-place processing or crushing. Over the filter, or choked zone, is a frost protection layer, which may also double as a root or growth zone, especially if the proper materials are selected to favor plant growth.

The uppermost layer is a gravel mulch or soil covered with a thin veneer or rock, the gradation of which is selected to enhance vegetation and simultaneously provide adequate erosion resistance. As noted in Figure 5.4, alternative components may be used to meet other site-specific requirements and constraints.

This cover is discussed for these reasons:

- It provides an upper bound of maximum functions, complexity, minimum infiltration, and a rational compromise between vegetation and erosion control.

- It has been successfully used on the Durango, Colorado, UMTRA Project disposal cell.

Cover 5: The Erosion Resister

Figure 5.5 shows the details of the erosion resister cover. This cover from the waste up incorporates the following components:

The radon barrier, overlain by the frost protection layer, offers the advantage that a thin radon barrier may be used. In practice, it may be feasible to place only the least practical constructible layer of 150 mm. Any inadvertent differences would be compensated for by the significant additional flux attenuation afforded by the thick frost protection soils.

The infiltration barrier, a geomembrane, is placed within the zone of freezing and thawing. It is protected by the overlying drain and bedding layer.

A layer of coarse gravel or cobbles is placed over the drain. The first function of this layer is biointrusion -- to impede burrowing animals, prevent inadvertent human intrusion, and possibly hinder the downward growth of roots.

A second and vital function of this layer is to prevent erosion of the cover in the event of erosion of a gully formation in the overlying vegetation supporting zone.

This uppermost layer, for optimum vegetation growth, is soil. In that this uppermost layer is not intended to be the primary erosion barrier, it need not be designed to be invulnerable to erosion in the event of the design precipitation. The upper surface of the erosion barrier gravel must be choked to prevent piping of the overlying soil.

In that the erosion barrier rock has to be sized to resist erosion by flow in a gully in the overlying soil, the thickness of the soil should be kept to a minimum. The thicker the soil, the deeper the gully, the larger the rock required for stability.

A reasonable alternative to pure soil is a layer of gravelly sandy silt and clay with adequate fines to support a reasonable stand of vegetation. The gravel percentage and size may be selected to provide erosion resistance against annual or somewhat greater (up to a 100-year recurrence interval) precipitation. This minimizes erosion potential and minimizes maintenance, yet provides an aesthetic perspective. In the event of precipitation that exceeds these lesser events, this upper layer may erode, but the cover and the cell are protected from erosion by the erosion barrier rock of the underlying layer.

While we believe that these covers encompass the full range of potential alternatives, many variations may be possible. The key to varying cover design is recognition of the many components that may be used in a cover. For this reason, we define in the next section the checklist cover and proceed to use it to examine the role of each cover component, and hence to establish a rational procedure to define a new cover for a new site.

COMPONENT FUNCTIONS

General

The following is a description of the main components that are common to the covers just discussed. This description is intended to set the stage for the subsequent discussion on the design of covers for specific situations; in effect, cover selection.

The Radon Barrier

The radon barrier's function is to reduce the flux of radon gas from the encapsulated wastes to no more than 20 pCi/m^2s. This is usually achieved by placing a barrier of soil or other natural materials over the waste. The barrier functions to impede or slow the diffusion of the radon gas through the material. The reduced diffusion velocity and the relatively short half-life of the radon as compared to the travel time through the barrier reduces the radon surface flux to or below the mandated limit.

In order to perform satisfactorily for such a period, the material of the radon barrier must remain intact (uncracked and unpenetrated by roots or animals). Cracks or other holes through the radon barrier allow the gas to pass rapidly from the waste to the environment. Furthermore, the material of the radon barrier should remain relatively moist, for soil's radon attenuation capacity is significantly affected by soil moisture content due to the water occupying some of the pore spaces of the soil.

To achieve and ensure these performance requirements, the radon barrier must be protected from erosion, root penetration, inadvertent human and animal intrusion, desiccation, and freezing and thawing. This is done by placing additional layers of soil, rock, etc., over the radon barrier. In order to prevent cracking of the radon barrier by deformation of the wastes over which it is placed, the cell foundation and the wastes must be selected, or engineered, to preclude unacceptable total or differential settlement. The radon barrier material should also be chosen such that the cracking potential of the material itself is minimized.

Four alternative radon barrier designs are listed and evaluated in Table 5.1. The radon barrier with the most significant advantages is a layer of compacted clay/silt at least 450 mm thick. This approach is easy to design and construct. It is robust, well-established, proven, and accepted. Provided the appropriate materials are placed over the soil, it is protected from erosion, desiccation, freezing and thawing, and root, animal, and human intrusion.

The only disadvantage is the need to select an appropriate soil and to control its placement density and moisture content. Cracking due to waste or foundation deformation may be controlled by selection of a suitable material, cell geometry (shape and thickness), waste and radon barrier placement techniques, and a sufficient overburden thickness.

TABLE 5.1

ALTERNATIVE RADON BARRIERS EVALUATION

EVALUATION CRITERIA	ALTERNATIVE	ADVANTAGES/ DISADVANTAGES
• Technically Effective: Proven and demonstrated performance on other sites. Established by measurement to control radon.	• Compacted clay/silt	• Advantages - Conventional and proven - Simple to construct - Easy to design and analyze • Disadvantages - Largest material volumes - Strict quality control required
• Simple Construction: Easy to place and construct using standard construction procedures.	• Bentonite amended sand	• Advantages - Dual function as low permeability infiltration barrier • Disadvantages - Susceptible to drying and cracking
• Dual Function: The component may or does function also as an infiltration barrier, or biointrusion control.	• Bentonite mat	• Advantages - Simple to construct • Disadvantages - Unproven - Thin and nonrobust
	• Waste placement sequence	• Advantages - Economic - Robust - Long-term effective • Disadvantages - Complex construction - Infiltration barrier still required

Waste placement sequencing as a function of radium-236 concentration to reduce the required radon barrier thickness should be done if practical and in compliance with the principle of reducing radon flux or exposure to as low as reasonably achievable (ALARA). Even if the waste placement does provide a layering sequence which is calculated to require no radon barrier, or a radon barrier thinner than 450 mm, standard practice is to place at least 450 mm of soil. This is done as a matter of conservative radon control and because a layer of low permeability soil no thinner than 450 mm is generally required at any rate for infiltration control, and because of the practicality of constructing a uniform layer of soil.

Other materials such as asphalt have been considered for the radon barrier. Asphalt has, however, been universally rejected because it is difficult to establish that it will not creep, crack, or slide off the waste over 1,000 years.

The Infiltration Barrier

The infiltration barrier's function is to reduce the flux of water to the encapsulated wastes to no more than 10^{-7} cm/sec. This is usually achieved by placing a geomembrane or a layer of low permeability clay over the waste. The barrier functions to impede or slow the passage or flow of moisture through the material. A soil infiltration barrier will also function as a radon barrier and is frequently referred to as a radon/infiltration barrier.

Table 5.2 lists five alternative infiltration barrier designs. These range from a geomembrane to a compacted silt and clay layer to a combination of both. The infiltration barrier with the most significant advantages is a combination of compacted clay/silt of at least 450 mm thick, bentonite mat, and a geomembrane.

Bentonite-amended sand may be used as an infiltration barrier where sufficient quantities of clay/silt are not economically available. It is difficult to mix and place, although, if this

TABLE 5.2
ALTERNATIVE INFILTRATION BARRIER EVALUATION

<table>
<tr><th>EVALUATION CRITERIA</th><th>ALTERNATIVE</th><th>ADVANTAGES/ DISADVANTAGES</th></tr>
<tr><td rowspan="5">• Technically Effective:
Proven and demonstrated performance on other sites

• Simple Construction:
Easy to place and construct using standard construction procedures

• Dual Function:
The component may or does function also as a Radon Barrier.

• Longevity:
Will function throughout the 1,000 year design life of the disposal cell

• Redundancy:
Provides back-up protection in the event of a failure</td><td>• Geomembrane</td><td>• Advantages
- Simple construction

• Disadvantages
- Limited life</td></tr>
<tr><td>• Bentonite Mat</td><td>• Advantages
- Simple construction
- Self repairing

• Disadvantages
- Thin and nonrobust</td></tr>
<tr><td>• Compacted Clay/Silt</td><td>• Advantages
- Conventional and proven
- Simple to construct
- Easy to design and analyze

• Disadvantages
- Largest material volumes
- Strict quality control required</td></tr>
<tr><td>• Bentonite Amended Sand</td><td>• Advantages
- Duel function as Radon Barrier

• Disadvantages
- Susceptible to drying and cracking</td></tr>
<tr><td>• Multicomponent Infiltration Barrier

- Geomembrane
- Geomat
- Compacted Clay/Silt</td><td>• Advantages
- Robust
- Conventional and proven

• Disadvantages
- Relatively expensive</td></tr>
</table>

is done properly, it yields a low permeability layer. Both a clay/silt radon barrier, and a sand amended with bentonite radon barrier may provide a dual function as an infiltration barrier.

A bentonite mat would theoretically impede infiltration and reduce it to standard's level. Alone, however, this is a relatively thin layer of naturally occurring material, and it is difficult to establish unequivocally that it will function as desired for the required performance period.

A geomembrane has the advantage of being easy and inexpensive to place, but its performance over the long term is questionable due to its limited life.

The Drain

The drain/bedding layer has two functions. The first is to be the zone through which water seeping through the overlying soil is shed off the cover so that infiltration to the wastes is controlled. In this role the drain must be sufficiently permeable to provide rapid water flow and complete shedding. Also, the drain must prevent erosion of the underlying radon/infiltration barrier by water flowing in the drain.

The second function of the drain is to serve as a bedding layer that prevents the overlying gravel and cobbles of the biointrusion layer from punching, pitting, or deforming the surface of the underlying radon/infiltration barrier soils.

Selection of a suitable soil for the drain or bedding layer accordingly involves balancing the potentially conflicting demands for using (a) coarse, permeable materials to increase flow or shedding rates, thus lowering total infiltration, versus (b) fine soils to most effectively eliminate erosion of the underlying radon/infiltration barrier soils. Table 5.3 lists three alternative drain layer designs. They range from allowing the biointrusion barrier to serve

TABLE 5.3

ALTERNATIVE DRAINS EVALUATION

EVALUATION CRITERIA	ALTERNATIVE	ADVANTAGES/ DISADVANTAGES
• Technically Effective: Proven and demonstrated performance on other sites	• Coarse sand	• Advantages - Conventional and proven - Simple to construct - Acts as a bedding layer • Disadvantages - Significant quality assurance requirements
• Sample Construction: Easy to place and construct using standard construction procedures	• Geonet	• Advantages - Simple construction • Disadvantages - Longevity questionable
• Longevity: Will function throughout the 1,000 year design life	• Biointrusion barrier serving as drain	• Advantages - Dual function • Disadvantages - Potential erosion of infiltration barrier - No bedding layer effect

a dual function as a drain, to naturally occurring materials (sand), to a geosynthetic. Criteria for evaluating the alternative drain layers are listed in the table.

The material with the most significant advantages is a layer of coarse sand at least 150 mm thick. This approach is easy to design and construct and is also well-established, proven, and accepted. Another advantage is that it serves a dual function as a bedding layer. The major disadvantage to this type of drain is that it necessitates a significant level of quality assurance to ensure proper placement and, ultimately, satisfactory performance.

The geonet is a plastic grid sandwiched between two layers of geotextile. The geotextile acts as a filter, trapping small particles, and the plastic grid channels the water off the cover. It is easy to place, but its performance over the entire design life of the cell is questionable; it will probably deteriorate within the first 50 to 200 years of the cell's life.

The biointrusion barrier would theoretically serve a dual function as a drain, but would not protect the fine-grained soils of the underlying layers from erosion or pitting. Such pitting could result in depressions in which water could pond and subsequently seep into the infiltration barrier. Also, the increased flow in the large voids of the biointrusion barrier may erode the underlying soil of the radon/infiltration barrier.

A layer of coarse sand is selected as the preferred alternative drain layer. This is because such a drain barrier has a significant history of regulatory and industry acceptance and will also serve as a bedding layer. It will also remain intact (in an appropriately designed cover) for the required performance period, and is demonstrated to expedite the flow of water off the cover.

The Biointrusion Barrier

The biointrusion barrier's function is to reduce two types or modes of biointrusion: plant intrusion, and animal including human intrusion. This is usually achieved by placing a layer of cobble/gravel, thick soils, or herbicide-imbedded geogrids over the waste.

If the biointrusion barrier is a layer of cobbles, the material gradation must prevent the passage of finer-grained soils into the rock voids; this is required to preclude development of soils that support root penetration.

Table 5.4 lists three alternative biointrusion barrier designs. These vary in type from naturally occurring materials to man-made geosynthetics. The biointrusion barrier with the most significant advantages is a layer of cobble/gravel. This approach is conventional and proven. It also may serve a dual function as a drain. The disadvantage of this type of biointrusion layer is the possibility of the voids filling with silt. This can be avoided by placing or forming a filter of finer-grained material above the biointrusion layer to filter and stop the migration of particles into this layer.

Thick soils may be used as a biointrusion barrier, where sufficient quantities of appropriate soils are available. The thickness of the soils necessary to provide proper protection from deep-rooting species will be such that large quantities of soil will be needed to construct the cover, which in turn will increase the height and volume of the disposal cell.

A geogrid with time-released herbicides would theoretically preclude plant growth. Alone, however, it is difficult to establish that it will function unequivocally as desired for the required performance period. Its effectiveness in deterring burrowing animals is also in question. A geogrid may be used in conjunction with another biointrusion material (i.e., cobble/gravel).

TABLE 5.4

RESULTS OF MODIFIED VALUE ENGINEERING: BIOINTRUSION BARRIER

EVALUATION CRITERIA	ALTERNATIVE	ADVANTAGES/ DISADVANTAGES
• Technically Effective: Proven and demonstrated performance on other sites	• Cobble/gravel	• Advantages - Conventional and proven - Dual function as a drain • Disadvantages - Possible filling of voids - Quality rock required
• Simple Construction: Easy to place and construct using standard construction procedures • Dual Function: The component may or does function also as a Radon Barrier.	• Thick soils	• Advantages - Simple to construct • Disadvantages - More expensive - May encourage root penetration - Does not provide for animal intrusion control
• Longevity: Will function throughout the 1,000 year design life of the disposal cell	• Geogrids with time-released herbicides	• Advantages - Simple to place • Disadvantages - Will degrade over time - Layer to protect against burrowing still required

Erosion Barrier

The erosion barrier's functions are to reduce the potential for erosion of soil from the cell cover, prevent the development of gullies, and protect the underlying cover components from erosion. To this end, the erosion barrier materials should not be eroded by run-off from the design precipitation (i.e., the probable maximum precipitation). Furthermore, the material of the erosion barrier must remain intact. If rock is used, it must be durable. If vegetation is used, it should be self-sustaining.

Table 5.5 lists five alternative topslope erosion barrier designs. These range from vegetated soil to rock. The barrier with the most significant advantages is the clump grass in vegetated gravelly soil. Provided the percentage of gravel in the mixture is appropriate, the material provides erosion resistance and supports the vegetation. The disadvantage of this combination is that it requires periodic maintenance.

The vegetated soil is aesthetic and is relatively inexpensive, but provides limited erosion resistance and requires significant upkeep. The gravel mulch is also relatively inexpensive and limits infiltration. This material performs poorly in the areas of erosion protection and aesthetics. The rock with soil-filled voids and the rock top components both provide good erosion protection with the rock erosion barrier requiring the lowest maintenance of all the materials under consideration. The rock with soil-filled voids has a high construction cost. The disadvantages of the rock layer are its poor aesthetics and inadequate function as an infiltration barrier.

Table 5.5

ALTERNATIVE EROSION BARRIER EVALUATION

EVALUATION CRITERIA	ALTERNATIVE	ADVANTAGES/ DISADVANTAGES
• Technically Effective: Proven and demonstrated performance on other sites • Simple Construction: Easy to place and construct using standard construction procedures	• Vegetated soil	• Advantages - Good aesthetics - Limits infiltration - Low cost • Disadvantages - High maintenance requirements - Limited erosion protection
• Longevity: Will function throughout the 1,000 year design life of the disposal cell	• Clump grass vegetated gravelly soil	• Advantages - Erosion protection - Good aesthetics - Limits infiltration • Disadvantages - Moderate maintenance requirements
• Maintenance: Low-level of upkeep involved	• Gravel mulch	• Advantages - Relatively inexpensive - Limits infiltration • Disadvantages - Poor erosion protection - Poor aesthetics
• Aesthetics: Blends in well with surrounding environment	• Rock with soil-filled voids	• Advantages - Erosion protection • Disadvantages - Expensive - Moderate maintenance requirements
	• Rock	• Advantages - Erosion protection - Low maintenance requirements • Disadvantages - Poor aesthetics - Poor infiltration barrier

SELECTION OF COVERS: THE CHECKLIST COVER

Checklist Approach

The checklist approach involves a process of successive elimination for specific reasons of one or more components from the so-called "checklist cover." Figure 5.6 shows the checklist cover that includes all reasonable components. Table 5.6 lists the purpose and function of each component.

Checklist Cover Objectives

The checklist cover, as shown on Figure 5.6, incorporates all reasonable components possibly required at a site to:

- Control erosion
- Limit infiltration
- Provide freeze/thaw protection
- Inhibit radon emanation
- Drain or shed precipitation
- Control biointrusion
- Be self-renewing and adaptable to climatic changes if vegetation is used.

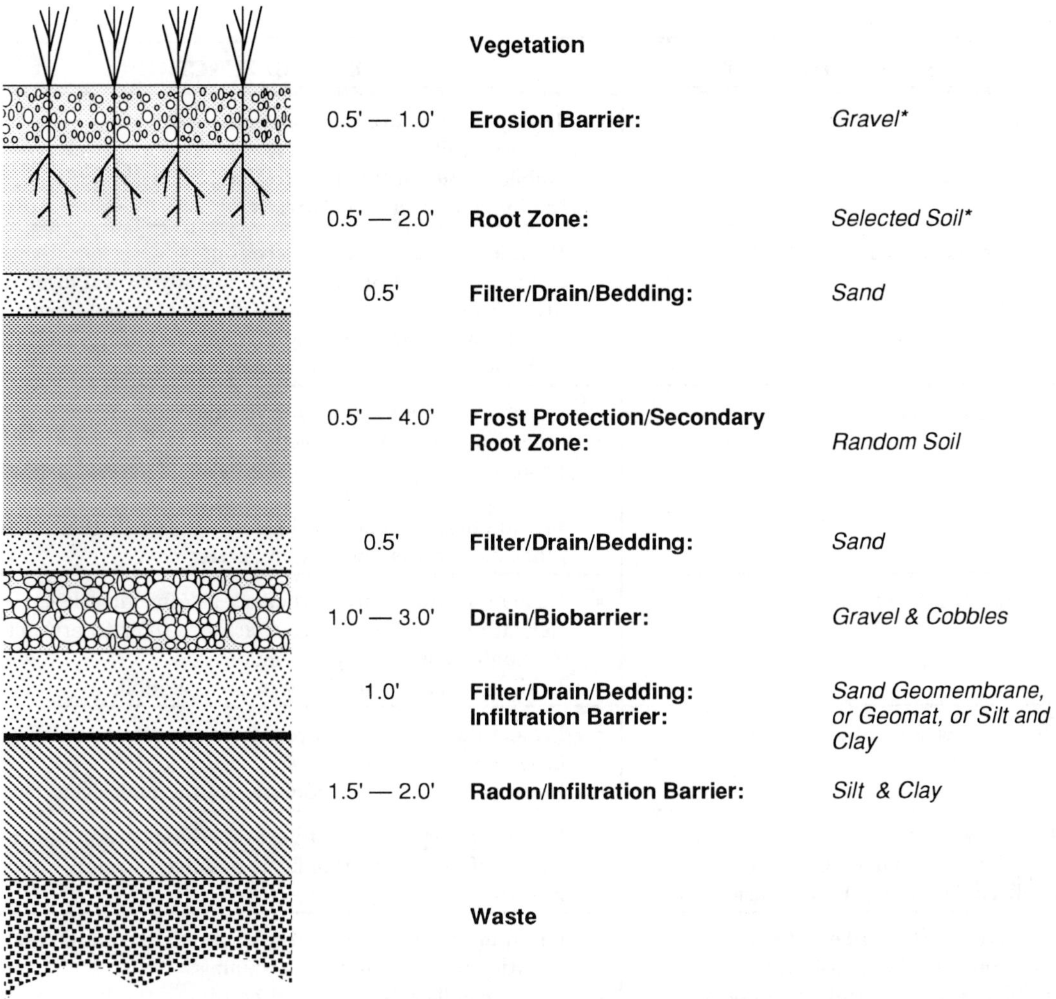

* In some instances, these two components may be inverted or mixed

This is not an actual cover. It is a compilation of all the feasible components that may be used in a cover.

FIGURE 5.6

THE CHECKLIST COVER

TABLE 5.6

COMPONENT ELIMINATION CRITERIA

COVER COMPONENT	PURPOSE AND FUNCTION
Vegetation	• Transpire moisture that enters the soil. • Reduce infiltration. • Stabilize soil and reduce erosion. • Minimize impact of rainsplash.
Erosion-barrier (durable rock) layer above	• Provide protection against erosion. • Reduce evaporation rates from underlying soil layer in drier environments - preclude drying of radon barrier. • May be mixed with rooting medium to form gravel mulch with dual (though compromised) functions.
Rooting medium	• Provide rooting medium for vegetation. • Store water for plant growth. • Protect underlying biointrusion layer of surface exposure. • Provide frost protection. • Provide host medium for burrowing animals that might otherwise burrow into radon barrier.
Bedding/Filter/Drain (Sand)	• Beneath erosion barrier rock act as a bedding layer. • Beneath erosion barrier rock act as a filter to prevent erosion of the underlying random soil. • Drain to expedite shedding of precipitation.
Frost protection (random fill)	• Protect the underlying layers from the effects of frost heave and frost penetration. • Preserve the physical properties of the underlying layers.
Filter (layer of sand or pea gravel overlying layer of coarse aggregate. Alternatively choked rock of biointrusion layer.)	• Prevent piping of soil into erosion/biointrusion barrier. • Drain infiltration as rapidly as possible to retard root growth.
Biointrusion/Drain layer (500 to 1000 mm of cobbles with a low coefficient of uniformity to prevent biointrusion)	• Drain infiltration as rapidly as possible to retard root growth. Impede burrowing animals. • Act as capillary break at the bottom of the layer to prevent upward movement of water and downward unsaturated flow (enhances moisture storage capacity). • Control topslope erosion if vegetation and topsoil eroded.
Filter/Bedding/Drain (150 to 300 mm layer of pea gravel, sand, or silt)	• Protect the underlying liner system from displacement and rock penetration. • Drain water laterally off the pile to limit infiltration.
Infiltration barrier (a liner system, or high-percentage bentonite mix [with silt or sand] or geomembrane)	• Intercept moisture. • Control infiltration. • Inhibit infiltration while mature vegetation community is establishing or after severe disturbance of the vegetation.
Radon barrier (clay/silt)	• Inhibit radon emanation. • Limit infiltration.

Application of the Checklist Top Cover Approach

In order to determine the appropriate components to be incorporated into the top cover or the side cover at a particular site, proceed as follows:

1. Obtain site-specific data
2. Examine relevant characteristics of the natural landscape (gullies, vegetation, and the like)
3. Examine the checklist cover and eliminate components on the basis of the component elimination criteria list in Table 5.7
4. Compile the final cover as a composite of the remaining components.

The checklist approach to the design of a cover is simplistic in that each component tends to be viewed in and of itself. In reality, there is considerable interaction of the various components. It has been said that the various components, properly selected, form a functional-synergistic entity. For example, in theory, the use of a bentonite mat as the only operational infiltration barrier is reasonable. Demands such as stability constraints dictate relatively flat slopes for the bentonite mat. Thus, there may be little gravity-induced run-off or shedding of precipitation through the drain above the bentonite mat. A hydraulic head could build up above the thin bentonite mat; the result would be an increased gradient and hence increased percolation through the infiltration barrier. To reduce the potential for buildup of water in the drain, it is prudent to place a soil layer (probably required anyway for frost protection) above the infiltration barrier, and to establish vegetation in the soil. The evapotranspiration of the vegetation will reduce the frequency and the amount of percolation reaching the drain, and therefore the need to rely on lateral shedding to remove water from the pile. The designer should always look for opportunities to enhance the interactive or synergistic effect of the various components in a cover.

TABLE 5.7

COMPONENT ELIMINATION CRITERIA

COVER COMPONENT	PURPOSE AND FUNCTION	RATIONALE FOR ELIMINATION (based on site-specific conditions)
Vegetation	• Transpire moisture that enters soil. • Reduce infiltration. • Stabilize soil and reduce erosion. • Minimize impact of rainsplash. • Provide shallow-rooting grass cover that competitively excludes deep-rooted trees and shrubs.	• Arid climate; vegetal productivity may be too low or undependable to provide a stabilizing cover. • Topsoil suitable to support vegetation is not available; topsoil is highly erodible because of physical structure or properties. • Large rock would be required to control gullies, and such rock is not available. • Rock quality would have to meet the same requirements as that for frequently saturated conditions -- scoring 65% or better. • It is possible to show that a significantly thick rock layer can be placed on the pile to inhibit the establishment of vegetation. • Vegetation not required to enhance cell aesthetics.
Erosion barrier: rock layer above filter, drain, or bedding sand	• Provide protection against soil erosion used in conjunction with vegetation. • Reduce evaporation rates within the underlying soil layer in drier environments -- preclude drying of the radon barrier. • Resist invasion by burrowing animals such as badgers and ground squirrels.	• Environment is wet/humid (semitropical). Vegetation will grow regardless of presence of rock and will constitute an unacceptable and onerous maintenance burden. • Suitable quality rock not available. • Unacceptable visual impact. • Adequate vegetal coverage to protect against erosion. • Flat enough slopes that additional erosion protection not needed. • Increased potential infiltration due to reduced evaporation not acceptable. • May favor invasion of deep-rooting trees and shrubs that can damage the underlying layers.

TABLE 5.7 (Continued)

COMPONENT ELIMINATION CRITERIA

COVER COMPONENT	PURPOSE AND FUNCTION	RATIONALE FOR ELIMINATION (based on site-specific conditions)
Rooting medium	• Provide rooting medium for vegetation. • Store water for plant growth. • Protect the underlying biointrusion layer from surface exposure. • Provide frost protection. • Host burrowing animals (reducing likelihood of their movement to deeper layers.	• Vegetation would not be used for any of the reasons stated in No. 1, above.
Frost protection (random fill)	• Protect the underlying layers from the effects of frost heave and frost penetration. • Preserve the physical properties of the underlying layers.	• Regional frost penetration depth is insignificant; protection, if required, can be afforded by the erosion barrier or rooting medium (if included).
Filter (layer of sand overlying layer of coarse aggregate: alternatively choke the biointrusion rock)	• Prevent piping of soil into erosion/biointrusion barrier. • Drain infiltration as rapidly as possible to retard root growth.	• Biointrusion layer would not be used for any of the reasons stated in the section below. • Potential for slope instability exists, particularly on sideslopes.
Biointrusion (2-3 feet of cobbles with a low coefficient of uniformity to prevent biointrusion)	• Retard root advancement by remaining free of capillary moisture. • Impede burrowing animals. • Prevent inadvertent human intrusion. • Act as a capillary break at the bottom of the layer to prevent upward movement of water and downward unsaturated flow (enhances moisture storage capacity). • Control topslope erosion if vegetation and topsoil eroded away.	• "High" quality rock not available (frequently saturated conditions). • No regulatory or institutional need to inhibit biointrusion or inadvertent human intrusion. • Deep-rooted species may be excluded by shallow-rooted species via ecological competition.
Drain/Filter/Bedding (pea gravel, sand, or silt)	• Drain water laterally off the pile to limit infiltration. • Protect the underlying liner system from displacement and rock penetration.	• Do not have underlying liner system for any of the reasons stated below. • Potential for slope instability exists, particularly on sideslopes.

TABLE 5.7 (Continued)

COMPONENT ELIMINATION CRITERIA

COVER COMPONENT	PURPOSE AND FUNCTION	RATIONALE FOR ELIMINATION (based on site-specific conditions)
Infiltration barrier (Geomembrane or high percentage bentonite)	• Intercept moisture. • Control infiltration. • Inhibit infiltration while mature vegetation community is establishing or after severe disturbance of the vegetation.	• Saturated hydraulic conductivity of radon barrier, amended or not, is low enough to lead to groundwater compliance. • Potential for slope instability exists.
Radon barrier (clay/silt)	• Inhibit radon emanation. • Limit infiltration.	• Not required because flux is below standards.

Case History 2: Application of the Elimination Criteria

The following is a brief example of the application of the component elimination criteria to a site in the Midwest.

Vegetation: The main reasons for eliminating vegetation (i.e., a dry climate, absence of aesthetic considerations, unavailability of suitable soils or rocks) are not significant at a Midwestern site. Therefore, in general, vegetation may conveniently be used at the site on the disposal cell. The only reason to eliminate vegetation is its unreliability compared to rock in controlling erosion.

Erosion Barrier: One of the main reasons for eliminating a rock erosion barrier is the absence of an economically available durable rock. This is not the case at our site. Two possible reasons for eliminating rock as the uppermost component are the potentially unacceptable visual (i.e., aesthetic) impact, and the humid climate of the site which makes the growth of vegetation inevitable. Perhaps the most compelling reason to avoid using rock, except as a thin veneer to "mulch" the soil, is that the rock may be invaded by deep-rooting trees and shrubs which, if not removed, will damage the underlying layers. Upon their eventual death, the taller trees will topple and uproot, damaging the surficial layers of the cover. Only a grass-covered soil layer, or perpetual maintenance, can be relied upon to avoid this scenario.

Elimination of rock as the uppermost cover component for erosion protection does not, however, mean that durable rock may not be used to control erosion. The durable rock may be compatible with vegetation if sufficient soil is forced or allowed to migrate into the voids of the rock. This too would encourage deep-rooting trees and shrubs, requiring frequent maintenance. A small amount of soil filtered into the interstices of a thick rock layer does little to improve the ecological performance of the cover, but it does provide a surface that may be manicured and mowed, which is not an option for a purely rock cover. Alternatively, a small amount of rock may be added on top of the soil to form a gravel mulch or the rock may be covered by a soil layer. In the case of soil over rock, the rock is sized to resist erosion by flow that may occur in a potential (hypothetical) gully that may develop in the soil. Accordingly, for the site in the Midwest, it is concluded that there are no valid reasons to eliminate rock as a way enhancing the erosion resistance of the cover.

Rooting Medium: Reasons for eliminating a rooting medium are the same as the reasons for eliminating vegetation. As it has already been concluded that there are no valid reasons to eliminate vegetation, it follows that the rooting zone cannot be eliminated. (Note that the root zone may be a separate layer placed for the purpose.) Alternatively, the frost protection layer may double as a primary rooting zone. If a separate upper rooting zone is provided, the frost protection layer may serve also as a secondary rooting zone.

Frost Protection Zone: The frost penetration depth in the Midwest is significant. It is not feasible to eliminate the frost protection zone or to relax the need for frost protection of the radon and infiltration barrier. However, if other components have thicknesses sufficient to provide frost protection, the layer identified as such may not be required.

Filter Over the Biointrusion Barrier: This must be used if a biointrusion barrier is used to prevent migration of soil fines into the biointrusion barrier.

Biointrusion Barrier: The main reasons for eliminating the biointrusion barrier are judgmental (i.e., how likely and damaging is animal, plant, and human intrusion). RCRA covers do not incorporate robust biointrusion barriers, therefore it may be argued that this is not an overriding institutional concern. Another judgmental reason to eliminate the biointrusion barrier is the belief by some that loose-cobble biobarriers are ineffective against roots, or at least will not deter root advancement for the extended design lives of waste disposal facilities. Another consideration is the fact that the operative factor in preventing biointrusion may be the layer of rock, but rather the adequacy of water and nutrient relations in the overlying rooting medium. If sufficient moisture and nutrients are supplied to plants, the physiological pressure for downward root advancement is relieved. The likelihood of root penetration of deep layers (hence need for a biobarrier) is also reduced if shallow rooting species such as grasses are established as a permanent (climax) community on the cover. Biointrusion barriers do appear to discourage burrowing animals, but not all cells will attract these animals. For these reasons, the biointrusion barrier elimination is neither absolute nor unequivocal. A biointrusion barrier is included in some covers, and omitted from others.

Drain/Filter/Bedding: This is eliminated or retained as the specifics of the cover alternative dictate. Generally, the presence of a drain overlying the infiltration barrier reduces the head on the barrier, thus reducing the flux throughout the cover. Bedding material between the biointrusion rock and the radon/infiltration barrier is needed.

Infiltration Barrier: The main reason to eliminate a separate infiltration barrier, i.e., one that does not also function as the radon barrier, is that the other cover components are in and of themselves adequate to reduce infiltration to the levels required by all performance criteria. Thus, if the radon/infiltration barrier is of a low hydraulic conductivity soil, an additional infiltration barrier may be repetitive and redundant. The most frequent reason for including a separate infiltration barrier is that some regulations mandate its inclusion. All of the alternative covers include a geomembrane infiltration barrier, but this may be eliminated if the regulations are deemed inapplicable or the radon/infiltration barrier soils have a sufficiently low hydraulic conductivity.

Radon Barrier: This component may be eliminated if the radon flux is below standards. Even in such a case, however, it may be expedient or prudent to place a layer of clean, fine-grained soil to provide a working surface or bedding layer in subsequent cover components. It is important to keep plant roots out of the radon barrier because the plants may transpire radon to the surface in rates that will exceed the standards.

SELECTION OF A PREFERRED COVER

Different cover types or configurations are possible and appropriate for the different cell types and encapsulation facilities. Therefore, in designating a preferred cover type, it must be individually done for each cell and cell content type and configuration. Accordingly, this section describes two cover selection methodologies.

Cover Comparison by Multi-Utility Attribute Methods

In principle, the checklist cover approach should always, if used in an objective manner, be able to identify a preferred cover alternative. In practice, this is not so. The reason is that the identification of a preferred cover alternative involves subjective and unquantifiable factors. Standard decision-making approaches to selection of preferred alternatives may be used in such cases. Probably the most comprehensive of these procedures is the Multi-Utility Attribute (MUA) Decision-Making Methodology.

In brief, the MUA approach involves designation of a number of selection criteria, the weighing of each, and assignation to each alternative of a score in accordance with some predetermined scale. The sum of the weighted scores is the ranking determinant. The main advantage of the MUA approach is that it explicitly identifies the criteria that affect selection. Different decision-makers can apply their own weights to assess the effect of their own subjective preferences. Accordingly in this section, the basic MUA approach is used.

Cover Ranking Factors

For the five cover alternatives shown in Figures 5.1 to 5.5, the following ranking factors may be defined:

Compliance with Regulations: Simply stated, a cover either does or does not comply with a governing regulation. It is not considered that there are degrees of regulatory compliance. If it does not, then it may not be used. Accordingly, this factor is not generally considered in the comparison of alternative covers for disposal facilities. It may, however, be used as an elimination criterion.

Compliance with ARARs: While difficult to quantify, it is considered that there may be subjectively different degrees of compliance with ARARs. A cover may be functionally equivalent, if not fully conformant, to a specific ARAR. Accordingly, this may be used as a comparative and ranking criterion.

Infiltration: While the erosion protection material has a very large effect on the infiltration through the cover, the rest of the cover also has a significant effect. The location of drains and the presence of a geomembrane are both critical to infiltration. Clearly, infiltration should be used as a comparative and ranking factor.

<u>Frost Protection</u>: Standard design practice dictates that the radon/infiltration barrier be protected by overlying layers from freezing and thawing. This must be done as a matter of good practice for all covers. Therefore, it is not used as a comparative or ranking criterion. It is recognized that different cover configurations and cover materials may result in greater or lesser frost protection layer thicknesses, but that leads to different cover costs, and so is incorporated into that aspect of cover evaluation and comparison.

<u>Cover Cracking</u>: While cover cracking is to be avoided if possible, cover cracking as a comparative and ranking criterion is not incorporated for the following reasons. First, cover cracking potential is primarily a function of the cell content, placement details, and the geometry of the cell, not the cover configuration. Secondly, the properties of the soil of the radon/infiltration barrier is the second most important factor in controlling or eliminating cover cracking. An appropriate radon/infiltration barrier soil must be selected regardless of the details of the remainder of the cover. It is recognized that the addition of extra load onto a radon/infiltration barrier may reduce the likelihood of cracking of the radon/infiltration barrier. In general, however, all covers are amenable to an increase in thickness (and cost) if necessary for crack potential reduction.

<u>Slope Stability</u>: Covers without geomembranes are inherently more stable than covers with geomembranes. If stability is a desired criterion, the omission of the geomembrane is desirable. It is, however, possible to achieve the design stability for all covers discussed in this report, and therefore slope stability is not generally used further as a comparative or ranking criterion.

<u>Biointrusion</u>: Biointrusion is highly dependent upon the type of erosion protection used on the cover. The presence or absence of capillary breaks, thick frost protection layers, and loose cobble biointrusion barriers will influence the likelihood and consequences of biointrusion. For this reason, biointrusion is a highly weighted comparative and ranking factor.

Maintenance Requirements: Different covers will have different maintenance requirements. These maintenance requirements are primarily due to the type of erosion protection used, and therefore maintenance is not counted as a comparison or ranking factor.

Cost: Cost must not drive decision-making when it comes to environmental restoration for a long period. Conversely, it is imprudent to spend more money than is reasonably necessary to remediate a contaminated site. For these two reasons, cost as a comparative and ranking factor may be included but assigned a relatively low weight.

Scoring Scales: Table 5.8 shows scales that may be used to score the cover alternatives for the various comparison criteria. Our scales are of our own making; others may wish to compile their own scales and test the sensitivity of the ultimate ranking to different ranking factor weights and scale factors.

Cover and Erosion Protection Comparison

In this book, all aspects of the performance of covers are discussed, all the different possible components of a cover for a disposal cell are evaluated, and the integration of cover components into a number of different covers is discussed. The cover alternatives vary only from the bottom of the erosion protection material on the surface down to the top of the radon/infiltration barrier. Each cover alternative can be used with the most effective erosion protection system. Therefore, various erosion protection alternatives are evaluated separately from the five cover alternatives.

Erosion Protection Ranking Factors

Erosion protection on a cover is achieved primarily by the uppermost component. Alternative erosion protection layers and materials have been used. These include vegetated soil, vegetated gravelly soil, vegetated rock/soil matrix, vegetated soil over rock,

TABLE 5.8

COVER PERFORMANCE

CRITERIA SCORING SCALES

Score	Criterion
ARAR Compliance	
1	Non Compliance
2	Partial Compliance
3	Reasonable Compliance
4	Significant Compliance
5	Total Compliance
Infiltration	
1	Low probability that long-term infiltration will be less than 10^{-7} cm/sec.
2	Reasonable probability that long-term infiltration will be less than 10^{-7} cm/sec.
3	High probability that long-term infiltration will be less than 10^{-7} cm/sec.
4	Reasonable probability that long-term infiltration will be less than 10^{-8} cm/sec.
5	High probability that long-term infiltration will be less than 10^{-8} cm/sec.
Biointrusion	
1	No specific features for biointrusion control
2	Has significant rooting zone at shallow depth
3	In addition to rooting zone, has capillary break below the rooting zone
4	In addition to the above, has an additional capillary break
5	In addition to the above, has a biointrusion layer
Cost	
1	81 to 100 percent of cost of most expensive cover alternative
2	61 to 80 percent of cost of most expensive cover alternative
3	41 to 60 percent of cost of most expensive cover alternative
4	21 to 40 percent of cost of most expensive cover alternative
5	0 to 20 percent of cost of most expensive cover alternative

vegetated gravel mulch, and nonvegetated rock. The following factors may be used as ranking factors in a comparison and selection of a preferred erosion barrier configuration and material.

Erosion Protection: The degree of erosion protection proffered by the various alternatives varies. The degree of erosion protection can be most easily evaluated by looking at the design storm which the material will withstand. While maintenance procedures may minimize cumulative damage, an event of design magnitude or greater could cause extensive damage. For this reason, erosion protection is used as a ranking factor.

Aesthetics: One of the most important design considerations is that a disposal cell blend in with or complement the surrounding area. This is generally used as a ranking factor for both top and sideslopes and is given a high weighting.

Infiltration: The erosion protection system has a large effect on the infiltration of surface water through the cover. The erosion protection affects the amount of run-off that occurs and the amount of evaporation that takes place. The type of erosion protection also affects the type and amount of vegetation, which in turn affects the evapotranspiration of infiltrating water. Therefore, infiltration is a ranking factor and is assigned a high weighting.

Biointrusion: Biointrusion is influenced by the erosion protection material used on the slope, in that the material used for the uppermost layer determines the climax species that will grow. Larger particle sizes such as gravel and cobbles encourage more deeply-rooted shrubs and trees over shallow-rooted grasses. While an underlying biointrusion layer may resist root penetration, a dense growth of deep-rooted species will relentlessly probe and eventually penetrate this layer. A thick rooting medium also protects against biointrusion by serving as a favorable layer for plant roots, by reducing their inclination to extend downward. Because of the potential of biointrusion to damage the cover, it is a ranking factor with a high weight factor.

Maintenance Requirements: Maintenance requirements include both repair of erosion and mitigation of biointrusion. To avoid counting biointrusion twice, for the purpose of this exercise, only repair of erosion and age-related damage features will be considered a ranking factor. Because this is related strongly to erosion protection which has already been accounted for, it is given a lower weight factor.

Cost: Cost must not govern decision-making when it comes to environmental restoration for the long period. However, spending more money than is reasonably necessary to remediate a site is injudicious. For these reasons, cost is included as a ranking factor, but is assigned a low weight.
Erosion Protection Scoring Scales: Table 5.9 shows the scale for each ranking factor used to numerically compare the erosion protection alternatives. Our scales were determined by ourselves; others may wish to compile their own scales and test the sensitivity of the ultimate ranking to different weight factors and ranking scales.

Cover Rankings

Table 5.10 shows different criteria weights, a comparison, and ranking of the five alternative covers. The criteria weightings used on Table 5.10 are based on the idea that prevention of infiltration is a significant objective.

A brief examination of the table indicates that significant changes in both the weights and scale numbers are required to alter the rankings. Accordingly, it is considered that the system and its results are robust. We recognize the subjectivity of the criteria weightings and even the assignment of criteria scores, hence the sensitivity of the rankings shown in these tables. Others may, and probably will, disagree with our assignment of number. All we can do is invite them to assign their own numbers to the system or process and use it as a basis for logical argument and rational decision-making.

TABLE 5.9 EROSION CONTROL CRITERIA SCORING SCALES

EROSION CONTROL	
Aesthetics	
1	Contrasts sharply with surrounding area and has no screening
2	Contrasts sharply with surrounding area but has perimeter screening
3	Blends in well with surrounding area but has no perimeter screening
4	Blends moderately with surrounding area and has perimeter screening
5	Blends in well with surrounding area and has perimeter screening
Infiltration	
1	Minimizes evaporation and does not provide for evapotranspiration through vegetation; minimum run-off
2	Allows for some vegetation growth but minimized evaporation; moderate run-off
3	Allows for some vegetation growth and allows for some evaporation, moderate run-off
4	Maximizes evapotranspiration and allows some evaporation; maximum run-off
5	Allows for maximum evapotranspiration and evaporation; maximum run-off
Biointrusion	
1	No specific features for biointrusion control
2	Continuous or regular and significant maintenance required to control biointrusion, i.e., mowing more than twice each year
3	Regular maintenance required to control vegetation, i.e., mowing once each year
4	Intermittent maintenance required to establish and perpetuate climax vegetation, or to remove unplanned vegetation, i.e, removal of deep-rooted vegetation when observed
5	Designed to accommodate climax vegetation or preclude vegetation
Maintenance Requirements	
1	May require significant repair of gullies and reseeding
2	May require minimal repair of gullies after drought or fire
3	May require minimal repair of gullies under PMP conditions
4	May require minimal repair of gullies under PMP conditions when coupled with drought or fire conditions or if event occurs in early years of vegetation
5	Gullies will not form under PMP condition
Cost	
1	81 to 100 percent of cost of most expensive erosion protection alternative
2	61 to 80 percent of cost of most expensive erosion protection alternative
3	41 to 60 percent of cost of most expensive erosion protection alternative
4	21 to 40 percent of cost of most expensive erosion protection alternative
5	0 to 20 percent of cost of most expensive erosion protection alternative

TABLE 3.10

COVER SELECTION
INFILTRATION AS PRIMARY FACTOR

Ranking Factor Weight	ARAR Compliance 0.15	Infiltration 0.50	Biointrusion 0.25	Cost 0.10	TOTAL	RANK
COVER TYPE						
SIMPLE	1 / 0.15	2 / 1.00	1 / 0.25	4 / 0.40	1.80	5
RCRA	2 / 0.30	4 / 2.00	2 / 0.50	3 / 0.30	3.10	4
DOUBLE DRAIN	3 / 0.45	4 / 2.00	2 / 0.50	1 / 0.10	3.05	3
MULTICOMPONENT	5 / 0.75	4 / 2.00	5 / 1.25	1 / 0.10	4.10	1
EROSION RESISTER	3 / 0.45	2 / 1.00	2 / 0.50	1 / 0.10	2.05	2

LEGEND

SCORE → 4 / 2 ← WEIGHTED SCORE

Decision Tree Approach

Figure 5.7 shows a decision tree that requires the asking and answering of simple questions, which nevertheless have answers that are highly subjective:

- Is erosion control more or less important than biointrusion control?
- Is regular maintenance acceptable?
- Are aesthetics a significant design consideration?
- May different covers be used on different parts of the pile?

By following through the decision tree, different covers may be selected. The path of most interest to us, and that which involves the answers to the questions that are thought to be the most reasonable, is the following:

- Erosion control is more important than biointrusion control. The reason is that the consequences of failure of an unattended pile due to erosion are of considerably greater significance and detrimental impact to the surrounding environment than the consequence of failure by uncontrolled biointrusion.

- Regular maintenance is both acceptable and probably inevitable. The reasons for this approach are that engineering works are being constructed that are, however designed, unnatural. To maintain their design features and desired function, they will have to be looked after to ensure they behave as intended. And even if building a natural entity were successful, it is recognized that natural processes and responses would not result in the performance and environmental protection sought.

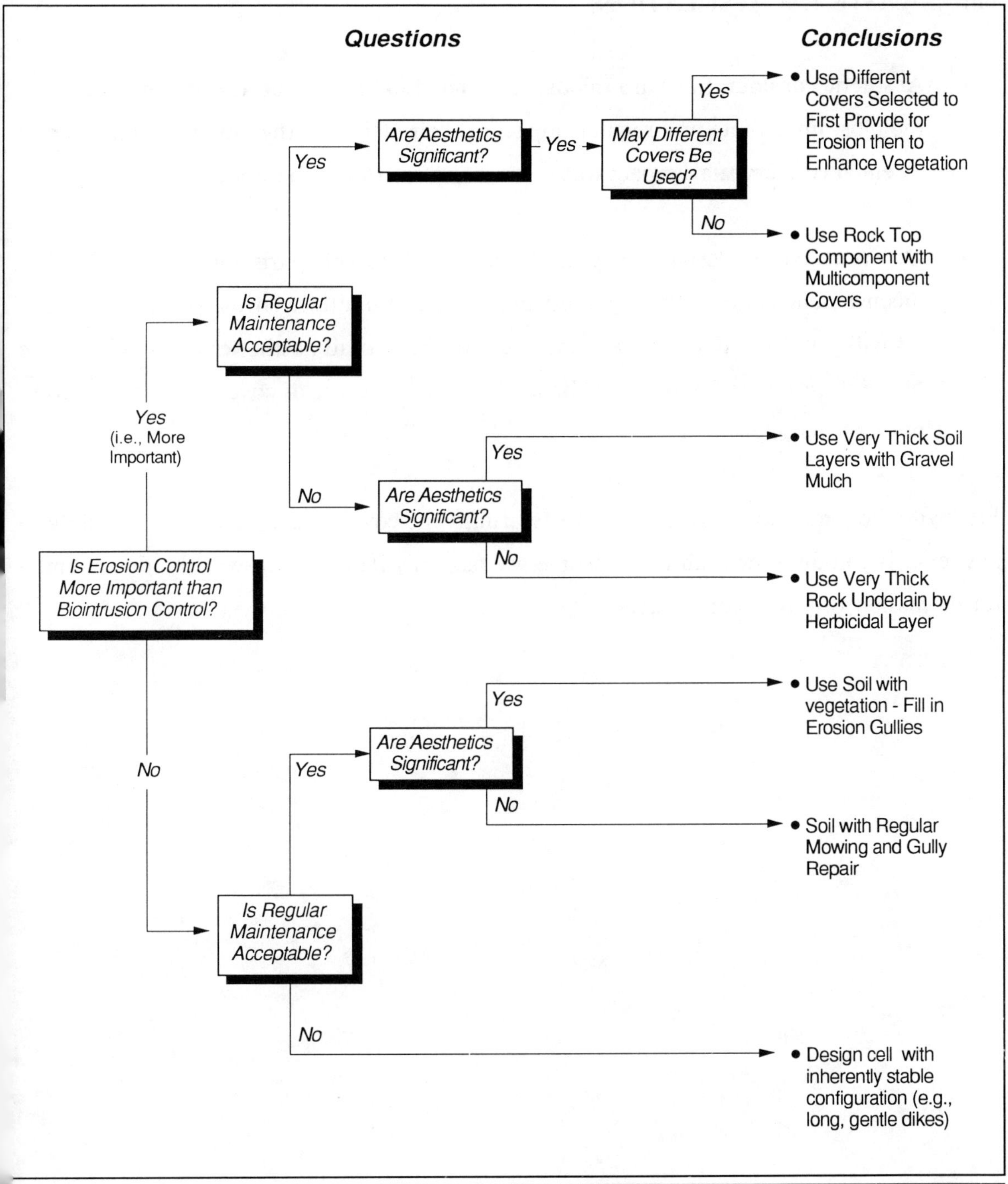

FIGURE 5.7

COVER SELECTION LOGIC CHART

- Aesthetic considerations are important. The visual impact of human construction is the most immediate and regular impact of their actions. Anything that can be done to enhance the visual impact to the environment should be done.

- Different cover designs may be used on different parts of the pile. The inconvenience and additional construction cost of multiple cover configurations are a small price to pay for the enhancement of the overall aesthetics and performance of the disposal facility that may be achieved by a variety of cover types on different parts of the pile.

We invite the reader to follow through to determine the cover design that results from these answers. If you disagree with the outcome, we ask only that your answers be documented for discussion with decision-makers.

Chapter 6

PERIMETER ENCAPSULATION SYSTEMS

ALTERNATIVES

The perimeter encapsulation system of a disposal cell may consist of a multicomponent cover composed of layers of soil and rock. Alternatively, the perimeter encapsulation system may be a clean-fill dike of compacted soil and rock (see Figure 6.1). If a multicomponent cover is used, the outer geometry of the waste is formed to a relatively flat slope, conventionally no steeper than five horizontal to one vertical (5H:1V). A cover is placed over the wastes. This cover may be any of the alternative covers discussed in the previous chapters.

A perimeter clean-fill dike is created by placing and compacting uncontaminated soil and rock. The outer slopes of the dikes may vary from 3H:1V to 10H:1V. The upper component of a clean-fill dike may be either rock or soil and gravel in which vegetation growth is encouraged.

SELECTION CRITERIA

Criteria for selecting a multicomponent cover or a clean-fill dike include: the desired cell performance (stability, erosion control, infiltration); the availability of clean or very slightly contaminated material; and the availability of a large enough site to accommodate the waste and additional material for the clean-fill dike.

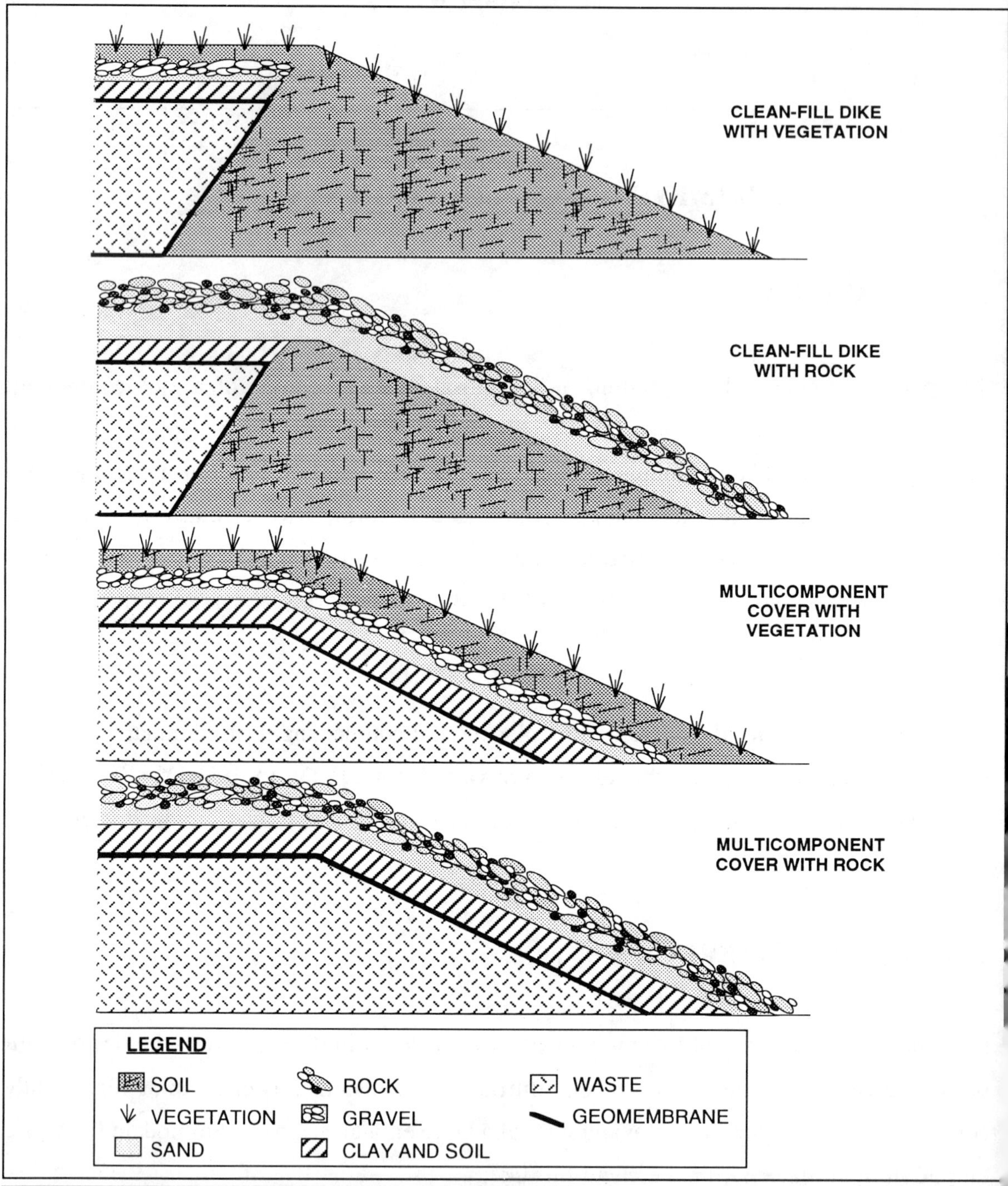

Alternative perimeter encapsulation systems include clean-fill dikes and multicomponent covers.

FIGURE 6.1

PERIMETER ENCAPSULATION ALTERNATIVES

Figure 6.2 shows the geometry of a range of clean-fill dikes. Figure 6.3 shows a vertical innerface clean-fill dike. Choice of the specific dike geometry is a function of the site constraints and the design and performance criteria established for the encapsulation system.

Figure 6.4 shows the results of a Value Engineering Workshop comparison of different perimeter encapsulation systems, specifically those shown in Figure 6.1.

EVALUATION OF DIKES

Clean-fill dikes score considerably higher than multicomponent covers for these reasons:

- Slope Stability: Clean-fill dikes have higher long-term static and dynamic factors of safety for both deep-seated and shallow surficial instability than multicomponent covers. The main reason is the presence of the geomembrane that is needed in the basal liner, and around and over the wastes; the geomembrane is a low-strength potential failure plane. Conversely, the clean-fill dike does not incorporate a geomembrane, either at its base or along the upper part of the outer slope. Potential failure is affected only by the strength of the soils and rocks of the dike and its foundation; no weak geomembrane failure planes reduce factors of safety.

- Biointrusion: Clean-fill dikes are considerably less susceptible to damage by biointrusion than are multicomponent covers. If vegetation induces surface and near-surface damage, the consequences to a clean-fill dike are considerably less than for a multicomponent cover. The shear bulk of the clean-fill dike provides ample space and volume for roots from vegetation growing on the perimeter encapsulation system to thrive without entering the waste or compromising the integrity of the cell.

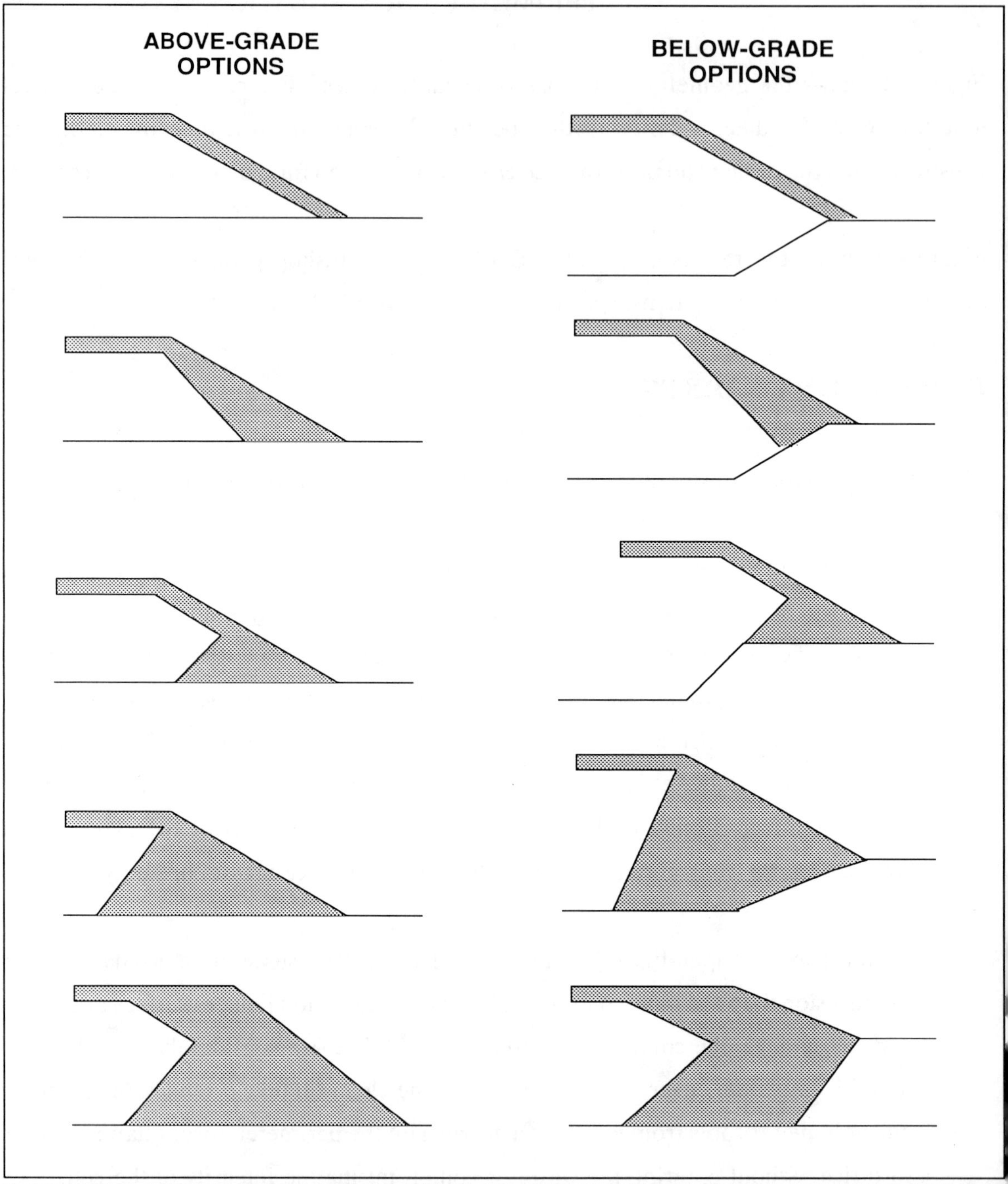

These are but some of the many different geometries that may be adopted for perimeter encapsulation systems.

FIGURE 6.2

DISPOSAL CELL SIDESLOPES

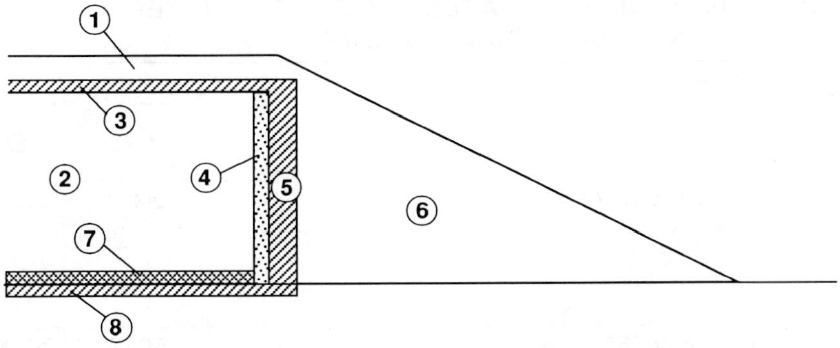

SECTION OF VERTICAL INNER FACE CLEAN-FILL DIKE

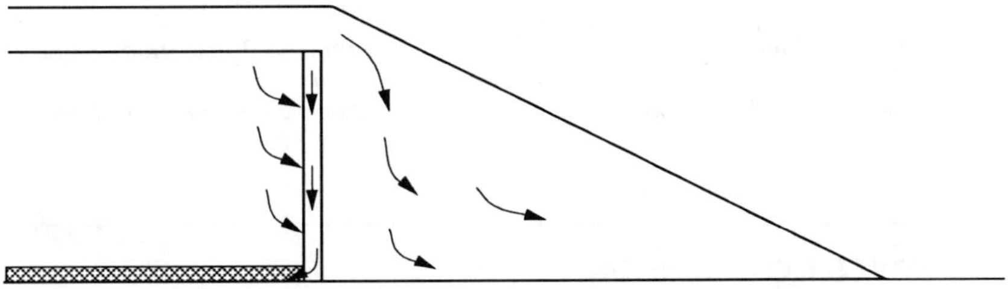

FUNCTION OF VERTICAL DRAIN

SEEPAGE FROM THE WASTE INTO THE VERTICAL DRAIN WILL FLOW DOWN TO THE BASAL LEACHATE COLLECTION SYSTEM. DIKE INFILTRATION WILL BE IMPEDED FROM ENTERING THE DRAIN BY THE LOW PERMEABILITY CLAY ZONE.

LEGEND

(1) COVER
(2) WASTE
(3) INFILTRATION BARRIER
(4) VERTICAL DRAIN
(5) LOW PERMEABILITY CLAY ZONE
(6) CLEAN-FILL DIKE
(7) LEACHATE COLLECTION SYSTEM
(8) BASAL LINER

To achieve all the advantages of a clean-fill dike, and increase cell capacity, the inner face of the dike may be vertical.

FIGURE 6.3

VERTICAL FACE CLEAN-FILL DIKE

ALTERNATIVES TO BE EVALUATED—WASTE PERIMETER ENCAPSULATION SYSTEMS

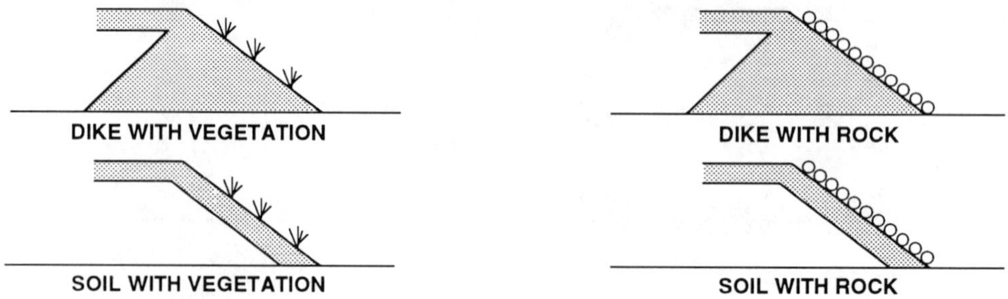

EVALUATION CRITERIA & WEIGHTINGS

Criterion	Weighting
Facilitate Construction	3
Facilitate Leachate Control	17
Increase Erosion Protection	22
Minimize Maintenance	8
Enhance Cell Performance	25
Enhance Public Perception	11
Minimize Worker Exposure	14

ALTERNATIVES COMPARISON

ALTERNATIVE	SCORE (% OF MAX)
Dike with Vegetation	**79**
Dike with Rock	**85**
Soil with Vegetation	**64**
Soil with Rock	**65**

CONCLUSIONS

- **DIKE WITH ROCK PROVIDES BEST PERIMETER ENCAPSULATION**
  Reasons – Robust: High Factor of Safety for Slope Stability
  – Substantial: Able to Accommodate Biointrusion to Wastes
  – Construction: Built Ahead of Waste; No Worker Exposure
- **ROCK PREFERRED TO VEGETATION**
  Reasons – Maintenance: Better Erosion Resistance
  – Stability: Durable Rock
  – Aesthetics: May be Publicly Acceptable

Value engineering used to compare alternative waste disposal cell perimeter encapsulation systems.

FIGURE 6.
VALUE ENGINEERING RESULTS AND SUMMAR

- Deformation: The deformation of a clean-fill dike depends only on the materials of the dike which may be carefully selected and placed. Conversely, the multicomponent cover rests on the waste, which may settle, deform, or even suffer deep-seated slope failure, hence impacting the cover. Less stringent waste placement control is feasible if clean-fill dikes are used.

- Construction: Clean-fill dikes facilitate construction because they may be constructed independently and ahead of the waste (see Figure 6.5). Dike construction may be scheduled so that no interfacing of the crews placing the dikes and the waste is necessary; if the waste is radioactive, this keeps exposures as low as is reasonably achievable.

- Erosion: In theory, there is no difference between clean-fill dike and multicomponent cover erosional resistance. Nevertheless, in the event of unplanned cover erosion, the consequences are less severe with the clean-fill dike with a multicomponent cover (see Figure 6.6). In brief, the wastes are closer to the outer erosion resistant layer with a multicomponent cover than with a clean-fill dike.

- Infiltration: Infiltration into and through a clean-fill dike does not affect leachate quantity or quality (see Figure 6.7). Hence overall leachate control is facilitated and reduced by using clean-fill dikes.

In conclusion, we note a considerable personal preference (engineer and ecologist, alike) for clean-fill dikes. They are robust, intuitively more stable for long-term cell performance, and more in accord with a total encapsulation philosophy.

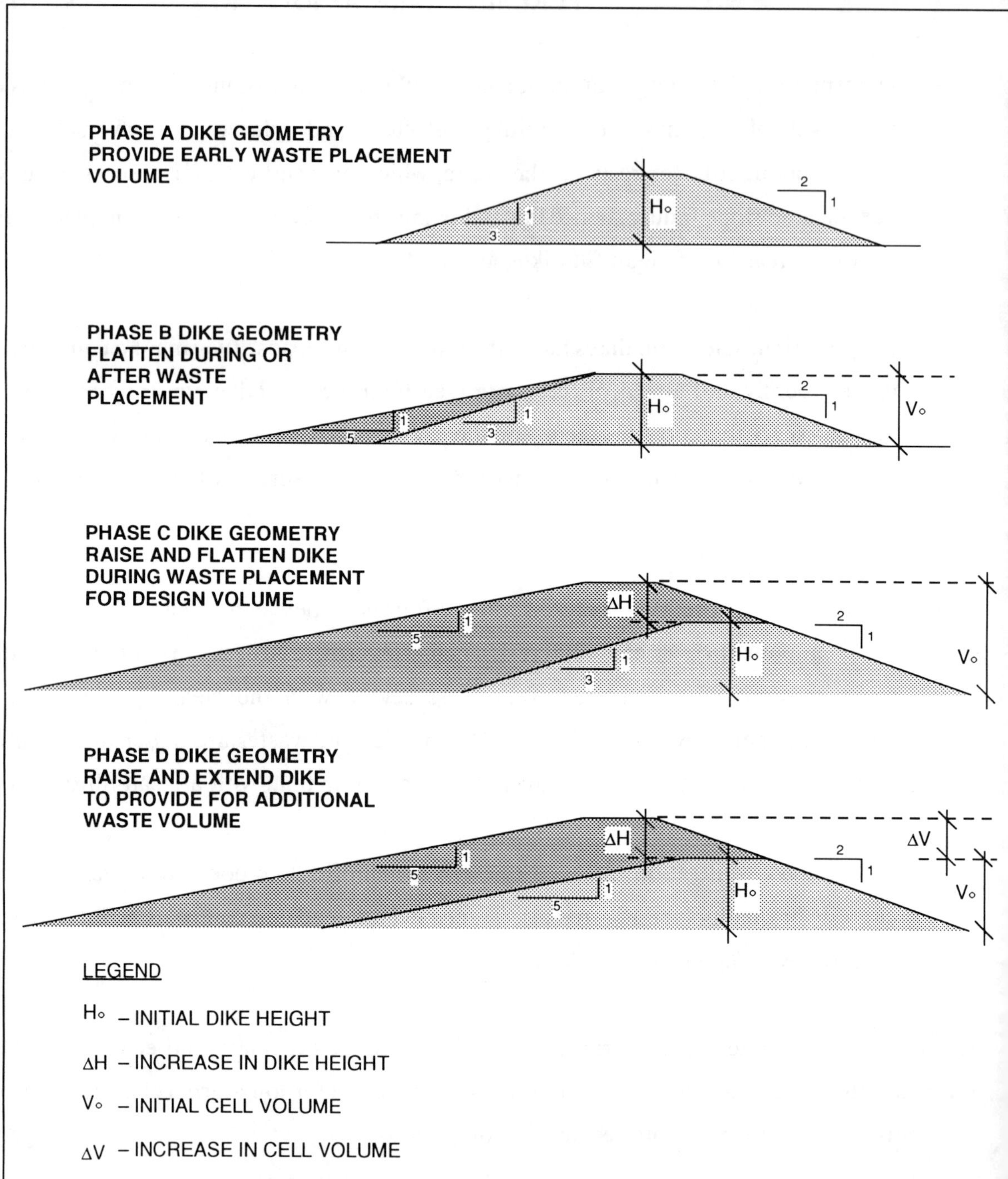

In order to provide for immediate waste volume, and possible increased waste volumes, the dike construction may be phased.

FIGURE 6.5

CLEAN-FILL DIKE CONSTRUCTION ALTERNATIVE PHASES

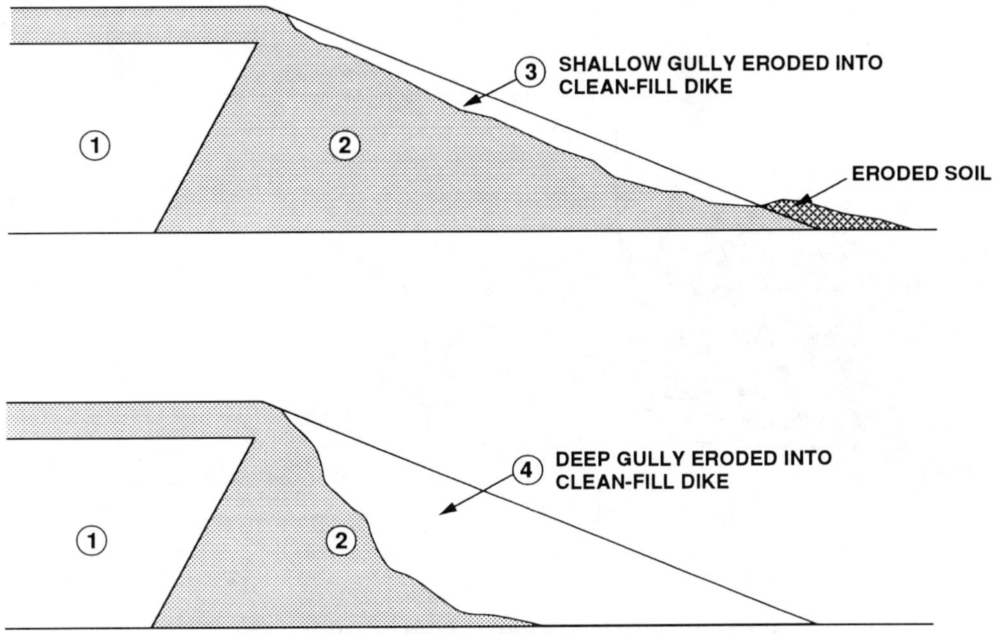

FACTORS THAT PRECLUDE THIS FAILURE SCENARIO

- Rock cover that resists erosion by probable maximum precipitation runoff
- Continued surveillance, monitoring, and maintenance

LEGEND

(1) WASTE
(2) CLEAN-FILL DIKE
(3) SHALLOW GULLY
(4) DEEP GULLY

Even significant gully formation by erosion of the clean-fill dike will not expose the encapsulated wastes.

FIGURE 6.6

CLEAN-FILL DIKE EROSION GULLY GEOMETRY

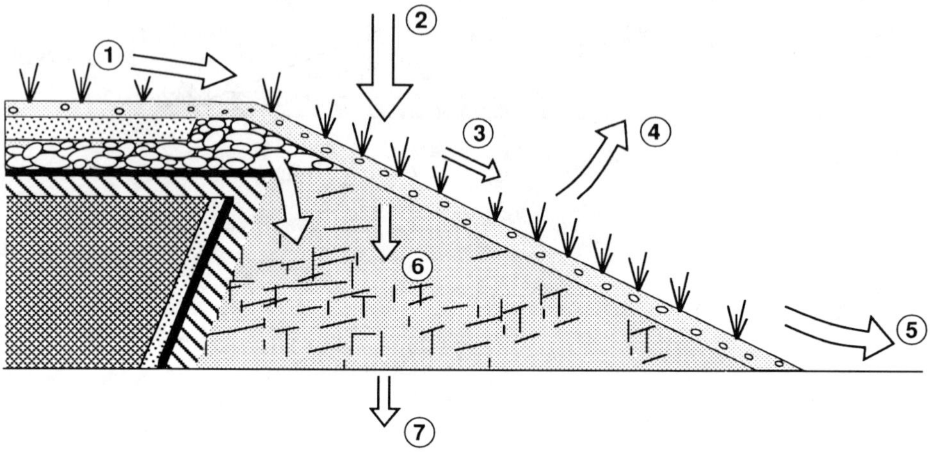

CLEAN-FILL DIKE SECTION WITH FLOW COMPONENTS

FLOW LEGEND

1. TOPSLOPE RUNOFF
2. SIDESLOPE INCIDENT PRECIPITATION
3. SIDESLOPE RUNOFF
4. SIDESLOPE EVAPOTRANSPIRATION
5. TOTAL CELL RUNOFF
6. DIKE INFILTRATION
7. SEEPAGE TO VADOSE ZONE

Vegetation on the dike reduces the quality of flow to the dike and groundwater.

FIGURE 6.7

CLEAN-FILL DIKE WATER BALANCE MODEL FOR VEGETATION COVER

Chapter 7

BASAL LINERS AND DRAINS

GENERAL

The long-term effectiveness of the encapsulation systems depends primarily on the covers and the perimeter encapsulation system. A basal liner and drain system may be used to achieve limited short-term performance objectives, primarily initial control of leachate from the wastes. The only valid long-term performance objective of a basal liner system is possibly the geochemical attenuation of the seepage from the waste. We believe that before wastes are encapsulated at a site, the designer must demonstrate with appropriate performance analyses that after the cessation of institutional controls, the cell performance will not result in unacceptable environmental impact. As described in a later section, this involves demonstrating both that the uncontrolled seepage from the drain outlets (if still functioning) will not impact surface or groundwater quality, and that seepage through a degraded liner (particularly a geomembrane) will not impact groundwater quality.

ALTERNATIVE BASAL ENCAPSULATION SYSTEMS

No Liner

An encapsulation system without a liner is standard practice if wastes are isolated in-situ. It is neither practical nor cost effective to pick up the wastes simply to place a liner. Liners are not used on the Uranium Mill Tailings Remedial Action Project disposal facilities. In both these cases, the philosophy is that the cover and perimeter dikes are the functional components that encapsulate the waste and that control infiltration and hence seepage from the cell.

I believe that the best approach to permanent encapsulation of wastes is the selection of a site where a liner is not required to effect groundwater protection. My logic is that in the long-term a liner will either fail, in which case the effect is identical to the no-liner case, as shown in Figure 7.1, or if it remains intact and functional, will simply divert the seepage to the perimeter of the cell where it will seep to the groundwater. At most, this represents a minor spatial diversion of the seepage.

Ideally, the disposal cell site should be underlain by soils of sufficient thickness and attenuative capacity to protect the groundwater from the potential consequence of contaminated seepage from the cell. If such a site does not exist in reasonable proximity to the waste, then the waste itself must be rendered innocuous, so that the seepage that is inevitable will not negatively impact the surrounding groundwater.

If the groundwater beneath the site is totally unsuitable for use, now or in the future (Class III), then this is another way that a site may be suitable for a cell without a liner.

Two approaches to the demonstration of groundwater protection by the site foundation soils, in the absence of a liner, are the travel time argument or the water quality argument. The travel time argument involves demonstrating that the time it will take for seepage from the cell to reach and impact the groundwater is in excess of the disposal facility design life. I personally find this approach offensive. I cannot concur with an approach that says in effect: we will not pollute the groundwater for at least 200 years, but thereafter all bets are off. This is "acceptable" only because the regulations say the cell must last for at least 200 years, but do not dictate what happens thereafter. The second, and I submit only ethically acceptable approach, is to demonstrate by means of a comprehensive, steady state, cell and

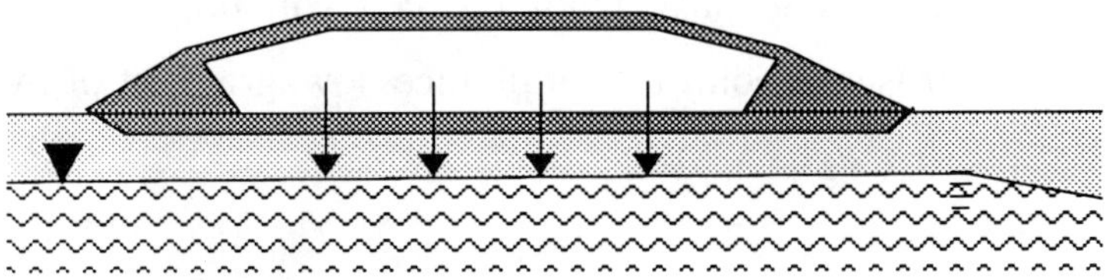

THE BASE LINER SEEPAGE MODEL

DESIGN DETAILS

- Block drain outlets
- Ensure seepage from cell through liner

GROUNDWATER PROTECTION MODEL

- Attenuation in foundation and vadose zone
- Dilution in groundwater

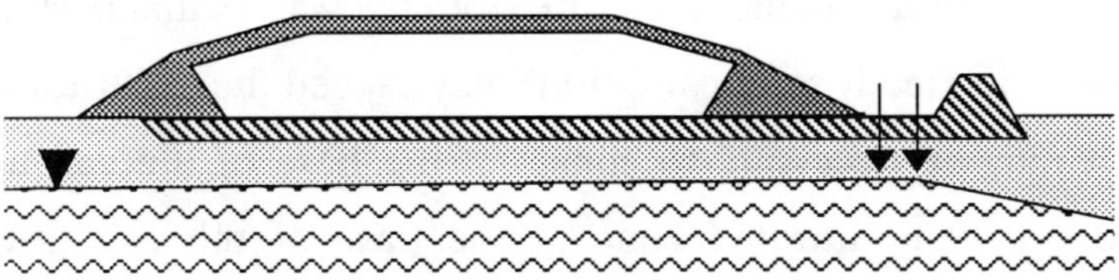

THE GEOCHEMICAL SUMP MODEL

DESIGN DETAILS

- Provide geochemical sump at drain outlet

GROUNDWATER PROTECTION MODEL

- Geochemical attenuation by clay sump liner

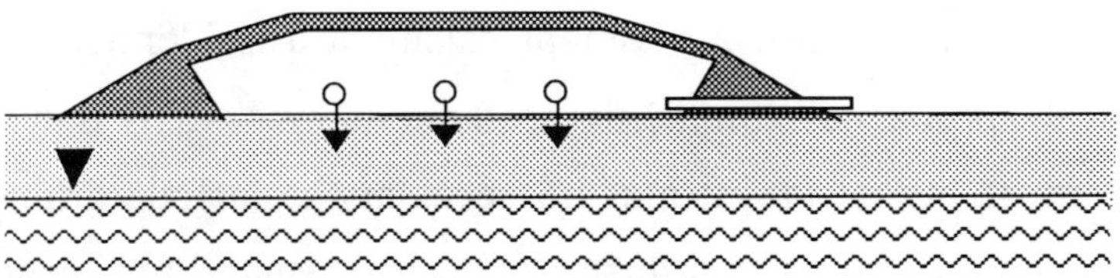

THE PERIMETER MULTIPLE DRAIN MODEL

DESIGN DETAILS

- Provide multiple drain outlets around cell perimeter

GROUNDWATER PROTECTION MODEL

- Dispersion of leachate over wider area
- Attenuation in vadose zone
- Dilution in groundwater

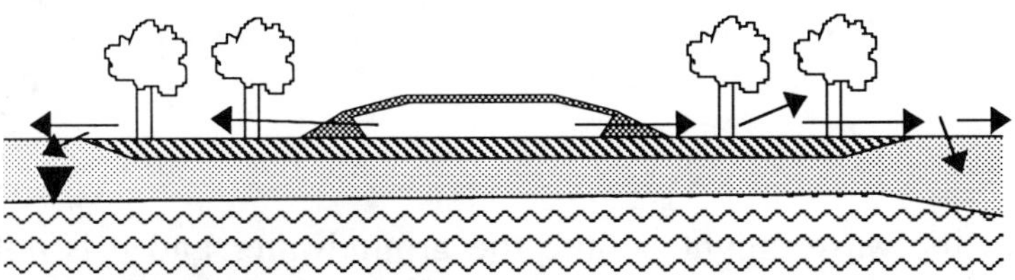

THE PERIMETER SPREAD AND INFILTRATION INHIBITOR MODEL

DESIGN DETAILS

- Grade site away from cell
- Construct low permeability layer around cell perimeter
- Provide multiple drain outlets

GROUNDWATER PROTECTION MODEL

- Leachate dispersed over wide area
- Evapotranspiration and rainwater dilution reduces seepage potential
- Low permeability inhibits infiltration

After cessation of institutional controls, groundwater protection is achieved primarily by infiltration minimization and quality attenuation.

FIGURE 7.1

LONG-TERM GROUNDWATER PROTECTION: ALTERNATIVE MODELS

groundwater performance assessment that the seepage from the cell will not result in exceedance of groundwater quality standards at a point of compliance downgradient of the cell.

The ideal site for a cell without a liner probably has a foundation soil of low permeability. Because the transient drainage (i.e., the initially higher flow rate from the wastes that occurs before the waste moisture content is at its steady state condition) may exceed the infiltration capacity of the foundation soils, it may be necessary to construct the base of the cell to provide a reservoir for the build up of a saturated zone at the base of the cell (see Figure 7.2). The increased hydraulic head at the base of the cell increases the infiltration rate. The trick is to ensure that there is adequate capacity in the "reservoir" to preclude overflow of the transient pool. The performance assessment calculation is identical to standard reservoir routing technical approaches commonly used for dams. If it is not possible to deal with transient drainage in this way, it may be necessary to include drains at the base of the cell to collect the excess transient drainage. This is discussed further in a subsequent section.

Geochemical Barrier

If the quality of the seepage from the waste might pollute the groundwater and it is not possible to alter the waste to moderate seepage quality, a geochemical barrier at the base of the cell may be required. A geochemical barrier is essentially a layer of selected soil with or without additives, such as carbon or lime, that is placed at the base of the cell to alter the quality of the seepage from the cell before it enters the foundation soils and rocks and mingles with the groundwater.

There are no case histories that I am aware of where a geochemical barrier is the sole basis of the encapsulation system's groundwater protection strategy. It is an integral part of the

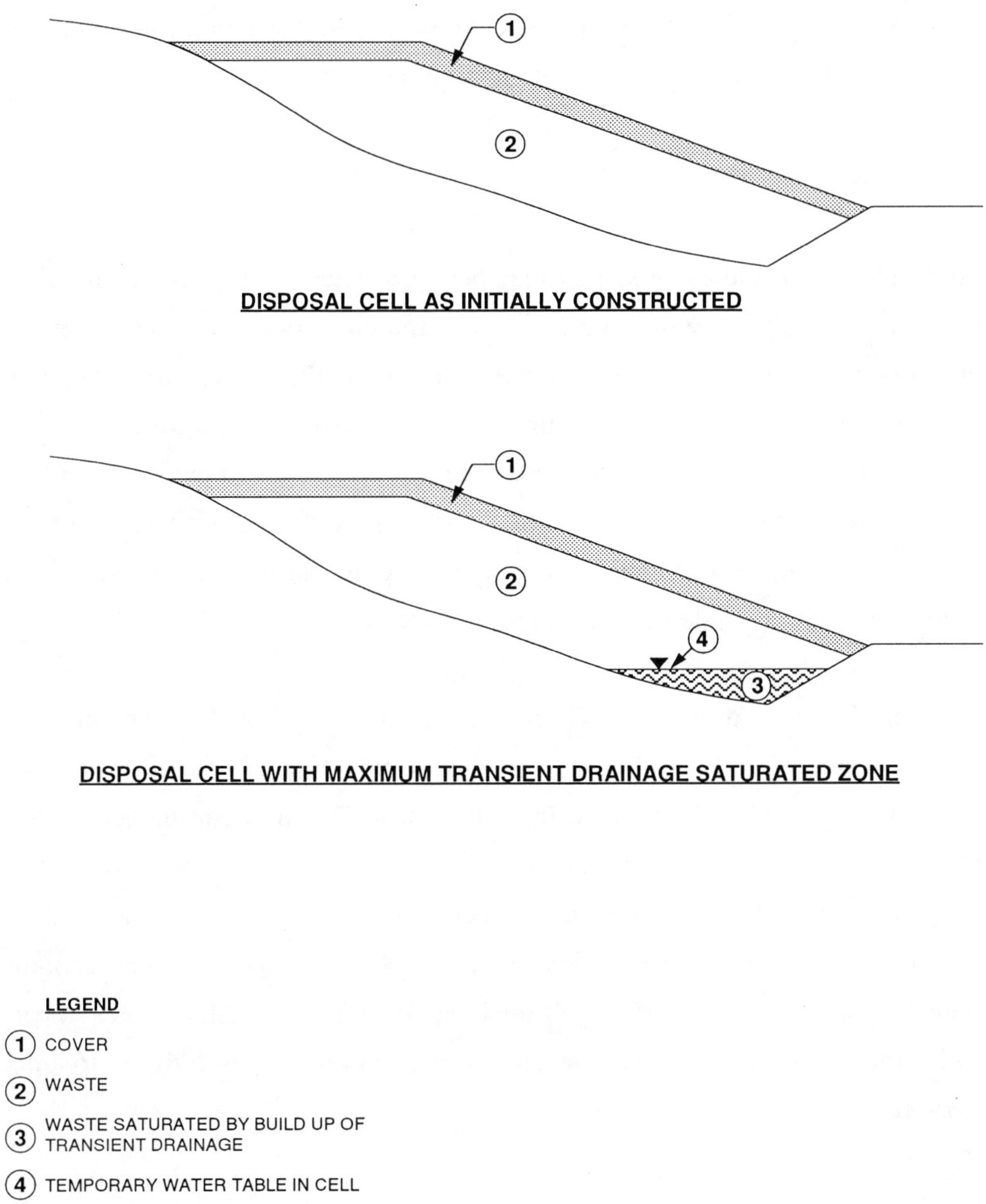

At placement the waste contains excess water, which, transiently, seeps to the base of the cell. If the foundation permeability is less than the transient seepage rate, a saturated zone may form in the waste. With time, all water will drain away.

FIGURE 7.2

TRANSIENT SEEPAGE POND DEVELOPMENT IN CELL WITH LOW PERMEABILITY FOUNDATION

design of the cell at sites such as the UMTRA Gunnison facility, where it is used in concert with seepage control and foundation seepage quality modification.

Liners and Drains

A single or double liner and drain system may be constructed at the base of the disposal cell. There is considerable literature on the design and construction of such systems, and we do not repeat any of that here. Rather we concentrate on the criteria for including such systems in an encapsulation system. The first and best reason to incorporate a liner and drain in a disposal cell is to control and collect excess transient drainage. Provided the period of transient drainage is less than the feasible period during which institutional controls may reasonably be invoked, the drains are a practical design detail to intercept contaminated flow and preclude impact on groundwater quality.

Many regulations for waste disposal incorporate the need for a double liner and drain. The lower drain is considered to be both a second backup drain and a leak detection system. Most of these regulations are formulated for cells that will encapsulate innocuous wastes with a relatively short period of potential contamination. The only federal law that deals with near-surface disposal of wastes that are unequivocally long-lived and noxious is the Uranium Mill Tailings Radiation Control Act of 1978. Neither the law nor the regulations, subsequently promulgated by the EPA, call for liners or drains. I believe this is a correct recognition of the futility of drains and liners for encapsulation systems designed to function for the long-term.

Unfortunately the laws that call for double liners and drains are considered to be potentially applicable or relevant and appropriate for many waste disposal facilities. If this is the case, the designer has no option but to obey the law, design and build liners and drains, and show with performance assessments that when, in the long-term, the drains clog, the liners deteriorate, and the facility is abandoned, the inevitable seepage at the perimeter of the cell (as shown in Figure 7.3) or through the base of the cell (as shown in Figure 7.4) will not negatively impact human health and the environment.

Liners and drains may be constructed of natural or synthetic materials. If the primary purpose of the drains and liners is control of transient drainage, synthetic materials are appropriate. If the design calls for long-term assured functioning of the drains and liners, then only natural materials should be used. Selection of the drain material and layout details to preclude clogging or blocking is the paramount design consideration. Selection and placement of the liner material to preclude deterioration is the primary design consideration.

Pipes may be placed in the drain material. Figures 7.5 and 7.6 show alternative layouts for drains in a leachate collection system. I believe that pipes should be viewed simply as an extra, and not as the functional water conveyance route. Pipes are too susceptible to breaking, deformation, and clogging to be relied on as a functional entity in an encapsulation system designed for long-term performance. Pipes are a valuable part of a drainage system if used as part of an overall performance monitoring network in a cell. The pipes may be located to collect seepage from the waste and the cover from discrete areas or parts of the cell. By monitoring the flow from individual drains, quantification of the hydrologic performance of areas of the cell is possible. If the cover fails in a specific area, increased flow from the subtending drain is a potentially sensitive indicator of localized failure.

Covers may be repaired or replaced during the period of institutional controls. Monitoring the flow from the uppermost drain is, therefore, reasonable. I cannot understand the point of monitoring flow from a lower drain, or leak detection system. I find it inconceivable that some day we will go in and pick up one or two million tons of material in order to place a

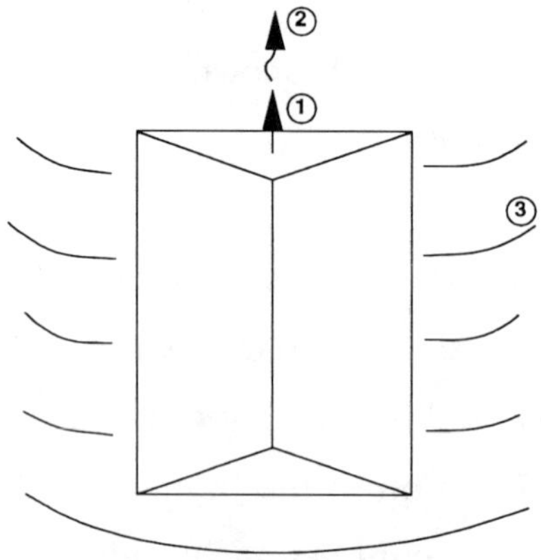

ABOVE-GRADE MODEL I
DRAIN OUTLET TO NORTH, CONCENTRATED SEEPAGE FLOWS OVER LAND TO DISTANT WATER COURSE, EVAPORATES OR SEEPS INTO SOILS

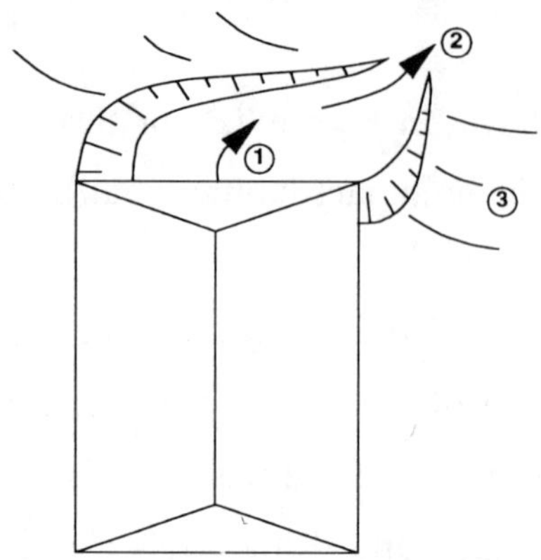

PARTIALLY BELOW-GRADE MODEL I
DRAIN OUTLET TO NORTH, CONCENTRATED SEEPAGE FLOWS THROUGH CHANNEL TO DISCHARGE TO NATURAL GRADE, HENCE TO FLOW OVERLAND TO DISTANT WATER COURSE OR SEEP INTO SOIL

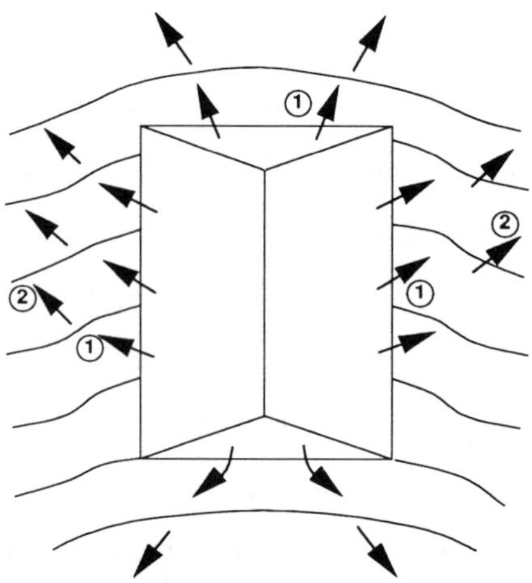

ABOVE-GRADE MODEL II
DRAIN OUTLETS AROUND PERIMETER DISSEMINATED SEEPAGE FLOWS AWAY FROM CELL; BECAUSE OF VERY LOW FLOW RATES MOST EVAPOTRANSPIRATES OR SEEPS INTO PERIMETER SOILS

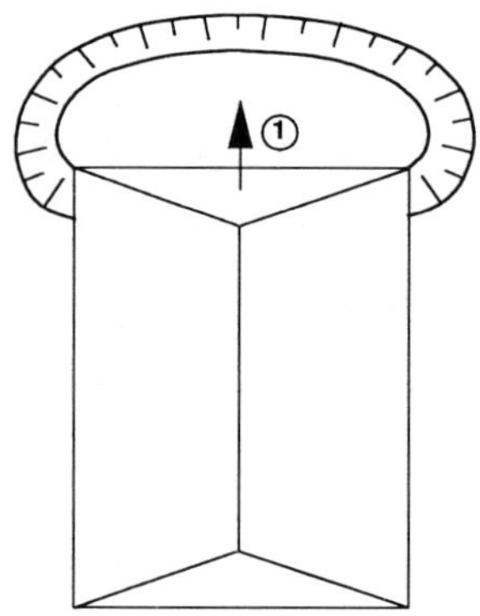

PARTIALLY BELOW-GRADE MODEL II
CELL SEEPAGE TO NORTH INTO SUMP FROM WHICH LEACHATE EVAPORATES OR SEEPS INTO SUMP FOUNDATION SOILS

LEGEND
(1) CELL SEEPAGE
(2) OVERLAND FLOW
(3) SURFACE CONTOURS

After cell abandonment, seepage will continue to emerge from the cell, and will flow overland to distant water bodies, be evapotranspirated or seep into the surrounding soils.

FIGURE 7.3

LONG-TERM DRAIN OUTLET AND SURFACE FLOW OPTIONS

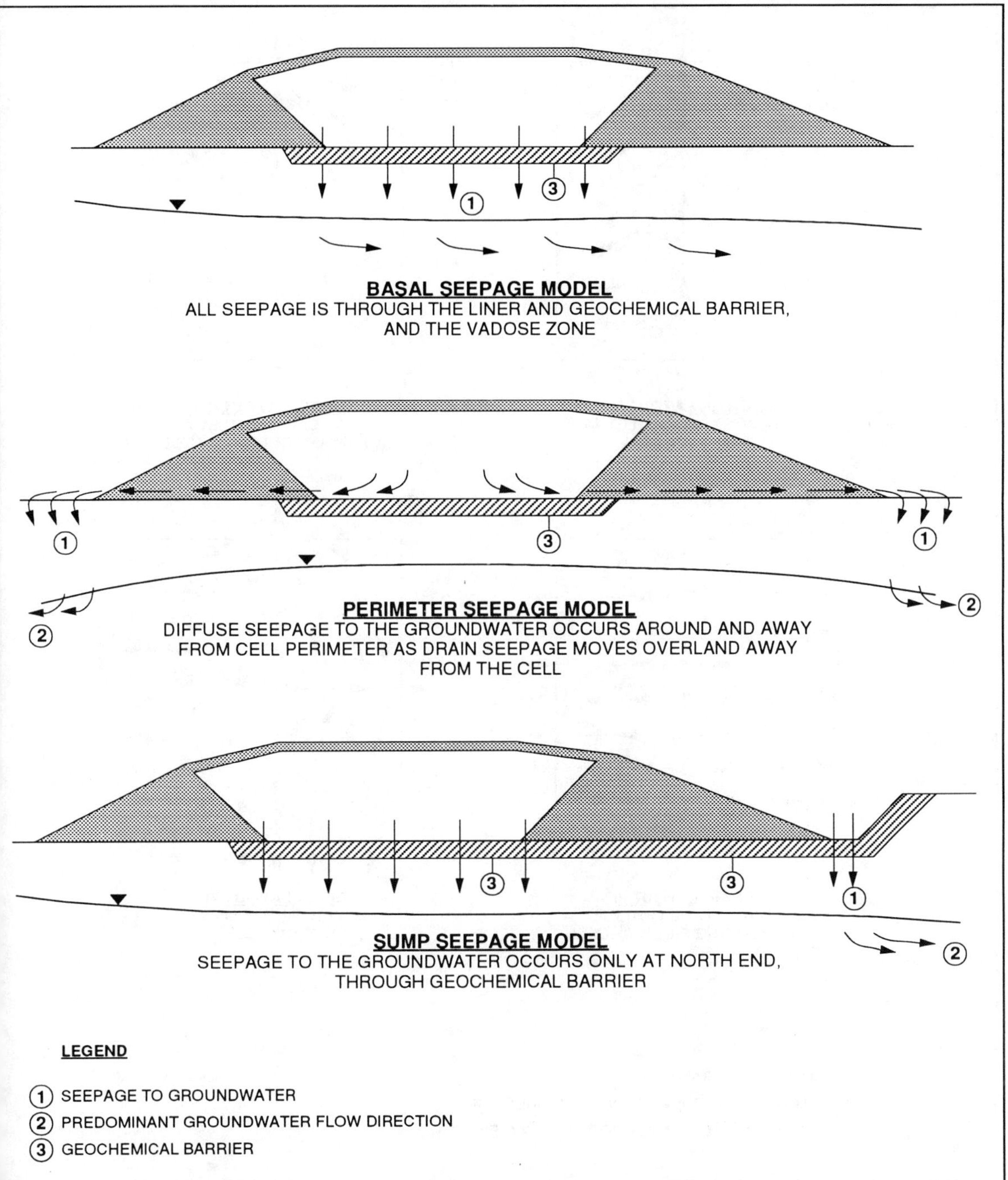

After cessation of cell operation and monitoring, minor seepage to the groundwater is feasible. The route and quantity depend on cell layout details, as shown in the alternatives.

FIGURE 7.4

SEEPAGE TO GROUNDWATER IN THE LONG-TERM

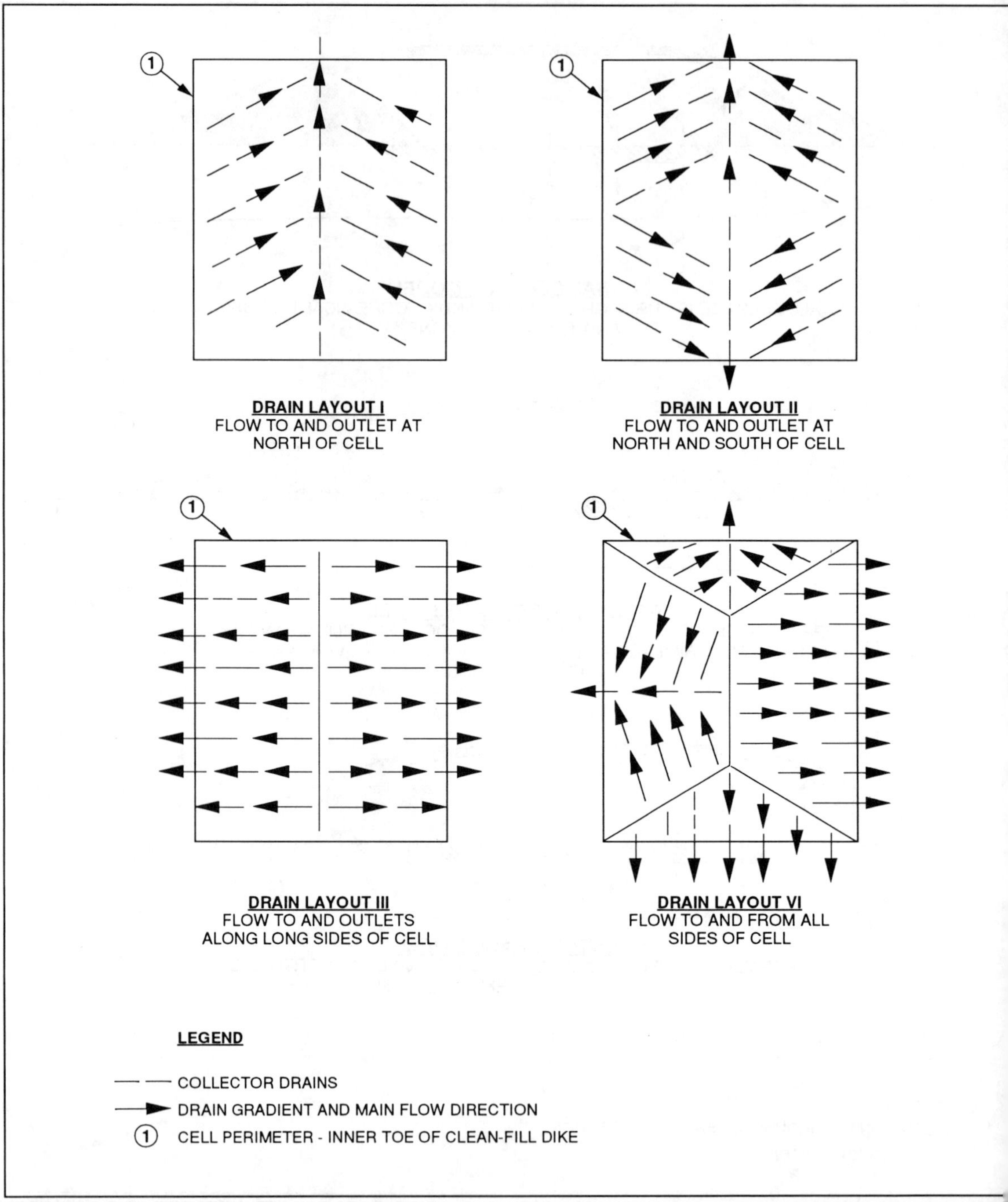

The leachate collection system drains may be arranged to focus or disseminate seepage from the cell.

FIGURE 7

ALTERNATIVE BASAL DRAIN LAYOU

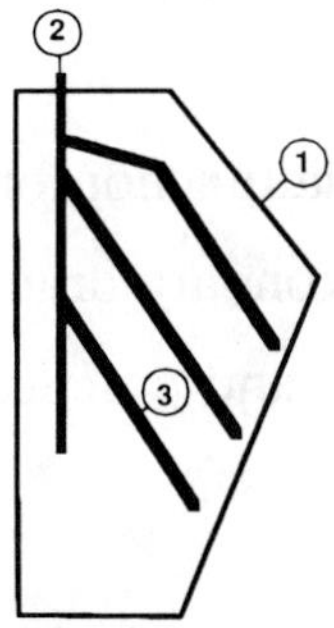

SINGLE DRAIN OUTLET

ADVANTAGES
- Easy Leachate Collection And Control
- Cost-effective Because Only One Dike Drain Required

DISADVANTAGES
- Vulnerable To Failure Of a Single Drain Outlet
- Complex Internal Geometry

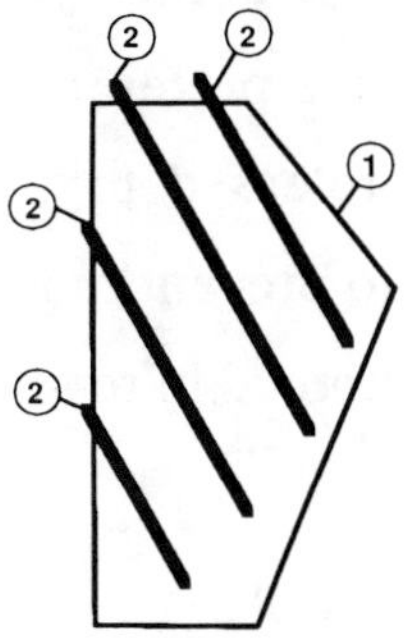

INDIVIDUAL DRAIN OUTLET

ADVANTAGES
- Sectional Monitoring Of Cell Flow Is Possible
- Simple Drain Geometry
- Less Vulnerable To Total Drain System Failure

DISADVANTAGES
- External Collection And Control Of Leachate Required

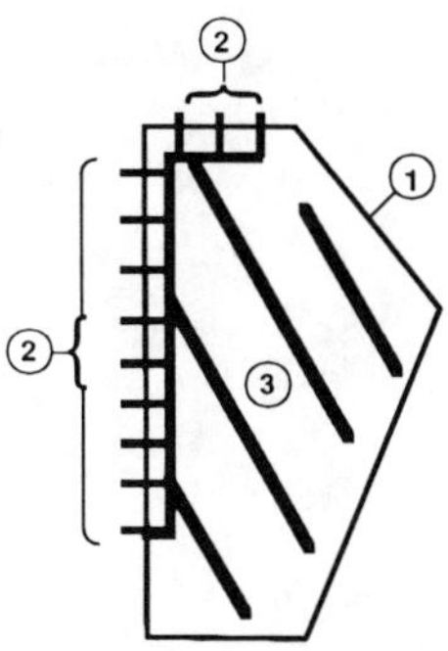

MULTIPLE DRAIN OUTLET

ADVANTAGES
- Highly Resistant To Single Component Failure
- Long-term Dispersal Of Seepage Reduces Potential Environmental Impact

DISADVANTAGES
- Complex To Build
- Multiple External Points For Leachate Collection & Control

LEGEND

(1) DISPOSAL CELL INTERNAL PERIMETER
(2) DRAIN OUTLETS
(3) INTERNAL DRAINS

One or multiple leachate collection outlet drains may be provided. The most reliable is an interconnected multiple outlet scheme.

FIGURE 7.6
ALTERNATIVE DRAIN OUTLET SCHEMES

new primary liner simply because the leak detection system indicates malfunction of the primary liner. If I am correct, the lower drain is thus reduced to a secondary drain, or redundant seepage control component; this in and of itself is reasonable and practical, if indeed this is recognized and acknowledged as such by all concerned.

A concern in the encapsulation of radioactive and mixed waste is that liners, drain layers and leachate collection and removal systems may constitute pathways for the preferential migration of radon gas. If the system daylights, then radon may be released to the environment in significant concentrations. Filters or traps may be needed to prevent radon migration. Once the transient drainage has been collected, grout may be pumped in to seal off potential pathways for radon migration.

Chapter 8

ECOLOGICAL SYSTEM

THE ECOLOGICAL APPROACH TO CELL DESIGN

The previous sections have outlined several systematic and objective approaches to cell design, including methods for selecting and perimeter encapsulating elements. These reflect the engineer's reliance and emphasis on the deductive approach to decision-making. Designs and decisions are deduced from logical thinking, calculations, and the results of performance assessments. It is this kind of structured creativity that has led to the engineering marvels of our technological heritage, cited proudly by our senior author in the book's opening chapter.

For the most part, even the most impressive engineered structures have required maintenance, or they have gradually failed in their purpose, integrity, and, with notable exceptions, beauty. The time is nigh for new confidence in our ability to engineer for perpetuity, because this is the challenge that must be met for our long-lived wastes. Maintenance is an unacceptable concession for systems to encapsulate waste whose longevity of hazard extends generations into the future. The ecological perspective is that, to the extent technically achievable, cells must be maintenance free.

How is this to be done? The systematic, deductive process described previously is the only solution, the only framework for designing a complex, multicomponent facility such as a disposal cell. But the process must be bolstered by a dose of intuition; inductive thinking that integrates observation, judgment, and understanding of nature into the design process.

The cell components closest to nature are the cover and perimeter encapsulation systems. For this reason, the concluding portion of this section on cell design -- on the ecological approach to cell design -- addresses these uppermost and outermost features of a cell. The ecological approach to cell design spelled out in the following section is then elaborated and substantiated in the sections on performance assessment, especially as related to vegetation (Chapter 17) and biointrusion (Chapter 18).

DESIGNING A VEGETATED COVER

There have been numerous references up to this point about the merits of the ecological perspective in general, and of vegetated covers in particular. It is now appropriate to summarize once and for all the definitive attributes of vegetated covers and to explain how such covers may be designed.

A vegetated cover is a cover that totally accommodates plants, using them to full advantage to maintain water balance, resist erosion, and prevent biological damage to the underlying layers. A soil cover does not qualify as a vegetated cover unless its bio-hospitality has been maximized. A rock cover cannot by definition be a vegetated cover, although rocks may be integrated into vegetated covers to enhance erosion resistance as long as they do not unduly depress biological vigor. Rocks may also be used to define the lower boundary of the rooting medium and thereby protect the underlying layers from biointrusion.

Definitive Attributes

The definitive attributes of a vegetated cover are simple and few. Specifically, a vegetated cover is a cover that:

- Fully accommodates organisms, using them to enhance the performance and longevity of a cell

- Does not try to resist biological colonization, because perpetual resistance is futile in the habitable portions of our living planet

- Relies on natural materials in more-or-less naturally occurring densities and stratigraphies to achieve the desired encapsulation

- Incorporates in its upper layers an adequate thickness of favorable soil to capture almost all infiltrating moisture, store nutrients, and sustain natural near-surface ecological processes

- Includes a transition beneath the favorable layer into an unfavorable zone that discourages roots and soil organisms from advancing downward

- Responds to environmental stresses such as climate change in a biologically predictable way; for instance, by increasing or decreasing the amount of vegetation to transpire the available moisture

- Repairs itself after minor damage by erosion, human intrusion, or other factors that perturb the surface

- Integrates into the native landscape with a gently graded topography and natural, self-sustaining plant community.

A final, summary attribute is that a vegetated cover should be permanent, at least on the scale of a human lifetime. The best vegetated covers will grow, thickening themselves with captured carbon and intercepted windborne matter. If the cover degrades or erodes, then the process must occur only gradually, in concert with similar processes that perpetually transform the surrounding native landscape. The design should be inherently fail-safe, uncompromised by potential failure scenarios that are tolerated only because it is deemed impractical or uneconomical to do otherwise. Long-term change in a cover may be inevitable; if erosion is unavoidable, then it must be gradual and predictable, releasing or exposing contaminants to the environment only very slowly, at a rate where dilution, attenuation, or contaminant transformation -- for instance, through radioactive decay, denaturation, or biodegradation -- renders the situation nonthreatening to the environment or to public health.

What this all means is that the design of a vegetated cover must be sound with respect to all facets of a multidisciplinary review. If any credentialed member of the design team, or of a review team -- be they ecologist, hydrologist, climatologist or engineer -- can legitimately challenge the ability of the cover to perform as intended in the face of reasonably foreseeable circumstances, then the design is unsuited to its higher purpose. The design must stand up to the rigors of thorough review, whether mathematical or intuitive, just as the cover itself will have to accommodate the varied affronts of nature.

Topslopes, Sideslopes, and Clean-fill Dikes

Vegetated covers are the covers of choice for the topslope of any cell, if indeed the purpose of that cell is to provide the longest possible, maintenance-free isolation of the wastes within. Topslopes directly overlie waste, so failures in cover performance bear immediately on the cell's ability to prevent contaminant release. It should be possible in almost every case to grade the topslope sufficiently flat as to obviate the need for large rock as a defense against heavy weather.

A simple soil cover belongs on a topslope only if the intent is to provide nothing more than temporary protection, and if underlying geosynthetics are to be the operative agent for water balance. A rock cover belongs on a topslope only if perpetual and frequent maintenance in the form of plant removal is desired; for instance, to provide everlasting employment for those who live around the cell.

The argument for vegetated covers on sideslopes of disposal cells is perhaps less compelling than for vegetated covers on topslopes, because erosion is an especially formidable hazard on the long or steep slopes that may be required to give the cell its required shape, size, and drainage characteristics. Without doubt, the attributes of a vegetated cover will be totally nullified if the cell is breached by a gully that only riprap would have prevented. Having granted this, I must nonetheless emphasize that if the sideslope indeed overlies waste, then degradation due to infiltration, drain clogging, and biointrusion will eventually impair cell performance. Once again, if a rock cover is to overlie waste, then the designer might as well include a multi-generational maintenance contract in the bid specification for the cell.

The requirement for vegetated covers on clean-fill dikes is less compelling still. Moisture infiltration and root penetration into the soil of the dike will not directly or immediately impact the waste. Deep-rooted trees and shrubs will predictably colonize the surface of a rock-covered dike, but the only damage would involve occasional lateral excursions by roots or animals into the waste, or gradual degradation due to the excavations, soil inversion, or concentrated flow that results from trees that topple on the sideslope. The rock-covered dikes will gradually roughen, churn, and erode to a modified grade, but the consequences of cell performance will be greatly mitigated by the fact that the dike fill is clean and the primary encapsulation system remains unbreached.

For Every Surface, a Vegetated Cover

The best cell designs will accommodate a vegetated cover on every surface, with riprap used only to provide extra protection against erosion at the points of greatest potential stress. The cell should be graded into a form that emulates and integrates with the surrounding geomorphology. Analyses have shown that clean-fill dikes may have to be sloped at 10:1 or gentler if they are to be vegetated, as opposed to riprapped, and relied upon to provide adequate protection against erosion. Thorough studies of the local geomorphology and of cover soil properties will be needed if steeper dikes are to be considered. In any case, the extended dikes generally require extra space, soil, and sometimes expense, all of which constitute a worthwhile investment if the payoff is a sound and enduring cell.

Vegetated Covers in Deserts?

The argument has been made that rock covers are appropriate in highly arid settings because vegetation is naturally sparse and few plants are present to invade and damage a cell. Furthermore, the lack of plants on a vegetated cover will render the soil highly vulnerable to erosion by the infrequent, but torrential storms that typify desert weather. The problem with this perspective is that, even in the driest desert, rock covers will gradually accumulate moisture due to the one-way action of the rock, allowing precipitation to infiltrate, but preventing soil moisture from evaporating. Drought-toughened desert plants are opportunistic and aggressive, relentlessly seeking pockets of available moisture and, once finding these, fully extracting the resource with their deep and extensive roots. Unfortunately, each of the rock-covered disposal cells now dotting the western deserts of the United States is but a man-made oasis, awaiting discovery, utilization, and gradual degradation by saltcedars, Russian olives, mesquites, cottonwoods, and a host of other desert phreatophytes.

As to the difficulty of establishing a protective mantle of vegetation in the desert, the answer lies in the effective use of desert pavement. An appropriate veneer of gravel over a well-graded soil will both stabilize the soil and harvest moisture to sustain vegetation. The desert pavement blocks evaporation and thereby increases the water supply for plants. This in turn increases the transpirational capacity available to remove moisture and prevent drainage after a desert deluge. The difference between this cover and the rock cover mentioned above is that the moisture is retained in soil near the surface, apart from the waste, where water-seeking roots will not damage the cell. As will be emphasized later, the key to designing a vegetated cover in the desert, as in all habitats, is to seek and emulate an appropriate natural model.

DESIGN APPROACH

Ecological cover design begins in nature. The multidisciplinary design team, led in this step by the ecologist, should descend upon the environs of the proposed cell and thoroughly study the habitats that may be emulated in the design and construction of the cell and of its peripheral tie-in to the landscape. The most important question to be answered is: what native plant community should be used as a model for the vegetated cover?

The geographic scope of the survey should encompass the area over which local climatic and topographic conditions are reasonably similar to conditions where the cell is to be built. The survey must not extend so far as to consider habitats that could not be supported on the cell. The search area on level terrain might be quite large because of the relative uniformity of conditions, whereas the search area in mountains would be smaller due to the dramatic microtopographic/climatic variability. On the other hand, if the searcher is careful to consider only those habitats in microtopographic settings similar to the proposed cell's, then the search area in montane settings may be expanded considerably.

The model community will, in almost every case, be a grassland: a desert grassland, a prairie, or a meadow. Grasses tend to be shallower-rooting than trees or shrubs. When grasses do root deeply, the roots are fibrous, thinner, and potentially less damaging to a cell's deep layers than are the woody, suberized roots of trees and shrubs. Grass roots form a dense and interwoven fibrous network to bind the soil and heal it when damaged. Grasses concentrate their biomass close to the ground, building a protective mat that provides the highest degree of basal and foliar cover to protect the soil against wind erosion, rainsplash, and overland flow.

Under some circumstances, grassland communities sustain themselves, and resist invasion by deep-rooted trees and shrubs, through the process of ecological competition. By fully covering the soil and extracting near-surface moisture, grasses prevent the seed of potential invaders from germinating. The tree and shrub seedlings that do manage to germinate in a vigorous grassland community are subsequently choked out by the sod or deprived of moisture by near-surface water depletion. The perpetuation of the grassland may be assisted by fire or by the action of large herbivores, both of which can be fatally damaging to the seedlings of trees, shrubs, and forbs, but which are often beneficial and revitalizing to grasses. This is because grasses grow from basal meristems at the bottom of each plant, whereas most other plants grow from terminal meristems that are damaged or removed by fire or herbivory. Lawnmowers, too, make quick work of terminal meristems and, in many respects, serve as well-tamed and well-timed grazers.

Thus, it is not just a grassland that is desired for a vegetated cover; rather, it is a self-perpetuating grassland that can be counted on for generations to competitively exclude deep-rooters. This is a climax community, the final and most advanced stage in ecological succession. Ecological succession is the natural process whereby plant communities gradually replace themselves, beginning with the smattering of weeds that first invade a disturbed area and progressing toward the complex, fully developed community that maximally utilizes an area's resources. Forest and woodlands, dominated by deep-rooted trees and shrubs, are the pervasive climax communities for broad tracts of land in the temperate zone. Prairie is the only large-scale grassland climax in North America, although native prairie has been almost completely displaced by agriculture. Even though prairie is practically extinct as a native habitat, it is an excellent choice for vegetated covers because of its self-perpetuation, facilitated by fire; its rarity in nature enhances its value as a habitat well worth restoring.

Even where forests are the pervasive climax, self-perpetuating grasslands occur in patches. We call these localized variants "edaphic subclimaxes" because the soil conditions support a community that will not, under the present circumstances, progress to the pervasive regional climax. If a cell is to be built in a wooded or shrub-populated area, then an edaphic subclimax grassland is probably the community choice for the vegetated cover.

Once a community has been identified by the design team that will serve as a model for the vegetated cover, then the effort should be directed toward a full understanding of the ecology of that community and of the circumstances that explain its presence, composition, and self-perpetuation. Soil properties such as texture, bulk density, stoniness, organic content and nutrient concentrations should be characterized, because these will be emulated in the rooting medium on the cover. The dominant plants should also be studied for their composition, density, and rooting behaviors; this information will assist not only in the formulation of the seed mix, but in the performance assessment for the cell. Computer codes for erosion and water balance both require plant data as inputs, and the model community can provide these data. The model community can also be used for geotechnical and geomorphological investigations that may be required to substantiate cell performance.

The model plant community serves as a general target for the design effort. The community installed on the vegetated cover will differ in several practical respects; for instance, the cover may have a lower boundary or biointrusion layer to confine the biological cycles to the cell's upper layers. The soil may differ somewhat based on the type of borrow material available, and a rock mulch may be used to enhance erosion resistance. The species composition will be less diverse because it is impractical to harvest and apply seed for every native species in the model community. In other words, the model community will serve as an ideal, or concept to uphold in the development of the actual cover community. Practical constraints dictate that the man-made plant community on the vegetated cover will at first be but a crude approximation of the natural model; however, if the design is astute and the execution faithful, then the community on the cover will gradually succeed toward a condition where it is indistinguishable from the natural model from which it was conceived.

Two other practical activities involved in the development of a vegetated cover are (1) construction of test plots and (2) consultation with local experts in agronomy, range management, agriculture, forestry, and other disciplines where individuals have direct experience revegetating land. The test plots, such as are used in mine reclamation, should encompass several hundred square feet and should help answer the following questions:

- Will the additional rock required for erosion protection significantly depress the vigor and sustainability of the desired plant community?

- What are the optimum practices for seed bed preparation, seed application, fertilization, and irrigation?

- What will the vegetated cover look like during the first few years after placement, and how much short-term maintenance will be required until the cover is fully established?

The actual planning, design, and construction of the test plots will yield a wealth of practical experience toward the same activities as they will be applied to the vegetated cover. The plots will also provide additional data for the water-balance and erosion models used for performance assessment and design optimization.

Local experts have enormous insight to contribute toward the planning, both for the test plots and vegetated cover. These practical-minded folks can also help plan the restoration of the area around the cell. Engineers and consultants all too often commit hubris on the issue of revegetation, selecting seed mixes, formulating seed bed preparation plans, and specifying a planting regimen from their urban offices with virtually no consultation with local green-thumbed experts. Several such experts should be consulted, because each may have specific insights on such diverse matters as the germination eccentricities of a particular species' seed to the economics and reliability of local planters and vendors. The best understanding will come from a synthesis of the inputs and perspectives of many experts.

Another key ingredient in the successful design of a vegetated cover is the same as for all elements of cell design: continuous refinement based on multidisciplinary review and value engineering. This process will greatly benefit the design as long as the review and refinement teams never lose sight of the ultimate objective of developing a cover that emulates and integrates with the native landscape, and is unchallengeable with respect to damage by erosion, intrusion, or infiltration.

Having provided general advice on the entire design process for a vegetated cover, I now embark on a brief discussion, one component at a time, of the ecological considerations in the selection of materials and placement of each layer. These remarks apply to the multicomponent cover beginning with the radon barrier and progressing upward to the soil/plant/atmosphere continuum.

COMPONENT DESIGN

Radon Barrier

The radon barrier in most cover designs is placed as an abiotic layer and that's how it should stay. The barrier generally consists of soil high in clay or silt, and is densely compacted during placement. To perform its intended function of preventing the release of radon gas, or any other gaseous contaminants, it must remain compacted and uncompromised by cracks, channels, or conduits. The moisture content must also remain relatively high to intercept radon and prevent desiccation cracking. As mentioned earlier, plants rooted in the radon barrier may dewater the barrier and transport the radon to the surface along with other soluble contaminants.

The radon barrier may be a redundant agent for resisting moisture infiltration or biointrusion, but it should never have either of these roles as its primary mission. It is critical that the overlying layers be constructed in such a way as to preclude the advancement of roots or burrowing animals into the barrier because, once these organisms enter the barrier, bio-modification will relentlessly proceed in a direction that reduces the barrier's performance.

The key to protecting the radon barrier from biointrusion is to confine the biological cycles to the overlying layers. However, other measures are possible to provide additional assurance. For instance, the radon barrier may be conditioned with water to that renders it chemically inhospitable to many organisms. Saline water may be used to condition the barrier, resulting in an elevated sodicity or salinity that will partially or temporarily deter root and animal advancement. A layer of lime on top of the barrier may serve the same purpose by creating a zone of highly alkaline conditions. These measures alone would do little to block the invasion of organisms if the radon barrier were directly exposed to the surface, because the world is full of plants and animals adapted to almost every hostile chemical and physical condition. However, when buried beneath physiologically friendly material, a zone of chemically hostile soil will accentuate the transition between biologically favorable and unfavorable conditions and thereby increase the system's resistance to biointrusion.

Biointrusion/Drain Layer

This layer, too, is placed in abiotic condition and is intended to remain that way. The rocks of the biobarrier should be washed of capillary fines and should be placed in a way that prevents the filtration of capillary fines into the interstitial void spaces. The effectiveness of the loose cobble barrier in resisting biointrusion will depend in large part upon the absence of capillary surfaces therein. Entering gravitational water must be quickly drained away. Gaseous water, hence a humid atmosphere, may be unavoidable, but this will be much less inviting to roots than would an abundance of liquid water accessibly stored on capillary surfaces.

Early cell designers proposed that loose-cobble biobarriers be installed at thicknesses of one meter or more, which provides plenty of protection against biointrusion but renders the cover very expensive. A less expensive approach is to place a sufficient thickness of rock to ensure that every root hypothetically entering the layer will have to grow through at least two void spaces before reaching the capillarily friendly layer that underlies the rocks. Depending on rock sizes, this results in a barrier between one and two feet in thickness.

The rocks in the buried layer should be sized to resist mobilization or displacement by the burrowing mammals that might venture to the bottom of the biologically favorable layer. Some biologists prescribe that the rocks be sized to weigh between two and three times the weight of the mammals that are to be deterred. This is impractical if we are designing against badgers, bears, or giant sandworms, but is a good rule of thumb for keeping out highly abundant ground squirrels or prairie dogs.

A last consideration to remember about loose-cobble biobarriers is that most biologists believe that the rocks and interstitial void spaces are in and of themselves ineffective against biointrusion. Some scientists accordingly consider them a waste of money. The operative factor against biointrusion is the transition between favorable and unfavorable media which, when overlain by an adequate thickness of biologically hospitable material, will serve as a biodeterrent.

Rooting Medium

The rooting medium is the thickest and most important layer of a vegetated cover. This is the layer that provides such favorable conditions for roots and animals that they are disinclined from venturing deeper into the cover. As mentioned above, the selection of soil type should be based on a model community that represents the long-term target conditions for the vegetated cover. Loams and silt loams are generally best because these have optimal water relations, allowing plenty of moisture to infiltrate and absorbing that moisture onto capillary surfaces. The clay and silt content should be high enough to provide good water storage, but not so high as to make the absorbed water unavailable to plants.

Water balance models such as HELP and UNSAT can also be used to guide the designer in selecting the best soil type, and in determining the optimal thickness and placement density. All these factors can contribute to the formulation on paper of the ideal medium. Then comes the less gratifying task of compromising the ideal to accommodate such realities as the availability of borrow material, and the expense of transportation and blending.

Where the resultant material is suboptimal, the designer can attempt to compensate by adding soil amendments, such as lime to add structure or peat to increase moisture retention. One principle holds true: every generation of plants rooted in the layer will improve that layer's suitability to subsequent generations. Every generation adds organic matter, loosens the soil, and introduces faunal associates such as earthworms and fungi to assist with the bio-improvement of the medium. So it is compellingly important to get something going on the cover and, if dieback occurs, to intervene and establish plants anew. No matter how bad the starting conditions, vegetation will eventually improve the substrate to the point where plants and animals can reasonably grow and reproduce.

Vegetation

The crowning layer of a vegetated cover is the vegetation itself, comprised of a community of plants that will reproduce, self-repair, and gradually increase in diversity. Perhaps surprisingly, the revegetation of the cover is not the most important determinant of the long-term performance of the cell. This is because the characteristics of the cover's uppermost surface and of the rooting medium will ultimately determine the resident plant community. These factors are more important over the long haul than are the starting conditions for the plant community.

Nonetheless, it is good practice to establish a robust, self-perpetuating community of native species as quickly as possible. Until this is done, the soil surface is at its most vulnerable state in terms of erosion and invasion by unwanted deep-rooters. Ecological competition may be an important factor in excluding deep-rooters, and the desired shallow-rooting grasses should be given a headstart. It is also important to green-up the cell for the ribbon-cutting ceremony.

Thus, seeding should be performed as soon as possible after final grading of the cover's surface. Some delay may be warranted in the light of local growth and precipitation cycles, but a temporary nurse crop of annual barley or brome grass should be established to provide erosion protection. Alternatively, a mulch of straw, woodchips or comparable organic residue may be added to stabilize the soil for the short-term and introduce organic matter that will benefit the cover indefinitely. The long-term community should, in almost every case, consist of mixed, native perennial species. Reclamationists, agronomists, and local conservationists will provide good guidance on the recommended species.

The philosophy for fertilization and irrigation should be to provide no more than would naturally be available during favorable conditions; for instance, after a burn that releases nutrients or a season of unusually heavy precipitation. Excess fertilizer or irrigation will create plant dependency, resulting in major dieback, when the supplementation is exhausted or terminated. Some reclamationists avoid irrigation even in the driest climates because they have seen such dramatic diebacks of profuse stands with vulnerable shallow roots. My philosophy is to irrigate no more than would represent the natural rainfall during the rainiest year of a 10-year cycle. In arid and semi-arid climates, it is just such years that nearly all growth and reproduction occurs. Time the irrigation to encourage deep-rooting and efficient water use. Of course, excessive irrigation and drainage on a disposal cell would be unwise.

Finally, it must be noted that revegetation efforts are rarely fully successful on the first try. Bare patches result from any number of factors, ranging from over-compacted soil to a plugged up seed drill. Emerging stands must be closely monitored, and bare patches must be noted and diagnosed for the impediment to revegetation. Once the problem has been identified, the bare area can be seeded and incipient erosion fixed. One or two such interseedings usually suffice.

Summing up: the goal is to arrive at a climax community as quickly as possible. This is achieved by 1) providing a medium that will support the climax community and 2) establishing a plant community that is relatively far along on the successional pathway.

This concludes the practical discussion of vegetated covers and how to build them. Of the alternative cover designs described previously, the full-component cover is best suited for refinement onto a vegetated cover, although any system can be rendered such by including an appropriate thickness of favorable growth medium and selecting the appropriate plant community.

Additional detail on the design, performance, and potential failures of vegetated covers are provided in Chapters 17 and 18, which describe the ecological model in performance assessment. These sections also describe failure scenarios of alternative cover designs attributable to ecological processes such as biointrusion.

SECTION III

ENCAPSULATION SYSTEM PERFORMANCE MODELING AND ASSESSMENT

CHAPTER

Chapter 9

THE MODELING AND SYSTEMS APPROACH

The technical approach that we favor to describe and assess the performance of an encapsulation system is to conceive of the cell as a system that is comprised of subsystems that together interact with the environment to encapsulate wastes and ensure permanent isolation. For the total system and for each subsystem, conceptual and analytical models are postulated and analyzed. This makes it possible to quantify cell behavior.

In this section of the book we define and discuss three main encapsulation systems or models, each of which consists of many subsystems and components. The primary systems or models are:

- The Hydrological Model
- The Geotechnical Model
- The Ecological Model

The hydrological model describes the flow of water and contaminants into, through, and from the encapsulation system and its surrounding environment. Figure 9.1 shows the various subsystems into which the Hydrological Model System may be divided. These include climate, surface water, cover and dike infiltration, waste flowthrough, basal liners and drains, foundations, the vadose zone, and groundwater flow models. The main considerations in modeling are each summarized in Figure 9.2.

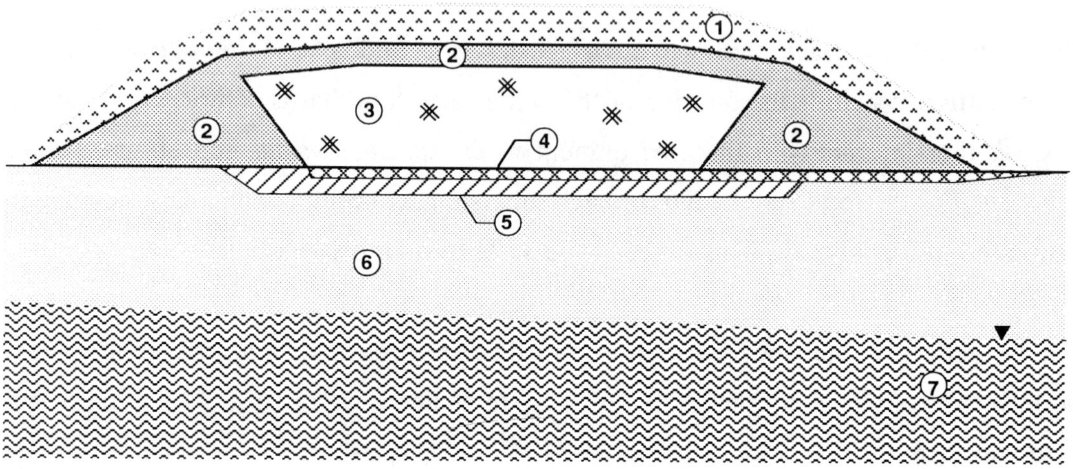

DISPOSAL CELL HYDROLOGIC SYSTEMS

LEGEND

1. SURFACE WATER FLOW SYSTEM
2. COVER AND DIKE FLOW SYSTEM
3. WASTE FLOW SYSTEM
4. BASAL LINERS AND DRAIN FLOW SYSTEMS
5. FOUNDATION FLOW SYSTEM
6. VADOSE ZONE FLOW SYSTEMS
7. GROUNDWATER FLOW SYSTEMS

These are the basic hydrological (and geochemical) systems that together compromise the disposal cell and waste encapsulation facility.

FIGURE 9.1

WASTER ENCAPSULATION FACILITY HYDROLOGICAL SYSTEMS

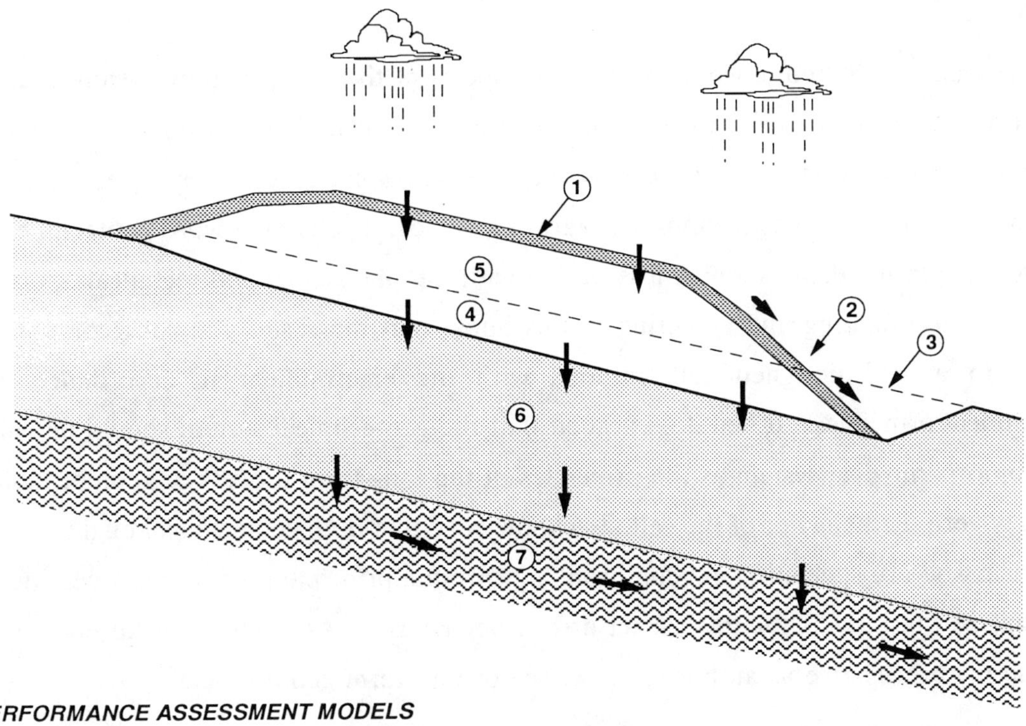

PERFORMANCE ASSESSMENT MODELS

Climate Model
- Precipitation
- Evaporation
- Temperature

Cover Performance Model
- Vegetation Performance
- Soil/Rock Water Retention and Through-flow
- Drain Lateral Shedding
- Infiltration Barrier Seepage

Vadose (Unsaturated) Zone Model
- Unsaturated Flow Rates
- Change of Seepage
- Water Quality

Groundwater Flow Model
- Regional and Site Flow Rates
- Background Water Quality
- Mixing, Dispersion and Attenuation
- Resultant Water Quality at Point of Compliance
- Groundwater Compliance Strategy

Tailings Through-flow Model
- Vertical Flow Rate
- Effect of Nonhomogeneity
- Change of Seepage
- Water Quality

LEGEND

(1) TOPSLOPE COVER
(2) SIDESLOPE COVER
(3) NATURAL GROUND LEVEL
(4) EXCAVATION ZONE
(5) ENCAPSULATED WASTE
(6) SOIL
(7) AQUIFER

The cell's hydrologic performance may be assessed by defining and quantifying models for subsystems and components.

FIGURE 9.2
PERFORMANCE ASSESSMENT MODELS

The Geotechnical System encompasses those aspects of the encapsulation system that are comprised of soil and rock and the soil-like wastes. We define and analyze models to quantify the interaction of these systems with the following forces: gravity, earthquakes, erosion by run-off, and site geomorphological processes. I select the word geotechnical as the overall title for this system as it is in line with the philosophy and technical approach that we are postulating and exploring in this book, that these soil and rock components should, in a well designed cell, interact with the environment in accordance with geotechnical principles. It could be argued that the geotechnical system should be called the geomorphological system as the principles of the long-term process of geomorphology should govern the behavior of the cell. I acknowledge that both geotechnical engineers and geologists who are specialists in geomorphology are the principal professionals who define and analyze the models for this aspect of cell behaviorist. As a geotechnical engineer, I exercise my prerogative as author to select the overall term geotechnical.

The Ecological Model is used to describe and quantify the performance of the cell as it is affected by interaction with living entities such as vegetation, animals, and humans. The ecological model also involves the incorporation of vegetation into the cell as a component to improve cell performance and improve the system's compatibility with the surrounding ecosystem.

Three modes of disposal performance are considered and evaluated:

- Normal, equilibrium performance
- Abnormal, or deviant performance
- Failure

Each of these modes of performance provide a basis for establishing cell performance monitoring data quality objectives and monitoring procedures as discussed in Section IV of this book.

Performance modeling is in reality no more than the old-fashioned process of collecting data about a site, the wastes, and the materials that constitute the encapsulation system, and compiling calculations that quantify the stability or the flow rate for the design layout of the proposed or actual cell. The technical approach of defining systems, subsystems, components, and then analyzing the behavior of each or collections of subsystems and components is also a standard engineering pastime for the average designer. Analysis may be done by solving equations that describe the system's behavior. Alternatively, we may set up and run simple or complex computer codes that model part of all of the cell or a specific system of the cell. Finally, we may simply go into the field and find a natural analog of the system, and by observing and explaining the behavior of the analog predict the performance of the cell system.

Because of these diverse procedures, some chapters of this section describe the concept or model that may be used to describe and analyze one or more system or subsystem, for example, the cover infiltration subsystem. Other chapters may concentrate on a specific technical approach or procedure, such as the complex issues associated with the topic of root biointrusion.

Chapter 10

THE HYDROLOGICAL MODEL: THE COVER

GENERAL

The cell hydrology model at its most general and simple may be conceived of as shown in Figure 10.1, which shows the major components of flow onto, off, into, through, and from the cell and the wastes. A simple one-dimensional model of all the components may be defined as shown in Figure 10.2. Examination of the components of the cell, as shown in the figure, and knowledge of the characteristics of the materials of the cell may enable the designer to compile a cell hydrological performance model, as shown in Figure 10.3, where quantities are derived primarily from the hydraulic conductivities of the materials and wastes that constitute the encapsulation system.

In what follows, I describe in considerably more detail each of the components and other possible models that may be used to assess and quantify the hydrological performance of a disposal cell.

THE CLIMATE

Climate is the prime factor influencing the hydrological performance of the cell. The precipitation component of the water flow model for climatic conditions must incorporate both the effect of short-term events of irregular occurrence, such as intense precipitation and average conditions, such as the annual rainfall and snowmelt quantities.

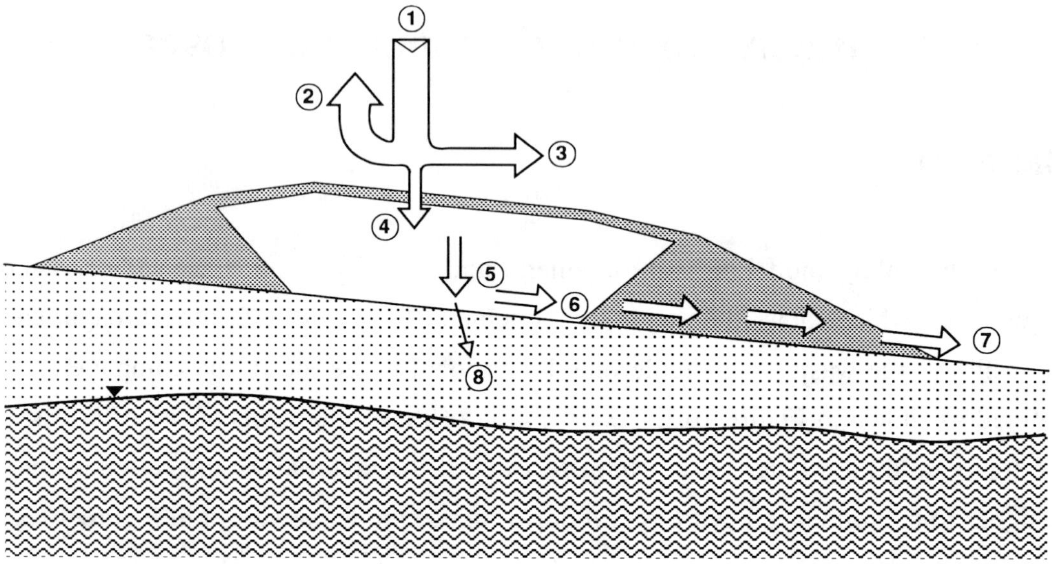

LEGEND

1. PRECIPITATION
2. EVAPOTRANSPIRATION
3. RUNOFF
4. COVER INFILTRATION
5. SEEPAGE THROUGH WASTE
6. LEACHATE COLLECTION SYSTEM FLOW
7. SEEPAGE FROM CELL
8. SEEPAGE THROUGH LINER

A small fraction of precipitation infiltrates the top cover, and leaves the cell via the drains, or seeps from the waste to the foundation and groundwater.

FIGURE 10.

DISPOSAL CELL HYDROLOGIC MODE

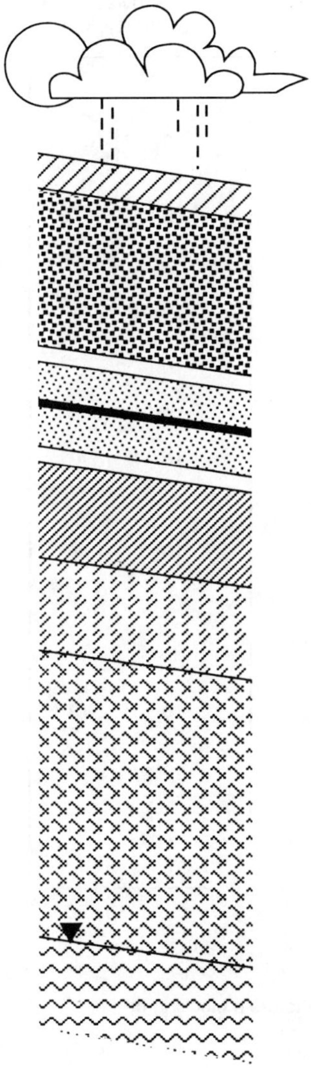

COMPONENT	ROLE AND FUNCTION
CLIMATE	• Source of water and infiltration
COVER	• Impede infiltration
WASTE	• Source of transient drainage (leachate) • Alters chemistry of infiltration water
DRAIN	• Leachate control
LINER	• Impede seepage
DRAIN	• Monitor leakage
LINER	• Impede seepage • Geochemically alter seepage quality
FOUNDATION SOIL	• Impede seepage • Geochemically alter seepage quality
VADOSE ZONE	• Geochemically alter seepage quality • Impede water seepage – ensure long travel time
GROUNDWATER	• Dilute seepage

For vertical seepage through the cell and its foundation, this model may be used to quantify flow rates.

FIGURE 10.2

ONE-DIMENSIONAL CELL SEEPAGE MODEL COMPONENT FUNCTIONS

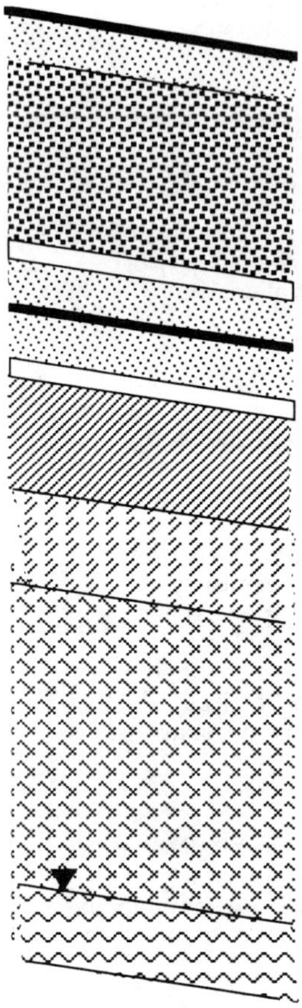

COMPONENT	HYDRAULIC CONDUCTIVITY OR FLUX (cm/s)	
	AVERAGE (SHORT-TERM)	RANGE (LONG-TERM)
COVER: GEOMEMBRANE	10^{-10}	10^{-12} to ∞
COVER: SOIL	10^{-8}	10^{-10} to 10^{-7}
WASTE	10^{-5}	10^{-7} to 10^{-4}
DRAIN	10^{-2}	10^{-3} to 0
GEOMEMBRANE	10^{-10}	10^{-12} to ∞
DRAIN	10^{-3}	10^{-4} to 0
LINER	10^{-7}	10^{-8} to 10^{-6}
FOUNDATION	10^{-8}	10^{-10} to 10^{-7}
VADOSE ZONE	10^{-7}	10^{-10} to 10^{-6}
GROUNDWATER	10^{-4}	

NOTES:

1. Short-term permeability based on design and construction values commonly achievable
2. Long-term permeability based on feasible material properties
3. Infinite permeability due to geomembrane degradation
4. Zero permeability due to drain blocking or outlet failure
5. 10^{-10} permeability for soils based on unsaturated hydraulic conductivity

The hydraulic conductivity and flux through the cell components may vary with time.

FIGURE 10.

ONE-DIMENSIONAL CELL SEEPAGE MODEL COMPONENT PERMABILITIES

Short-term events affect erosion potential, and hence the design of the cover. Average conditions affect such aspects of cell performance as infiltration and the groundwater regime.

Site evapotranspiration is a major component of the cell and immediate surrounding area water balance. Consequently, it impacts cell and subsurface infiltration. Evapotranspiration affects the growth of vegetation which is a major determinant of hydrological performance of the cover, the cell, and the immediate surrounding area.

Temperature affects evapotranspiration patterns and rates, influences vegetation growth, and controls snow melting patterns.

THE SURFACE RUN-OFF SUBSYSTEM

Cover Run-off: Soil, Rock, and Drain Performance

Figure 10.4 graphically describes the water balance of the top cover, particularly as it affects the clean water system of the cell. The clean water system encompasses the water that falls on the cell, evaporates from the cell, and runs off the cell to the surrounding environment. As this figure shows, most incident precipitation runs off, is evaporated, or drains away.

Figure 10.5 broadens this observation and depicts the surface flow routes impacting the cell surface hydrology model. Only a minimal amount of the moisture that falls on the cover will ever infiltrate the geosynthetic and low permeability infiltration/radon barriers. This component of the surface run-off system focuses on quantification of incident precipitation, evapotranspiration and run-off from the cover, particularly during storms and snowmelt.

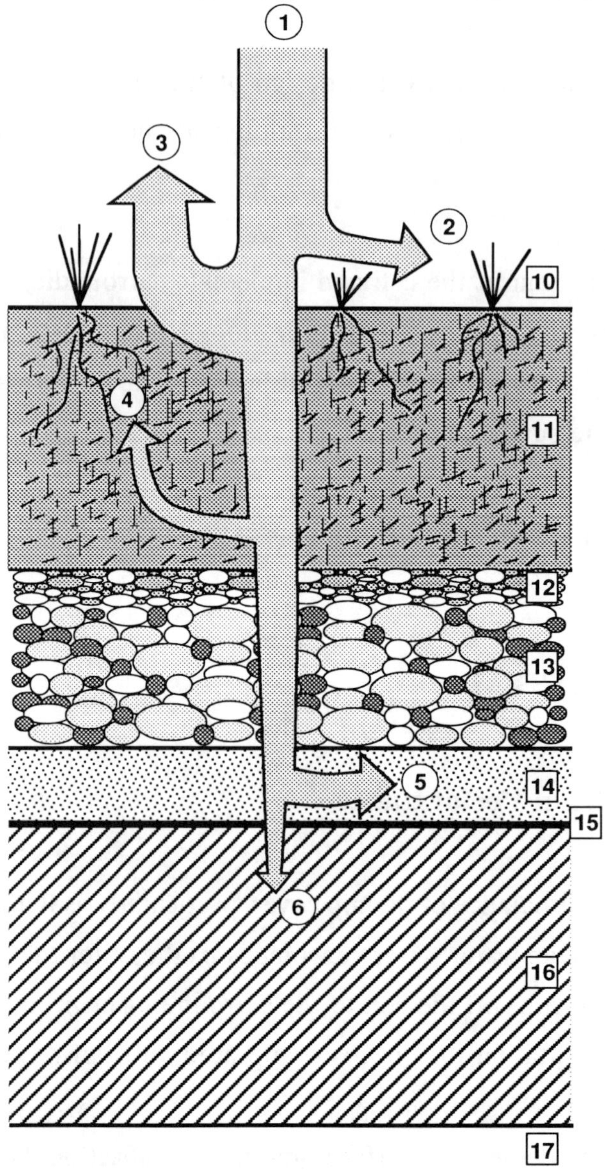

LEGEND

COVER WATER BALANCE
(1) Precipitation
(2) Surface Runoff
(3) Evapotranspiration
(4) Root Zone Moisture Removal
(5) Drain Flow
(6) Infiltration
COVER COMPONENTS
[10] Vegetation
[11] Rooting Soil: Sand, Silt, & Clay
[12] Choked Zone: Filter
[13] Biointrusion Barrier: Cobbles
[14] Drain: Sand
[15] Infiltration Barrier I (Geomembrane and Bentonite Mat)
[16] Infiltration Barrier II and Radon Barrier: Silt & Clay
[17] Waste

NOTE

SIZE OF WATER BALANCE ARROW AND THICKNESS OF LAYERS ARE NOT TO SCALE.

Most incident precipitation runs off, is evaporated, or drains away. Only a very small amount infiltrates the geosynthetic and the low permeability infiltration/radon barrier.

FIGURE 10.-

DISPOSAL CELL COVER WATER BALANC

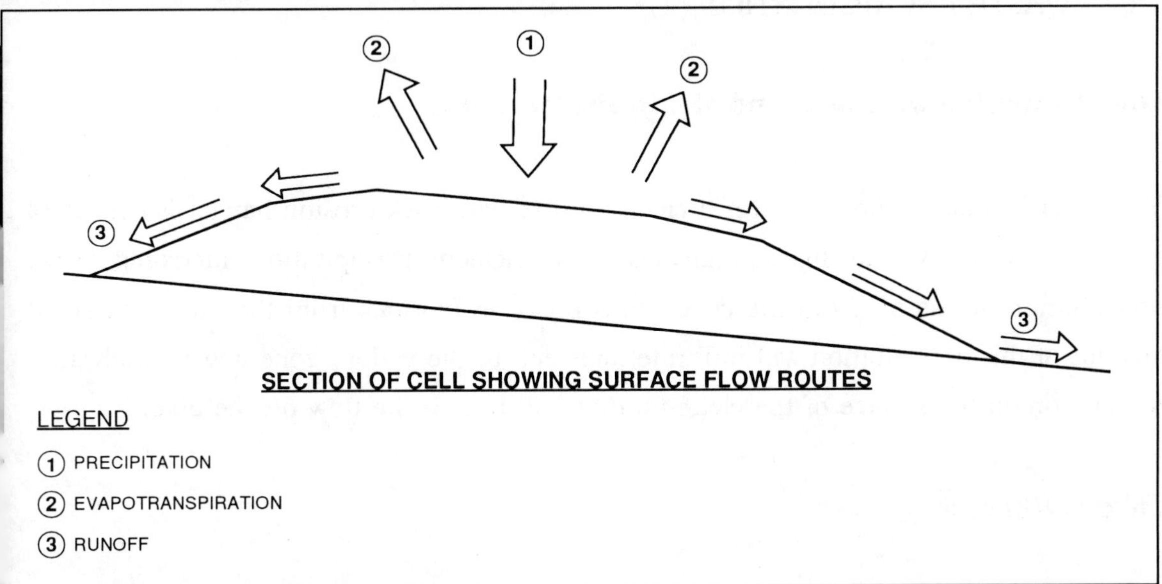

Most of the precipitation on the cell evapotranspires or runs off to the surrounding environment.

FIGURE 10.5

DISPOSAL CELL SURFACE HYDROLOGY MODEL

Dike Run-off: Soil, Rock, and Drain Performance

The water balance components for a clean-fill dike with rock erosion barrier is described in Figure 10.6. As this figure indicates, most incident precipitation, including direct precipitation or run-off from the cover, runs off or evaporates from the dikes. A small amount of this precipitation will infiltrate and seep to the vadose zone and groundwater. Vegetation on the surface of the clean-fill dike will reduce the flow off the dike.

Dike Infiltration

Infiltration of the clean-fill dikes is a part of the clean water surface system primarily because the infiltration to the dikes does not come into contact with the waste, but seeps through the dikes to the foundation soils and through the foundation soil to groundwater.

As shown in Figure 10.6, although most incident precipitation which falls or flows onto the clean-fill dikes will run off or evapotranspire, a small amount will infiltrate the dike and seep into the vadose zone and groundwater. Vegetation may slow run-off and thereby encourage infiltration; in most cases the plants will, however, evapotranspirate excess infiltration.

Site Surface Water Control: Run-off

Water flowing off the cover and sideslopes of the clean-fill dikes will mingle with precipitation on the perimeter buffer area to evaporate, infiltrate, or flow from the site. Figure 10.7 summarizes the components and concepts associated with this model for an above-grade cell.

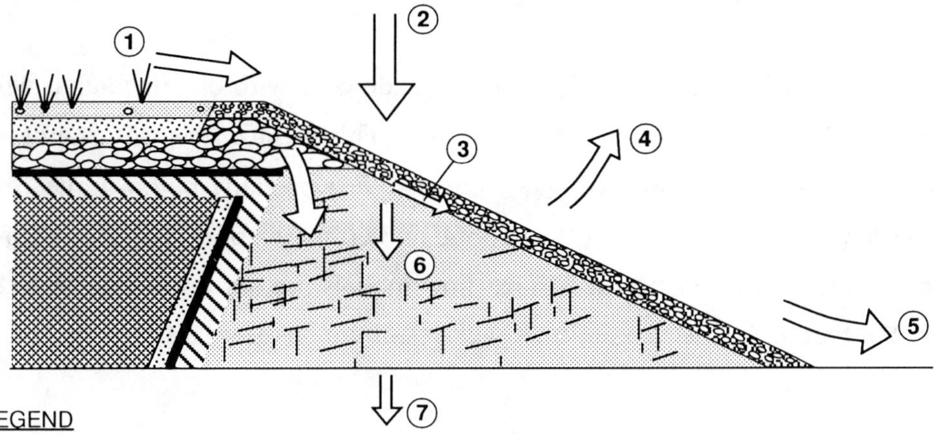

FLOW LEGEND

1. TOP COVER RUNOFF
2. SIDESLOPE INCIDENT PRECIPITATION
3. SIDESLOPE RUNOFF
4. SIDESLOPE EVAPOTRANSPIRATION
5. TOTAL CELL RUNOFF
6. DIKE INFILTRATION
7. SEEPAGE TO VADOSE ZONE

Some incident precipitation on the dike runs off and some infiltrates to seep to the groundwater.

FIGURE 10.6

CLEAN-FILL DIKE WATER BALANCE MODEL FOR ROCK COVER

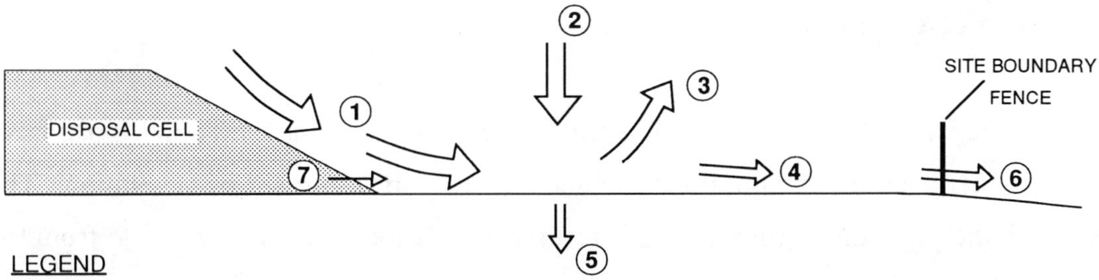

LEGEND

1. RUNOFF FROM CELL
2. PRECIPITATION
3. EVAPOTRANSPIRATION
4. OVERLAND FLOW
5. INFILTRATION
6. SURFACE FLOW FROM SITE - AT BOUNDARY FENCE
7. SEEPAGE FROM CELL DRAINS

Flow from the cover mingles with precipitation to evaporate, nfiltrate, or flow from the site (above-grade cell).

FIGURE 10.7

CELL PERIMETER SURFACE HYDROLOGY MODEL

In the long-term, above-grade or partially below-grade cells with open channels will allow surface flow and seepage to flow away from the cell. This surface flow will evaporate, seep into the soils, or continue to surface water bodies. For partially below-grade cells with outlet catch sumps, seepage and run-off will pond and either evaporate or seep through the geochemical layer. These long-term surface flow models are shown in Figure 10.8. It is unlikely that either of these possibilities will result in unacceptable geomorphological changes.

COVER WATER BALANCE MODEL: THE INFILTRATION BARRIER

General

This section discusses the water balance of the cover. It focuses on infiltration, i.e., the performance of the infiltration barrier and the quantity of seepage to the wastes.

Vegetation Performance

The role of vegetation in the water balance of the cover is complex. Some portion of infiltrating water is intercepted in the root zone and evapotranspirated by vegetation. The soil in which the vegetation grows acts as a reservoir for water which may migrate from the soil into the cell drainage system.

Drain

Another fraction of the moisture infiltrating the cover flows off the cell via the drain. As shown in Figure 10.9, water seeping from the overlying soil generally runs slowly through the drain and, depending on the rate of flow and the hydraulic conductivity of the infiltration barrier, soil may seep into the infiltration barrier.

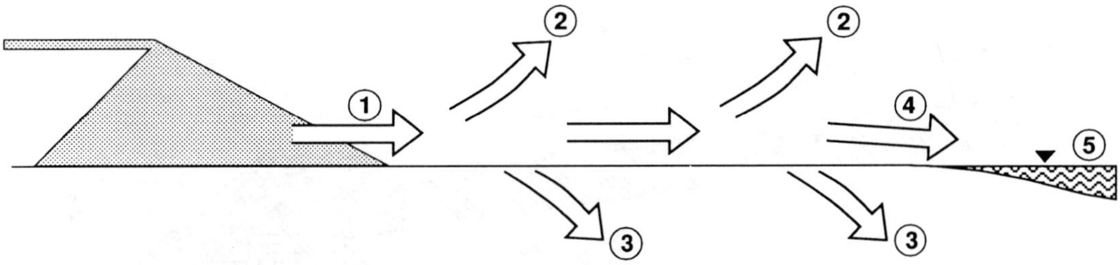

LONG-TERM SURFACE EMERGENT FLOW MODEL I

FOR ABOVE-GRADE OR PARTIAL BELOW-GRADE WITH OPEN CHANNEL: SURFACE SEEPAGE FROM THE DRAINS OR DIKES WILL FLOW OVERLAND AWAY FROM THE CELL TO EVAPORATE, SEEP INTO THE SOILS OR CONTINUE TO SURFACE WATER BODIES

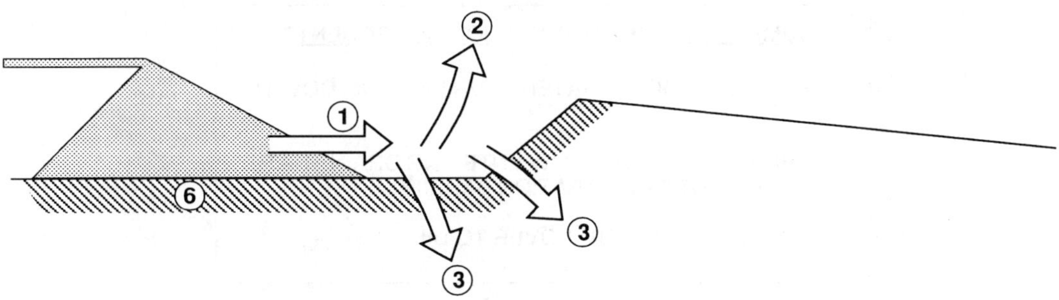

LONG-TERM SURFACE EMERGENT FLOW MODEL II

FOR PARTIALLY BELOW-GRADE CELLS OR CELLS WITH OUTLET CATCH SUMPS, SEEPAGE FROM THE DRAINS OR DIKES WILL POND AND HENCE EVAPORATE OR SEEP THROUGH THE GEOCHEMICAL LAYER

LEGEND

(1) SEEPAGE FROM CELL

(2) EVAPOTRANSPIRATION OF SEEPAGE

(3) SEEPAGE INFLOW TO FOUNDATION SOILS

(4) FLOW TO SURFACE WATER BODIES

(5) SURFACE WATER BODIES: PONDS, RESERVOIRS, CREEKS, RIVERS

(6) GEOCHEMICAL BARRIERS

In the long-term, after cessation of interception of seepage from drains or dikes, the surface flow from the cell will flow overland, either to surface water bodies, to be evapotranspirated, or to seep to foundation soils.

FIGURE 10.8

LONG-TERM SURFACE EMERGENT FLOW MODELS

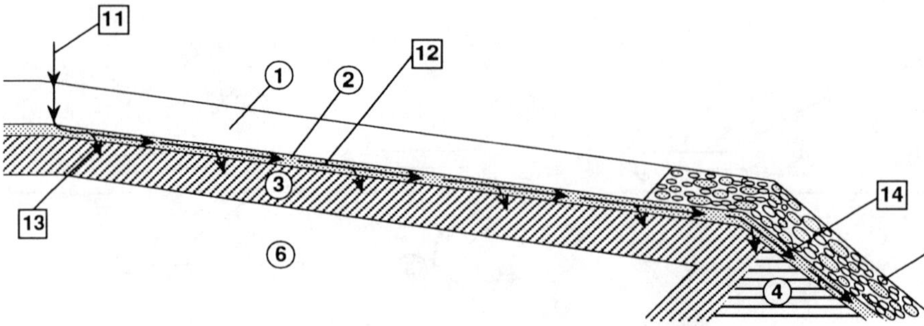

SECTION OF COVER SHOWING INFILTRATION AS WATER FLOWS DOWN DRAIN

SOME DRAIN AND INFILTRATION COMPONENTS

- TRAVEL TIME OF WATER IN DRAIN FROM COVER CREST TO DIKE
- PERCENTAGE OF FLOW ENTERING DRAIN THAT SEEPS INTO INFILTRATION BARRIER
- DRAIN FLOW FROM COVER TO DIKE

LEGEND

CELL COMPONENTS
(1) SOIL (2) DRAIN (3) INFILTRATION BARRIER (4) DIKE (5) EROSION BARRIER (6) WASTE
SEEPAGE
[11] FLOW THROUGH SOIL [12] DRAIN FLOW [13] CELL INFILTRATION [14] FLOW TO DIKE DRAIN

As water flows down the drain some will seep into the infiltration barrier.

FIGURE 10

COVER DRAIN LATERAL SEEPAGE

Infiltration Barrier

Each component of the cover plays an important role in protecting the infiltration barrier and limiting infiltration. Figure 10.10 illustrates the location and describes the function and nature of infiltration barrier components.

The Geomembrane

The geomembrane (if used) is a short- to medium-term infiltration inhibitor. In the long-term, deterioration of the geomembrane will probably render it ineffective. Figure 10.11 shows the distribution of the hydraulic conductivity of the geomembrane and other cover components. Also shown is the probable short- and long-term flux rates through the assemblage of components that together make up the infiltration barrier. Figure 10.12 shows feasible changes in cover flux over time as a result of decay of the geomembrane. It is likely that similar changes in the cover's resistance to infiltration over time will occur.

The Geomat

The geomat, if used, will provide additional inhibition of seepage into the cell. There are valid questions about the areal effectiveness of the mat, because of its relative thinness. Because of its bentonite composition, it will, however, be effective for the life of the cell.

The Soil Barrier

This layer of soil (usually clay and silt) serves the dual purpose of providing additional resistance to infiltration and providing a barrier to radon flux. Together with the bentonite mat layer, this compacted soil layer will provide a sufficient barrier to reduce infiltration to

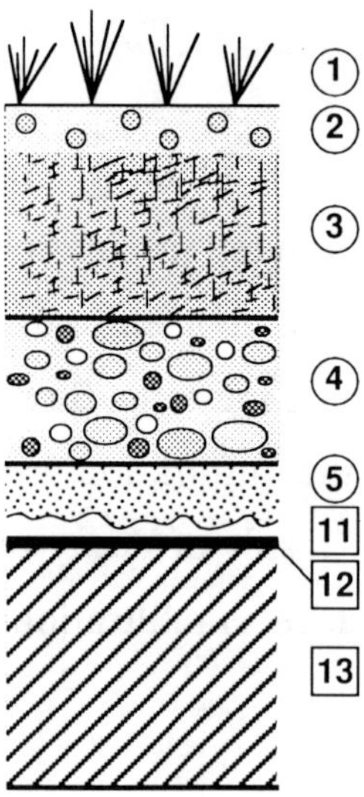

INFILTRATION BARRIER PROTECTION

(1) VEGETATION:	Control erosion and increase evapotranspiration
(2) GRAVELLY SOIL:	Control erosion
(3) RANDOM SOIL:	Freeze/thaw protection
(4) BIOINTRUSION COBBLES:	Animal, human, and root intrusion control
(5) DRAIN SAND:	Bedding to preclude cobble punching and deformation

INFILTRATION BARRIER COMPONENTS

[11] GEOMEMBRANE:	Very low permeability; may deteriorate with time.
[12] BENTONITE LAYER (IN GEOMAT):	Very low permeability; long-term durable natural material; also seals geomembrane pinhole leaks. Seepage may increase if water ponds over layer.
[13] RADON BARRIER CLAY/SILT:	Low permeability; seepage not sensitive to water ponding. Long-term efficacy may be affected by deformation, freeze/thaw, or dessication.

Each component of the cover plays a part in protecting the infiltration barrier and limiting infiltration.

FIGURE 10.1
INFILTRATION COMPONENT
THEIR NATURE AND FUNCTIO

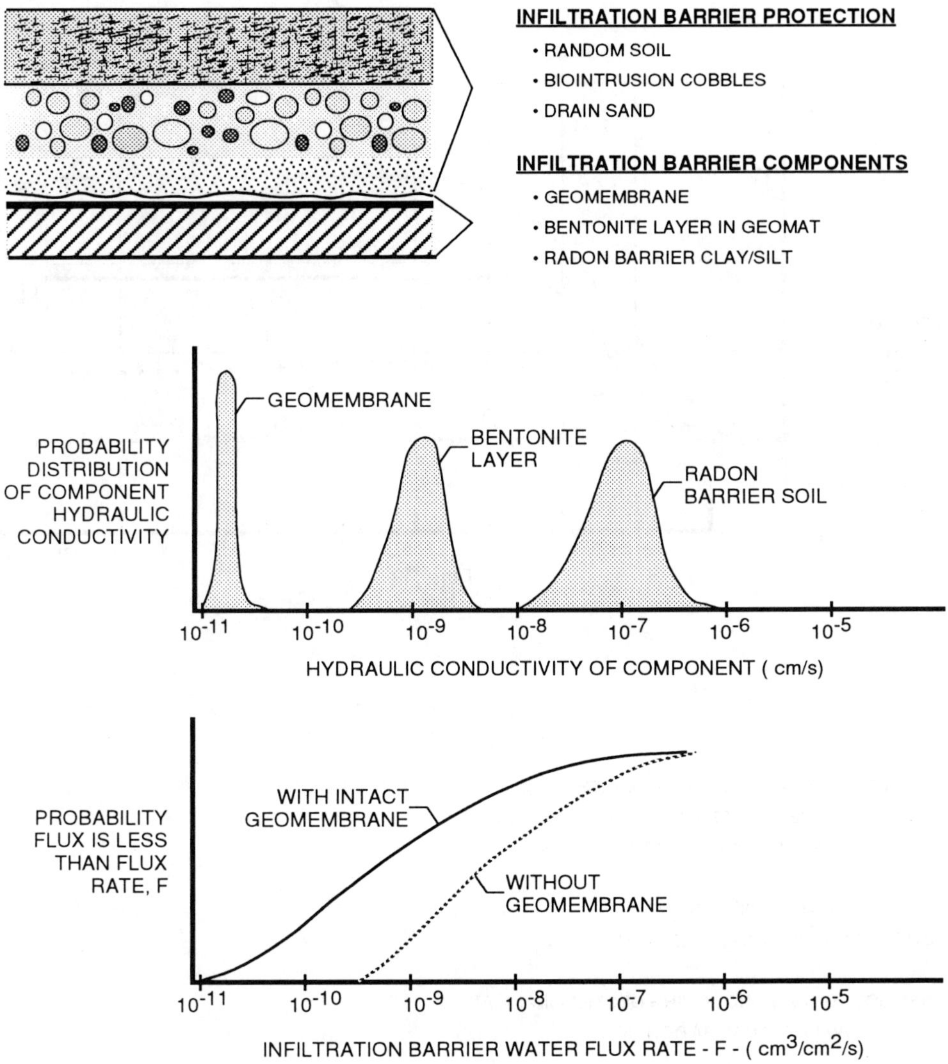

Material permeabilities vary about a normal distribution mean; the cover seepage rate is not an absolute value, but has a probability of being less than a given value.

FIGURE 10.11

INFILTRATION BARRIER COMPONENT PERMEABILITY AND FLUX VARIATIONS

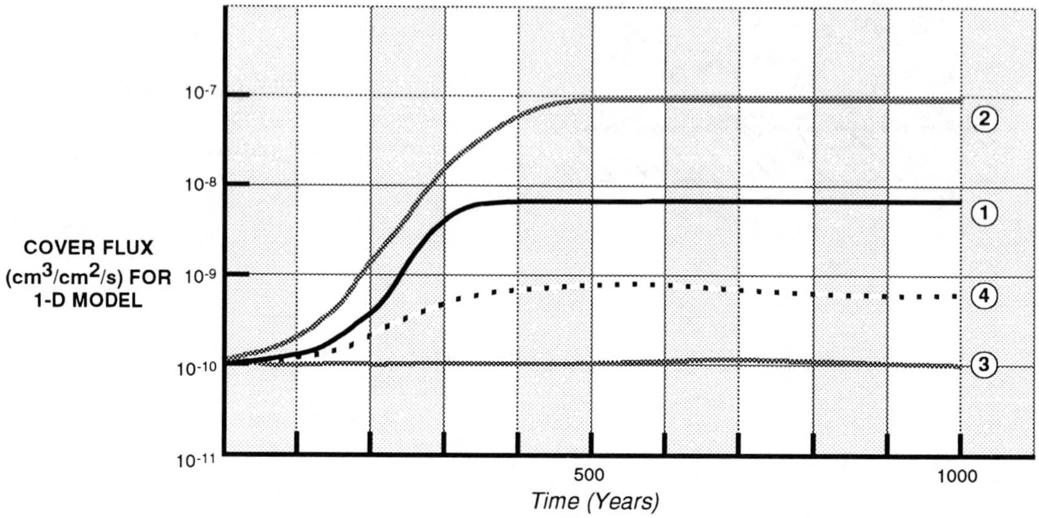

LEGEND

(1) MOST LIKELY AVERAGE COVER FLUX WITH DETERIORATED GEOMEMBRANE AND INFILTRATION BARRIER HYDRAULIC CONDUCTIVITY CONTROL

(2) UPPER BOUND WITH NON-FUNCTIONING GEOMEMBRANE OR BENTONITE LAYER

(3) LOWER BOUND FOR FUNCTIONING GEOMEMBRANE AND EFFECTIVE EVAPOTRANSPIRATION

(4) MOST LIKELY FLUX WITH BENTONITE MAT AS BACKUP INFILTRATION BARRIER

As time passes, and the geomembrane decays, the cover-flux will be governed by the infiltration barrier permeability and evapotranspiration.

FIGURE 10.1

COVER FLUX - VARIATION WITH TIM

below critical levels even in the absence of the geomembrane. Figure 10.13 describes the relationship between soil permeability and porosity and the travel time for water through the infiltration barrier. As shown in Figure 10.14, with the development of a saturated zone above the infiltration barrier, the seepage rate through the barrier may increase.

It should be noted that seepage through the soil barrier is also affected by climatic conditions which may result in a wide variation in component soil moisture content. Figure 10.15 summarizes this phenomenon. Material permeabilities also vary and as Figure 10.11 shows, seepage rate is not an absolute number, but is expressed as a probability. Consequently, no single absolute estimate of infiltration through the barrier is possible; however, estimates are produced by this model based on specific performance scenarios and physical parameters of components.

The soil barrier may be affected by the generation of heat during the stabilization process. Heat generation may affect seepage by increasing vegetated growth, drying or cracking the radon barrier, decreasing saturation of the infiltration barrier, softening and chemically destabilizing geomembranes, and reducing clay/geomembrane interface friction.

Infiltration Barrier Failure Scenarios

Potential cover failure scenarios that would increase the infiltration through the cover include: cracking or biointrusion of the infiltration barrier system, increase in soil permeability due to freezing and thawing, geomembrane deterioration, and cover drain blocking. As Figure 10.16 indicates, only if the water infiltration through the cover or into the leak detection system exceed maximum feasible values may component failure be said to have occurred.

Figure 10.11 shows the change in the feasible distribution of infiltration through the cover, with and without the geomembrane. The bentonite mat layer and the compacted silt and

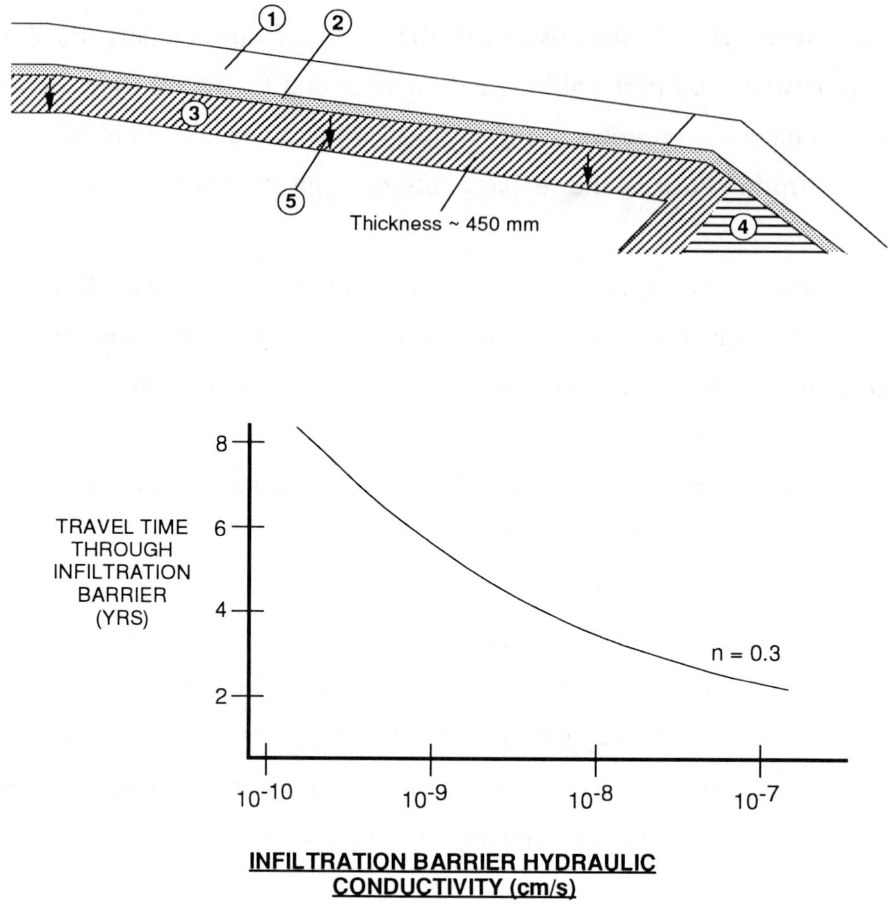

LEGEND

① SOIL
② DRAIN
③ INFILTRATION BARRIER
④ DIKE
⑤ SEEPAGE THROUGH BARRIER
n SOIL POROSITY

Travel time for water through infiltration barrier is controlled by soil permeability and porosity (n).

FIGURE 10.1
INFILTRATION BARRIER SEEPAGE TRAVEL TIME

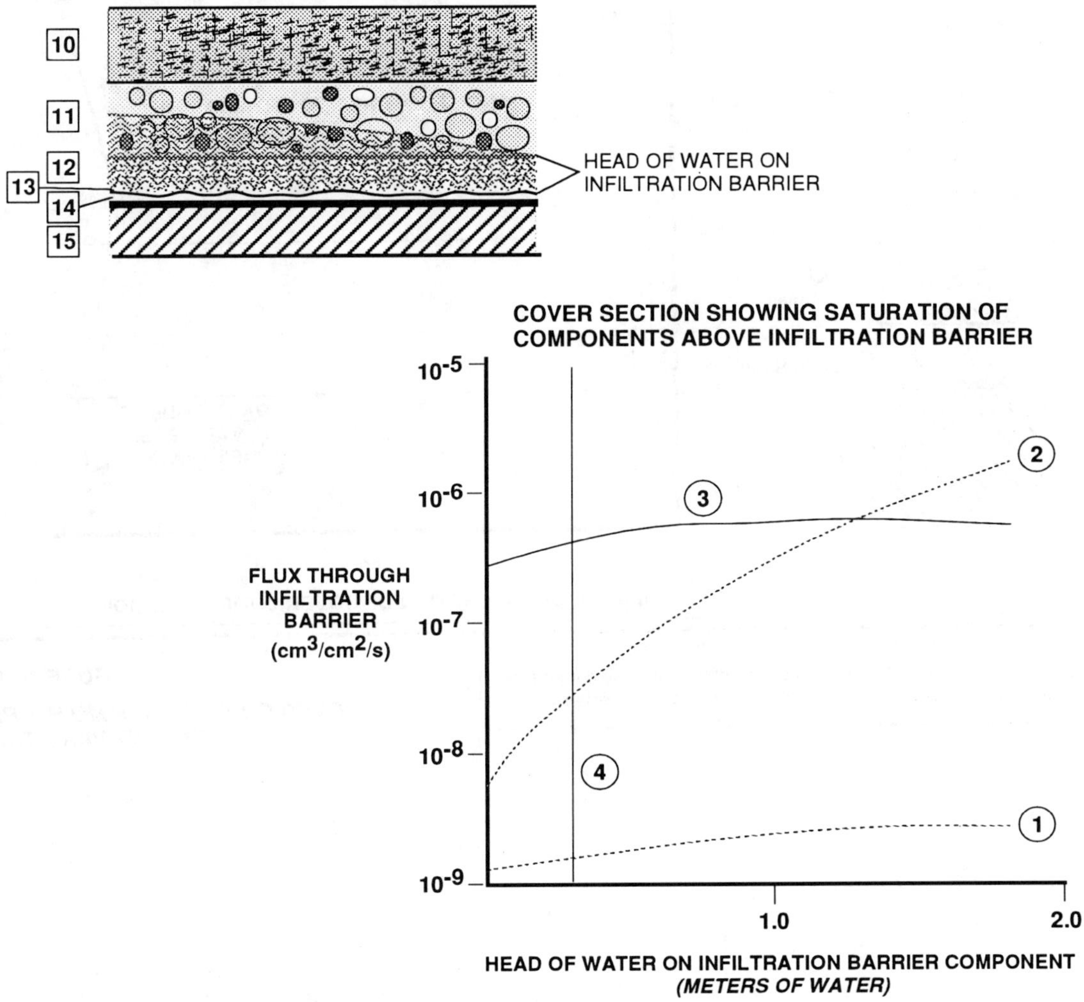

LEGEND

(1) INTACT GEOMEMBRANE
(2) BENTONITE LAYER (IN GEOMAT)
(3) RADON BARRIER CLAY/SILT
(4) FEASIBLE HEAD DUE TO DRAIN BLOCKING OR CLOGGING

[10] RANDOM SOIL
[11] BIOINTRUSION COBBLES
[12] DRAIN SAND
[13] GEOMEMBRANE
[14] GEOMAT WITH BENTONITE LAYER
[15] RADON BARRIER CLAY/SILT

Development of a saturated zone above the infiltration barrier could lead to an increase in the seepage rate through the infiltration barrier.

FIGURE 10.14

COVER INFILTRATION AS AFFECTED BY HEAD BUILD UP

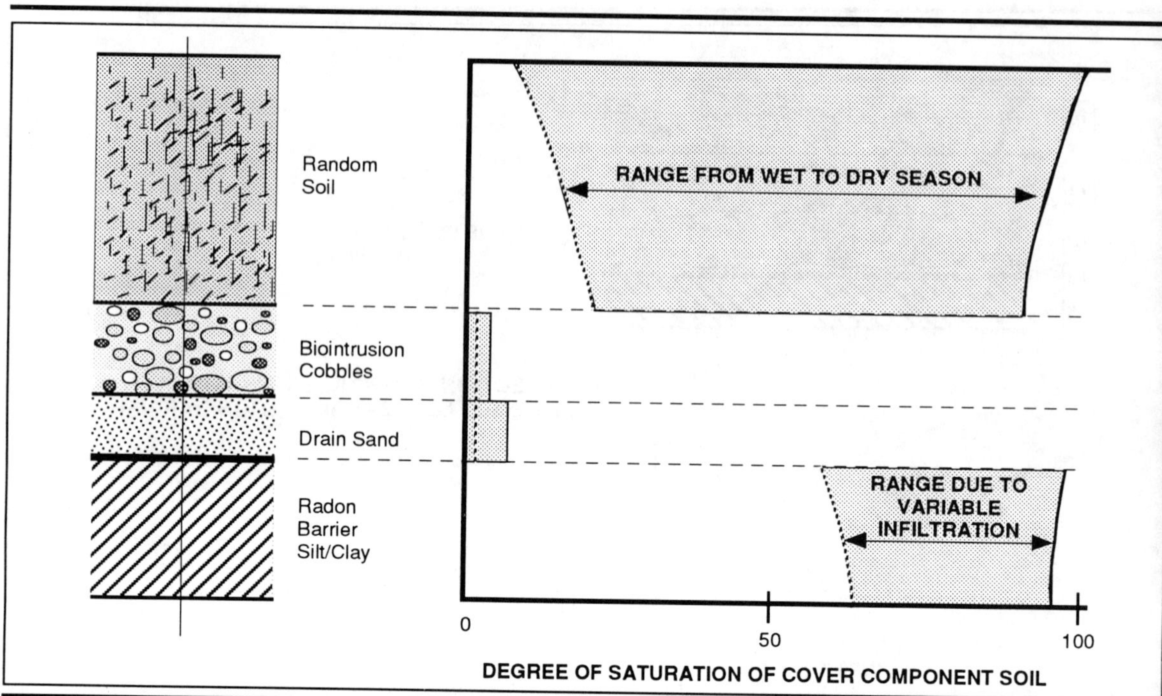

Climatic conditions and variations of cover infiltration may cause the cover component 's soil moisture content to vary over a wide range.

FIGURE 10.15

COVER COMPONENTS MOISTURE CONTENT VARIATION

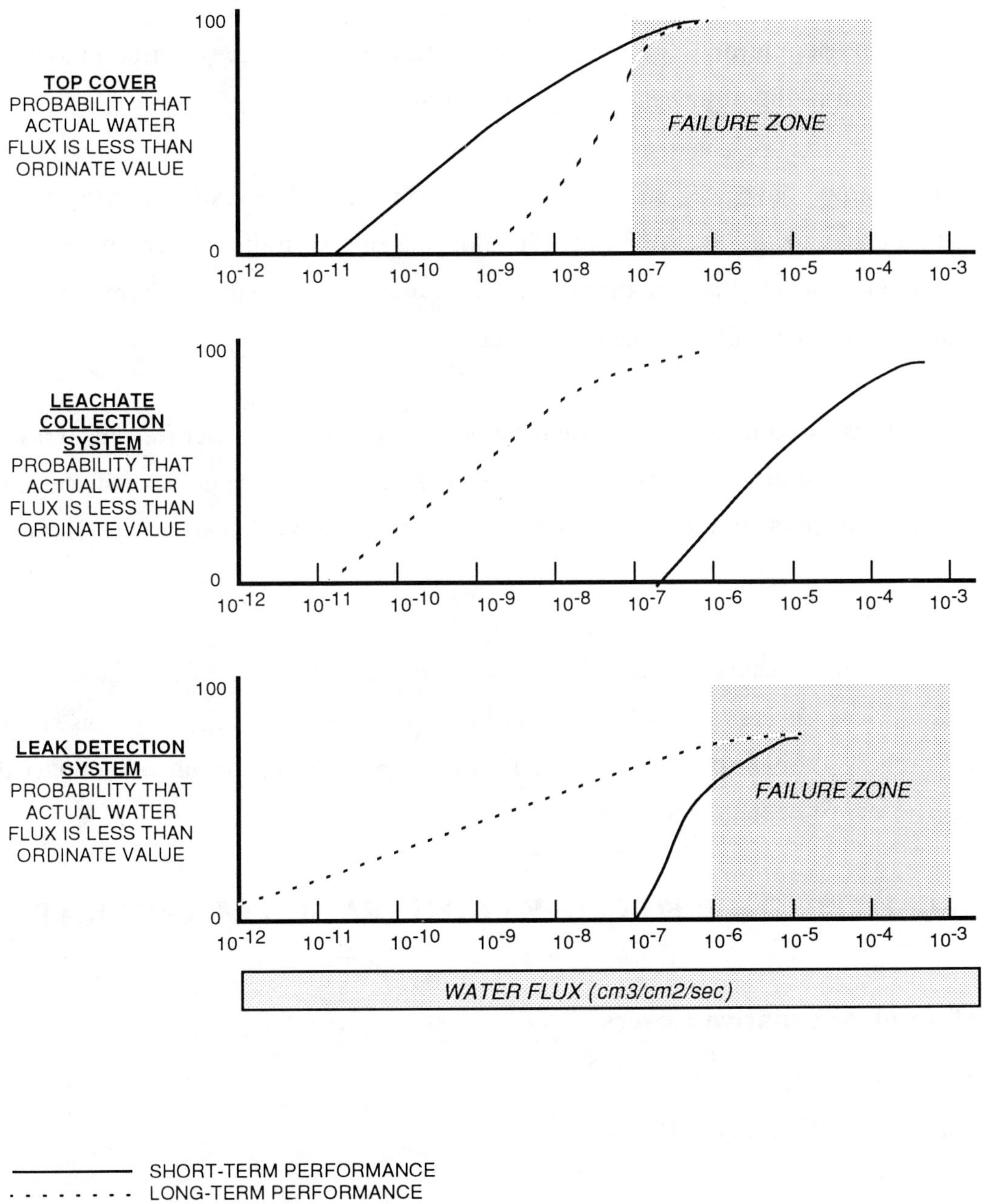

Only if the water flux through the cover or into the leak detection system exceeds maximum feasibility values may component failure be said to have occurred.

FIGURE 10.16

CELL HYDROLOGIC PERFORMANCE SYSTEM FAILURE ZONES

clay provide a sufficient layer of low permeability in the cover to reduce infiltration to below critical levels even in the absence of the geomembrane.

Figure 10.17 shows how seepage through the cover's soil layer will transfer to the biointrusion barrier in the event of complete clogging of the drain. As the biointrusion barrier is more permeable than the drain, this will have no effect on the performance of the cover with regard to lateral shedding of the water.

Figure 10.18 describes how infiltration (and radon flux) could increase if freezing/thawing damages the radon/infiltration barrier. Extreme freezing and thawing damage of the radon and infiltration barriers may be precluded by the thick overburden layers, as shown in Figure 10.19.

In addition to "natural" causes of cell failure, Figure 10.20 describes the potential effects of monitoring devices on cell performance. As the figure shows, boreholes or instrument monitoring tubes, through the cover into and through the cell, may negatively impact the desired performance of the disposal cell.

TECHNICAL APPROACH FOR COVER INFILTRATION QUANTIFICATION

Evaluation of Alternative Covers

Case History 3: Cover Infiltration

The infiltration through the five alternative covers defined in Section II was evaluated with the HELP model. Table 10.1 briefly describes the HELP computer code and compares it to an alternative code. Table 10.2 lists the various parameters used in the analysis. The topslope covers were evaluated assuming a two percent slope; the sideslopes were modeled

FEATURES THAT PRECLUDE MALPERFORMANCE

The very high permeability biointrusion cobbles are not susceptible to clogging. Loss of flow capacity in the drain can be augmented by cobble layer flow capacity.

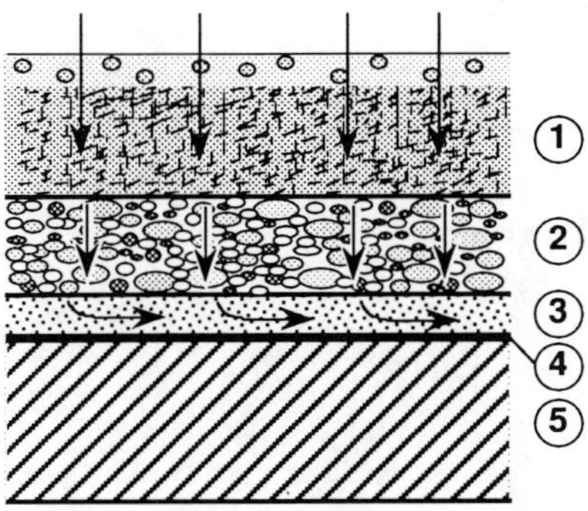

NORMAL PERFORMANCE

Seepage through soil and cobbles flows into drain sand and flows laterally in drain layer above infiltration barrier.

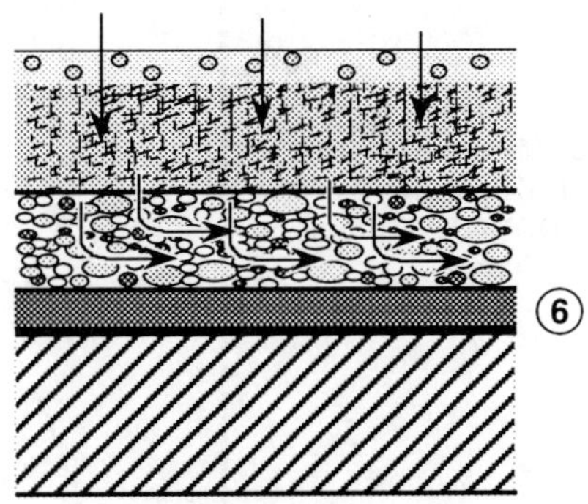

FAILURE SCENARIO: BLOCKED OR CLOGGED DRAIN

Seepage through soil and into cobbles flows laterally in cobble layer above clogged drain.

LEGEND

1. RANDOM SOIL
2. BIOINTRUSION COBBLES
3. DRAIN SAND
4. INFILTRATION BARRIER
5. RADON BARRIER
6. BLOCKED OR CLOGGED DRAIN

Blocking or clogging of the cover drain by physical or chemical processes will not impair the cover's ability to shed water.

FIGURE 10.17

COVER DRAIN CLOGGING SCENARIO ANALYSIS

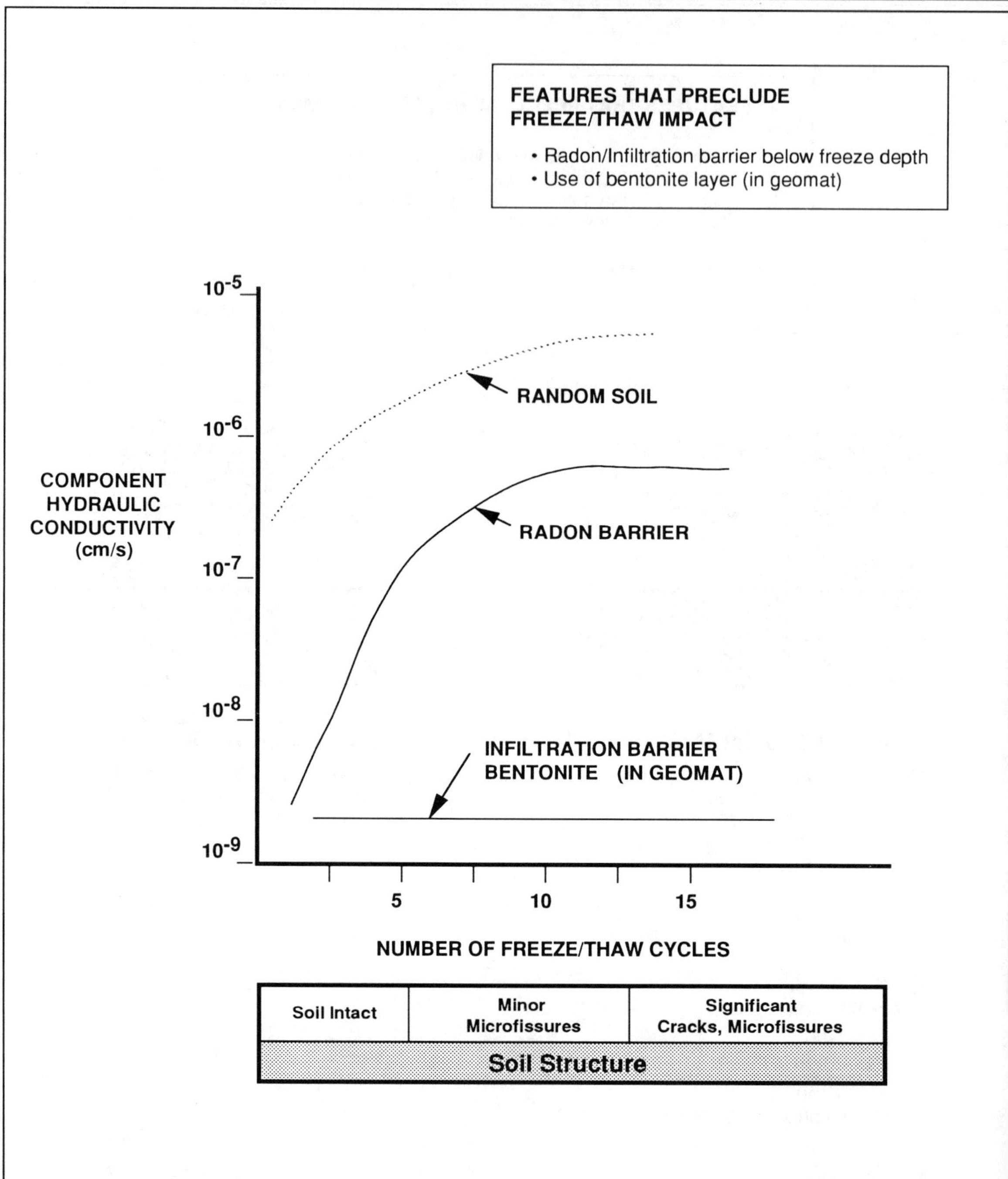

Infiltration (and radon flux) could increase if freezing/thawing damages the radon/infiltration barrier.

FIGURE 10.18

POTENTIAL IMPACT OF FREEZE/THAW ON SOIL HYDRAULIC CONDUCTIVITY

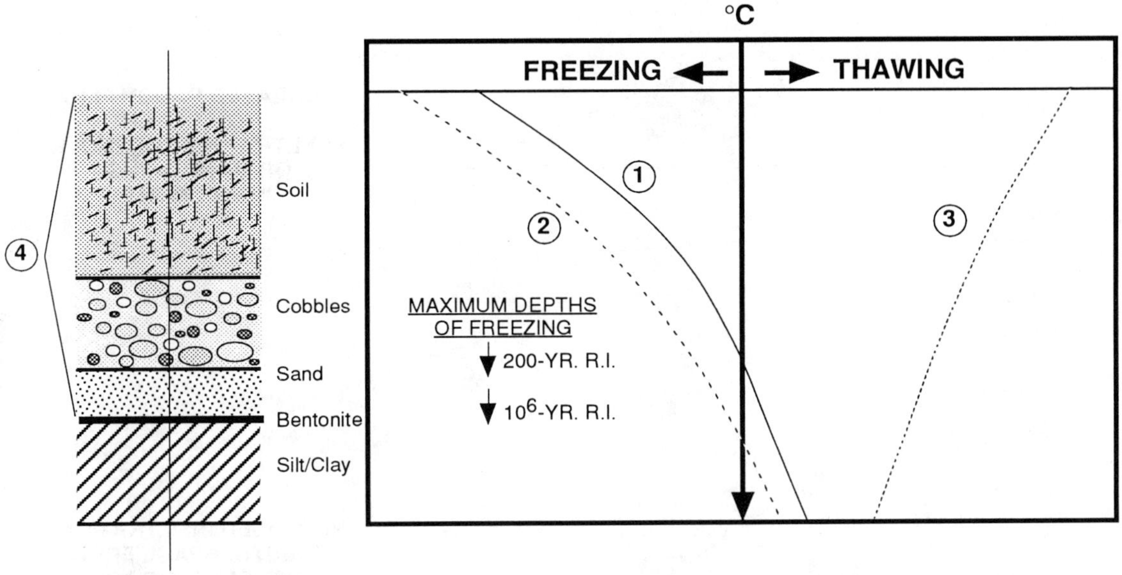

1. TEMPERATURE OF SOILS DUE TO 200-YEAR RECURRENCE INTERVAL FREEZING (PROBABILITY OF OCCURRENCE 1/200)
2. TEMPERATURE OF SOILS FOR MAXIMUM CONCEIVABLE FREEZING EVENT (PROBABILITY OF OCCURRENCE 1/10 ^{6})
3. TEMPERATURE OF SOIL FOR AMBIENT SUMMER CONDITIONS
4. FREEZE/THAW PROTECTION LAYERS

Extreme freezing and thawing damage of the radon and infiltration barriers is precluded by the thick overburden layers.

FIGURE 10.19

FREEZE/THAW TEMPERATURE PROFILES IN COVER

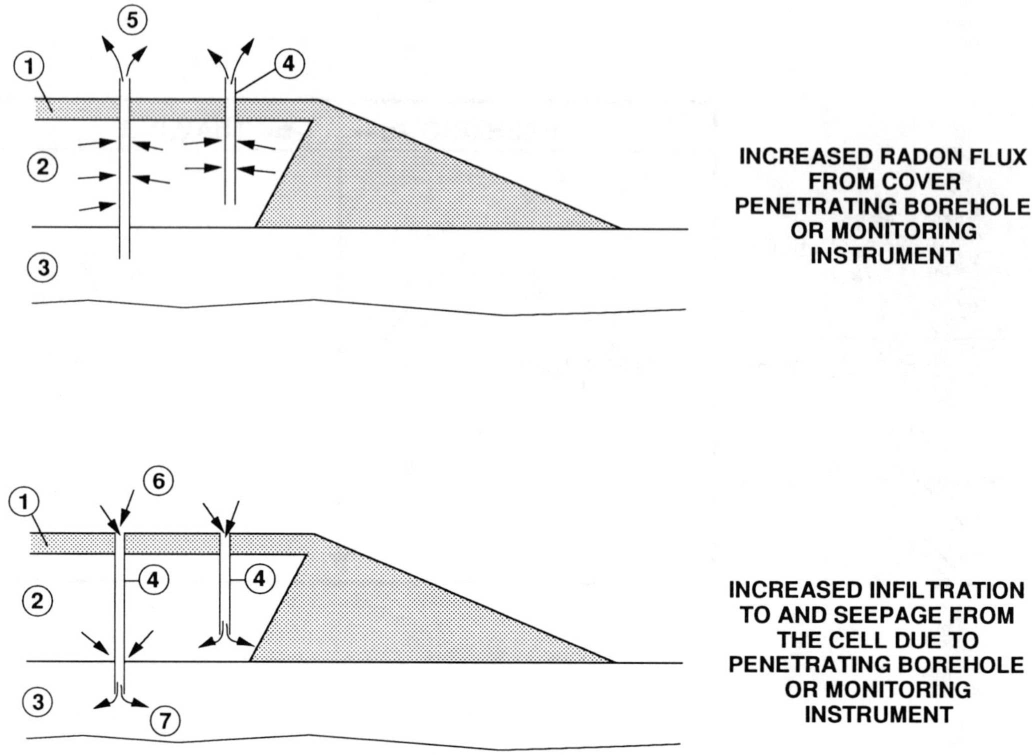

LEGEND

1. COVER
2. WASTE
3. VADOSE ZONE
4. PENETRATING BOREHOLE
5. INCREASED RADON ESCAPE
6. INCREASED INFILTRATION
7. INCREASED SEEPAGE

Boreholes or instrument monitoring tubes through the cover into and through the cell, may negatively impact the desired performance of the disposal cell.

FIGURE 10.2
COVER PENETRATING BOREHOLE NEGATIVE EFFECTS

TABLE 10.1

COMPARISON OF FEATURES OF SOME INFILTRATION MODELS WITH EXPLICIT VEGETATIVE TRANSPIRATION TERMS

	MODEL	
	HELP (Version 2)	**UNSAT-H**
Vertical Unsaturated Flow	Gross water balance-Darcian flow with free outflow (unit gradient), unsaturated hydraulic conductivity as function of water content (modified Brooks-Corey equation).	Numerical solution of general unsaturated flow equation, with hydraulic conductivity and matrix potential as functions of water content.
Run-off	SCS Run-off Curve Number method	Water in excess of what can be absorbed at the surface.
Lateral Drainage	Steady-state Boussinesq equation	None
Evapotranspiration	PET from modified Penman Equation Soil evaporation calculated from plant interception and snow accumulation Transpiration calculated from LAI given by vegetative growth and decay model	PET from Penman Equation Drying evapotranspiration calculated by Thornthwaite-Mather Method (as function of field capacity and wilting point) Vapor diffusion considered if soil surface dries out Plant roots treated as moisture sink, with root density specified by user

NOTES: Models selected from those previously applied to assessment of UMTRA Project or other DOE low-level waste disposal facilities.

Features as described in and/or inferred from model documentation.

TABLE 10.2

HELP MODEL SOIL PARAMETERS

MATERIAL PROPERTIES				
MATERIAL TYPE	**Porosity (vol/vol)**	**Field Capacity (vol/vol)**	**Wilting Point (vol/vol)**	**Hydraulic Conductivity (cm/sec.)**
Erosion Protection (Rock)	0.440	0.200	0.070	1.0
Vegetated Soil	0.500	0.370	0.220	0.001
Drain/Bedding	0.400	0.200	0.070	1.0
Frost Protection	0.290	0.200	0.100	0.00001
Biointrusion Rock	0.440	0.200	0.070	1.0
Radon/Infiltration (Barrier)	0.520	0.480	0.420	10^{-7}

using a twenty percent slope. The slopes were modeled with geomembranes overlying the radon/infiltration barrier. The EPA document "Quantification of Leak Rates Through Holes in Landfill Liners" was examined to determine how to model geomembrane leakage. This document provides leakage rates based on pressure heads and perforation size and type, and on the subgrade material. Generally, the higher the head, the higher the leakage rate. Round holes leak more than slits or elongated holes, and higher permeability subgrade raises the leakage rate. The HELP model allows a leakage rate to be input as a percentage of the water available from the layer overlying the geomembrane. Because the pressure heads associated with the alternative covers were very low, the EPA document did not provide guidance (being intended for liners which have much higher pressure heads). Therefore an arbitrary, but conservative, figure of one percent was used. Early in the analysis, some cover alternatives were run without geomembranes. Because infiltration through the clay infiltration barrier is also largely influenced by the head of water, it can be inferred that alternatives exhibiting lower infiltration with geomembranes will as well.

The results of the infiltration analysis are shown in Table 10.3. As seen in this table, the leakage through the covers varies from $3.87x10^{-9}$ cm/sec. to as little as $1.69x10^{-10}$ cm/sec. These modeling results indicate that there is a direct correlation between lower infiltration and the presence of vegetation, a thick rooting zone, lower permeability infiltration barriers, including geomembranes, and drainage layers overlying the infiltration barrier. Other studies (DOE, 1989) have shown that an increase in the cover inclination (or slope) and an increase in permeability of drains lead to a decrease in the infiltration into the cover.

Additional modeling runs were made with a "clogged" drain situation which was simulated by decreasing the permeability of the drains by a magnitude of 100, and by decreasing the porosity and field capacity. This was done in an attempt to observe the effect of plant roots partially filling the voids in the drains. This effect is more likely to occur in covers where

TABLE 10.3

HELP MODELING RESULTS

COVER		RUN-OFF (inches/yr)	EVAPOTRANS-PIRATION (inches/yr)	LATERAL DRAINAGE (inches/yr)	FLUX (inches/yr) (cm/sec.)
Simple Cover	Topslope	1.054	31.746	0.2402	0.0124 (9.98×10^{-10})
	Sideslope	0.366	32.094	0.5866	0.0159 (1.28×10^{-9})
RCRA Cover	Topslope	0.051	29.335	3.6121	0.0048 (3.86×10^{-10})
	Sideslope	0.079	29.834	3.1339	0.0031 (2.50×10^{-10})
Double-drain Cover	Topslope	0.000	28.098	4.9510	0.0036 (2.90×10^{-10})
	Sideslope	0.074	27.815	5.1613	0.0021 (1.69×10^{-10})
Multicomponent Cover	Topslope	0.051	29.336	3.5927	0.0117 (9.42×10^{-10})
	Sideslope	0.114	30.429	2.7996	0.0064 (5.15×10^{-10})
Erosion Cover	Topslope	0.000	28.107	4.4661	0.04807 (3.87×10^{-9})
	Sideslope	0.061	27.490	5.2370	0.2655 (2.14×10^{-8})

the drains are shallow and overlie soil that holds moisture. With an insufficient rooting zone near the surface, the plants roots will go deeper, particularly where a favorable medium exists. Deeper drain layers will provide some capillary break and, if they overlie a biointrusion layer or a geomembrane, are less likely to become clogged. Results of this effort are shown in Table 10.4.

Hence, for an encapsulation cell with the above considered variables, the preferred cover is that which has:

- The lowest achievable permeability infiltration barrier
- The most abundant vegetation stand
- The thickest, economically feasible rooting zone
- The greatest possible slope inclination
- The most and highest practical permeability drains, particularly below rooting medium materials.

The double-drain cover has the lowest average infiltration rate of 2.90×10^{-10} cm/sec. for the topslope, and 1.69×10^{-10} cm/sec. for the sideslope; however, when the clogged situation is examined, the RCRA cover is preferred.

INFILTRATION BARRIER

In order to evaluate the performance of the encapsulation cell, the implication of a cover infiltration range of 10^{-9} to 10^{-7} cm^3/cm^2/sec. should be evaluated.

TABLE 10.4

RESULTS OF DRAIN CLOGGING

COVER		FLUX (unclogged) cm/sec.	FLUX (clogged) cm/sec.	PEAK DAILY HEAD (unclogged) inches	PEAK DAILY HEAD (clogged) inches
Simple Cover	Topslope	9.98×10^{-10}	____	42.7	____
	Sideslope	1.28×10^{-9}	____	42.1	____
RCRA Cover	Topslope	3.86×10^{-10}	8.45×10^{-10}	0.8	15.1
	Sideslope	2.50×10^{-10}	4.19×10^{-10}	0.2	11.0
Double-drain Cover	Topslope	2.90×10^{-10}	1.09×10^{-9}	2.6	15.7
	Sideslope	1.69×10^{-10}	4.10×10^{-10}	0.6	13.0
Multicomponent Cover	Topslope	9.42×10^{-10}	9.50×10^{-10}	0.8	0.8
	Sideslope	5.15×10^{-10}	9.58×10^{-10}	6.0	15.3
Erosion Cover	Topslope	3.87×10^{-9}	2.32×10^{-7}	1.0	27.2
	Sideslope	2.14×10^{-8}	1.07×10^{-7}	0.0	11.6

In the event that an infiltration of 10^{-7} $cm^3/cm^2/sec.$ exceeds an acceptable performance limit, then an infiltration barrier soil that can be placed at an in-place hydraulic conductivity of 10^{-8} cm/sec. should be used.

Geomembrane Infiltration Barriers

In order to incorporate the influence of a geomembrane in the cover infiltration analysis, it is necessary for the design engineer to specify a leakage rate. Following is a detailed discussion of the technical approach to the selection of reasonable default geomembrane leakage rates.

The leakage rate through a geomembrane is dependent upon a number of variables, but is primarily dependent on the size and shape of the perforation, the liquid head on the geomembrane, and the hydraulic characteristics of the subbase. The leakage rate is nearly independent of the thickness of the geomembrane, the properties of the liquid, and the absence or presence of an underlying geotextile.

By placing the geomembrane over a low-permeability clay material, the potential for leakage is minimized. Therefore, it is only the size and shape of the holes in the geomembrane which are cause for concern. By using a thicker, roughened geomembrane, the possibility of random flaws is reduced due to the manufacturing process which bonds the two outer, textured layers to the middle layer. Construction seaming is the other way of introducing holes in the geomembrane. The caliber of the seam construction and the quality assurance program should be able to minimize leakage through seams.

Infiltration Barrier Design Guidance

In order to achieve the performance discussed above for the soil infiltration barrier, the following design guidance should be adopted in compiling a design of an infiltration barrier in a cover:

- The minimum required radon/infiltration barrier thickness should be 450 mm.
- If the RCRA regulations are ARARs, then the infiltration barrier thickness should be 600 mm.
- The minimum hydraulic conductivity of the infiltration barrier should be 10^{-7} cm/sec.
- If further analyses of the cell performance indicate the need for a lesser hydraulic conductivity, then bentonite amendment of the native clay materials may be necessary in the upper part of the radon barrier. Bentonite amended sand is preferred in the uppermost lift of the infiltration barrier not only to minimize infiltration, but to enhance slope stability.
- The minimum plasticity index of the soil of the infiltration barrier should be at least 15.

Chapter 11

THE HYDROLOGICAL MODEL: THE WASTES

GENERAL

This chapter deals with both the quantity and quality of seepage from the wastes both in the short- and the long-term. Various wastes may be placed in the disposal cell. Some may have been solidified and stabilized prior to placement into the cell.

TRANSIENT DRAINAGE

Transient drainage may occur if significant percentages of the wastes are placed in the cell with moisture contents in excess of the residual (long-term equilibrium) moisture content of the materials. Figure 11.1 describes how moisture in the waste at placement will seep from the waste at a decreasing rate until the waste is at a moisture content in equilibrium with that associated with steady state flux hydraulic conductivity moisture.

Alternative physical states of the waste (placed as a compacted soil as compared to a fully hydrated concrete or a vitrified mass) may affect the occurrence, rate, and timing of transient drainage. Figure 11.2 illustrates how the spacing of zones of high permeability wastes determines the period of transient drainage. The rate of transient drainage can be altered by changing the drainage path length (i.e., the longest distance a particle of water has to travel before it leaves the low permeability wastes and enters the higher permeability wastes.)

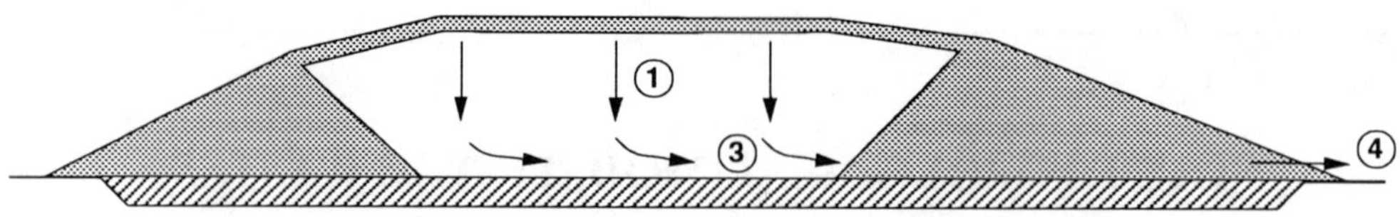

SOURCE AND ROUTE OF WASTE TRANSIENT SEEPAGE

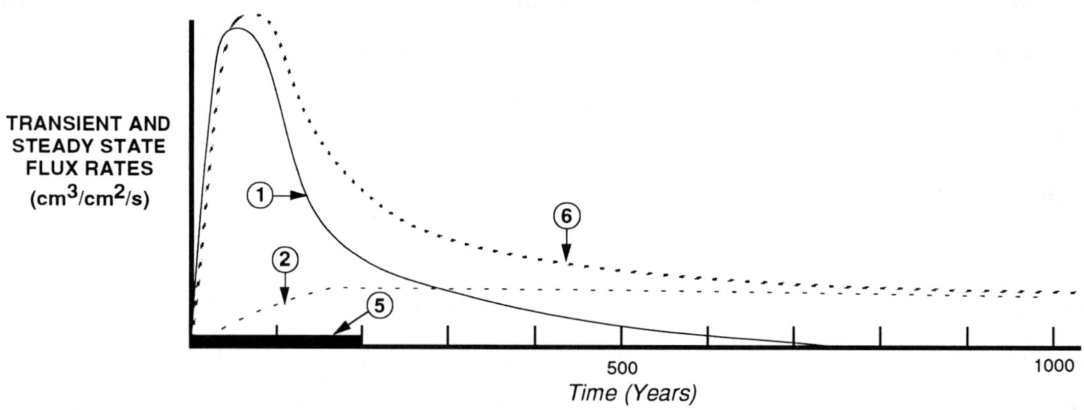

TRANSIENT AND STEADY STATE CELL SEEPAGE

LEGEND

1. TRANSIENT SEEPAGE RATE FROM WASTE
2. LONG-TERM STEADY STATE COVER INFILTRATION RATE
3. LEACHATE COLLECTION SYSTEM SEEPAGE
4. DRAIN OUTLET FLOW
5. PERIOD OF GEOMEMBRANE EFFICACY
6. TOTAL SEEPAGE FROM SINGLE DRAIN

Moisture in the waste at placement will seep from the waste at a decreasing rate until the waste is at a moisture content in equilibrium with that associated with steady state flux hydraulic conductivity moisture.

FIGURE 11.1

TRANSIENT SEEPAGE FROM THE WASTE

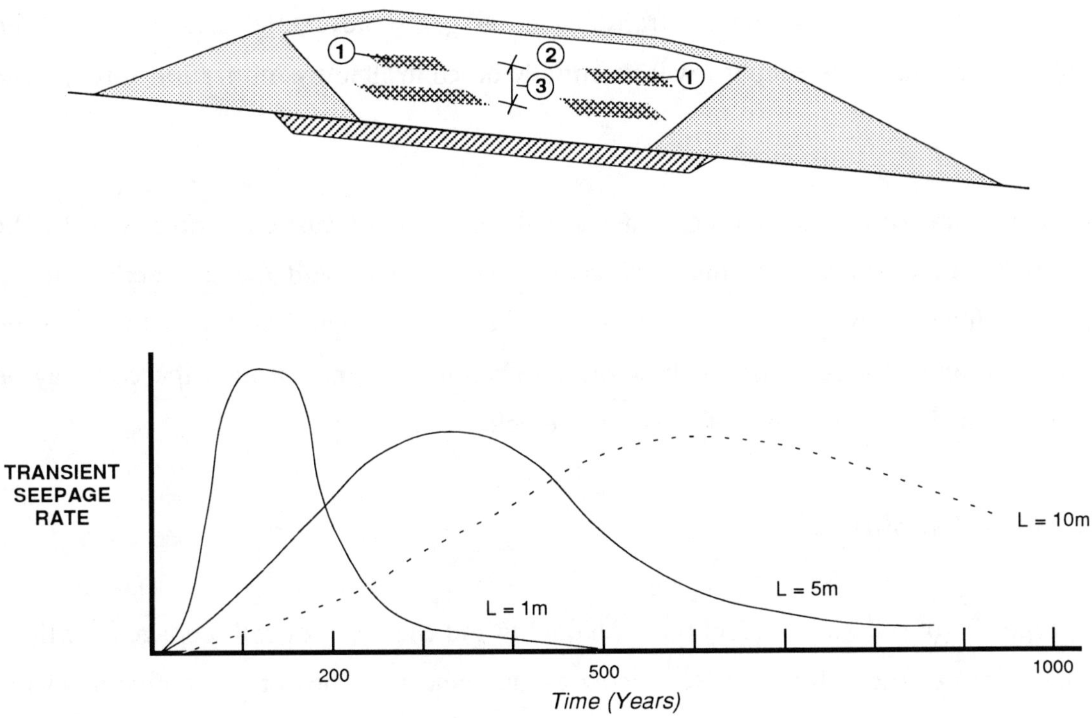

LEGEND

1. HIGH PERMEABILITY WASTE OR FILL
2. LOW PERMEABILITY WASTE
3. SPACING OF HIGH PERMEABILITY ZONES (L)

The spacing of zones of high permeability wastes determines the period of transient drainage.

FIGURE 11.2

TRANSIENT DRAINAGE RATES

STEADY STATE DRAINAGE

With the substantial cessation of transient drainage, steady state conditions will be established. Steady state seepage will essentially be controlled by infiltration through the cover.

Placement of the wastes (i.e., segregation or zonal placement of wastes of differing hydraulic conductivity) may control the routes of steady state seepage and the interaction of the seepage with wastes of differential capacity to affect the ultimate quality of the seepage. Figures 11.3 and 11.4 demonstrate how the infiltration into and through the cell may be dependent on the arrangement of waste in the cell.

SEEPAGE CHEMISTRY

The properties of cell wastes affect both the quality and chemistry of cell seepage over time. Passage of water through the wastes increases the concentrations of contaminants at the drain. The concentration of contaminants in the seepage may be reduced by geochemical attenuation in flowing through the liner and foundation soils.

Figure 11.5 presents one feasible procedure to modify leachate quality by passing all waste leachate through a geochemical sump before collection or free discharge.

Geochemical zonation beneath the clean-fill dike is illustrated in Figure 11.6. In this procedure, a layer of geochemically attenuating soil is placed at the base of the dike. This soil could be used to modify long-term leachate quality before free discharge.

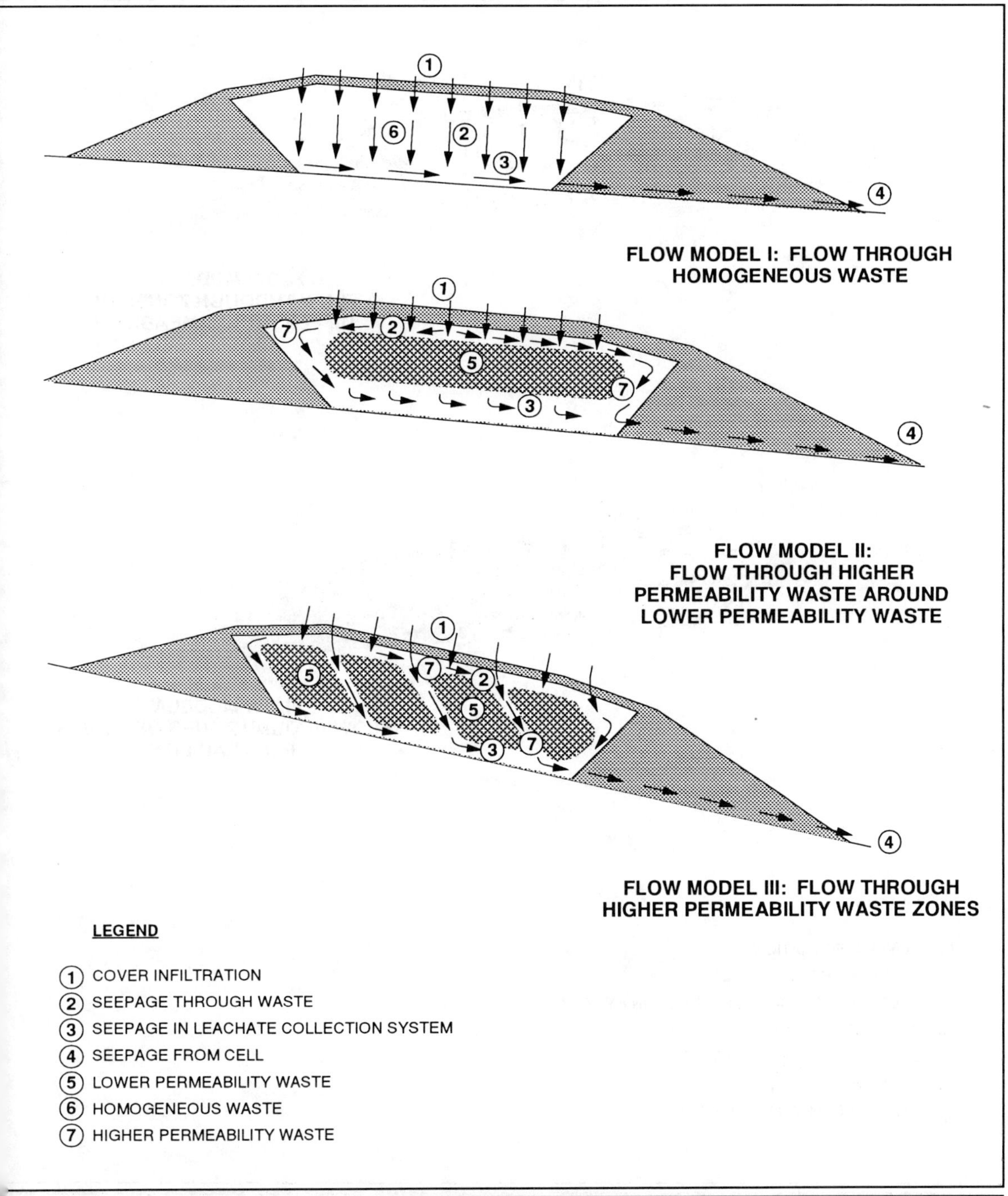

Depending on the arrangement of waste in the cell, the infiltration into and through the cell may be directed around or through specific waste zones.

FIGURE 11.3

WASTE THROUGH-FLOW MODELS I

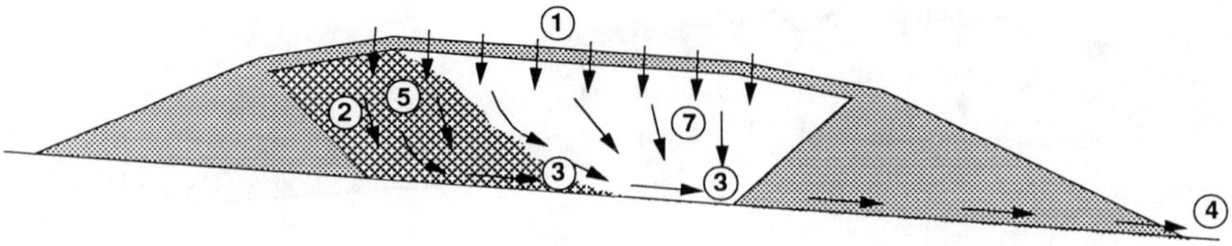

**FLOW MODEL IV:
FLOW THROUGH ZONES OF
DIFFERENT PERMEABILITY**

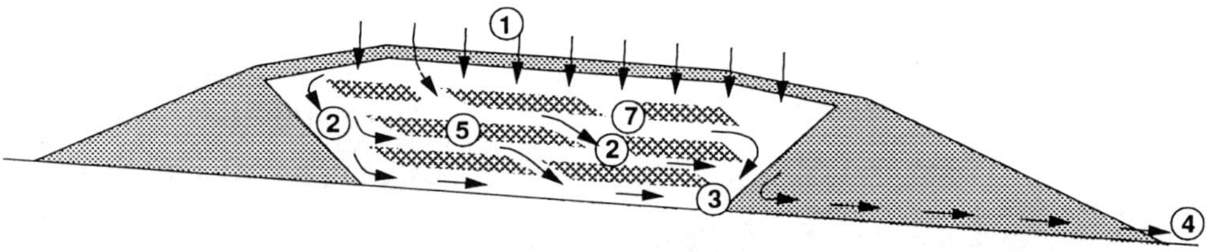

**FLOW MODEL V:
FLOW AROUND ZONES OF LOWER
PERMEABILITY**

<u>LEGEND</u>

① COVER INFILTRATION
② SEEPAGE THROUGH WASTE
③ SEEPAGE IN LEACHATE COLLECTION SYSTEM
④ SEEPAGE FROM CELL
⑤ LOWER PERMEABILITY WASTE
⑥ HOMOGENEOUS WASTE
⑦ HIGHER PERMEABILITY WASTE

Depending on the arrangement of waste in the cell, the infiltration into and through the cell may be directed around or through specific waste zones.

FIGURE 11.

WASTE THROUGH-FLOW MODELS

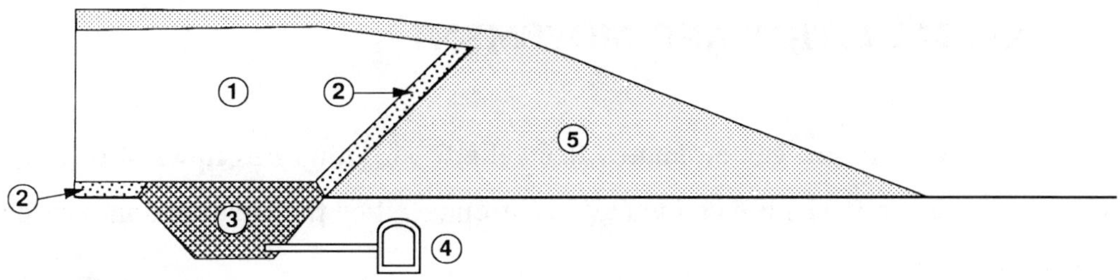

LEGEND

1. WASTE
2. DRAINS
3. ZONE OF GEOCHEMICAL ATTENUATION MATERIAL
4. ACCESS AND DRAIN COLLECTION GALLERY
5. CLEAN-FILL DIKE

A feasible procedure to modify leachate quality is to pass all waste leachate through a geochemical sump before collection of free discharge.

FIGURE 11.5

LEACHATE GEOCHEMICAL SUMP

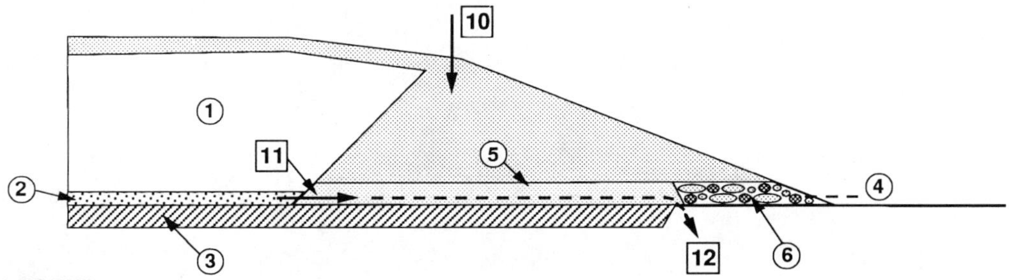

LEGEND

1. WASTE
2. DRAINS
3. BASAL LINER AND FOUNDATION
4. DRAIN OUTLET PIPE: BLOCKED AT COMPLETION OF TRANSIENT DRAINAGE
5. SELECTED GEOCHEMICALLY ATTENUATING MATERIAL
6. GRAVEL DRAIN TO CONTROL LONG-TERM SEEPAGE AT TOE OF DIKE

SEEPAGE QUANTITIES

10 DIKE INFILTRATION
11 DRAIN FLOW
12 FOUNDATION INFILTRATION

A layer of geochemically attenuating soil placed at the base of the dike could be used to modify long-term leachate quality before free discharge.

FIGURE 11.6

DIKE GEOCHEMICAL ZONATION

RADON GAS EMANATION AND MOVEMENT

Figure 11.7 shows how radon gas emanates from the waste, migrates upward through the waste and cell drains, and exits the cell via and is attenuated by the cover radon barrier and clean-fill dike soils.

HEAT GENERATION

A waste may be solidified and stabilized by the addition of cement before placement in a disposal cell. The hydration process may generate heat. As shown in Figure 11.8, cell components, including the soil barrier may be affected by the generation of heat during the stabilization process. Heat generation may affect seepage by increasing vegetated growth, drying or cracking the radon barrier, decreasing saturation of the infiltration barrier, softening and chemically destabilizing geomembranes, and reducing clay/geomembrane interface friction.

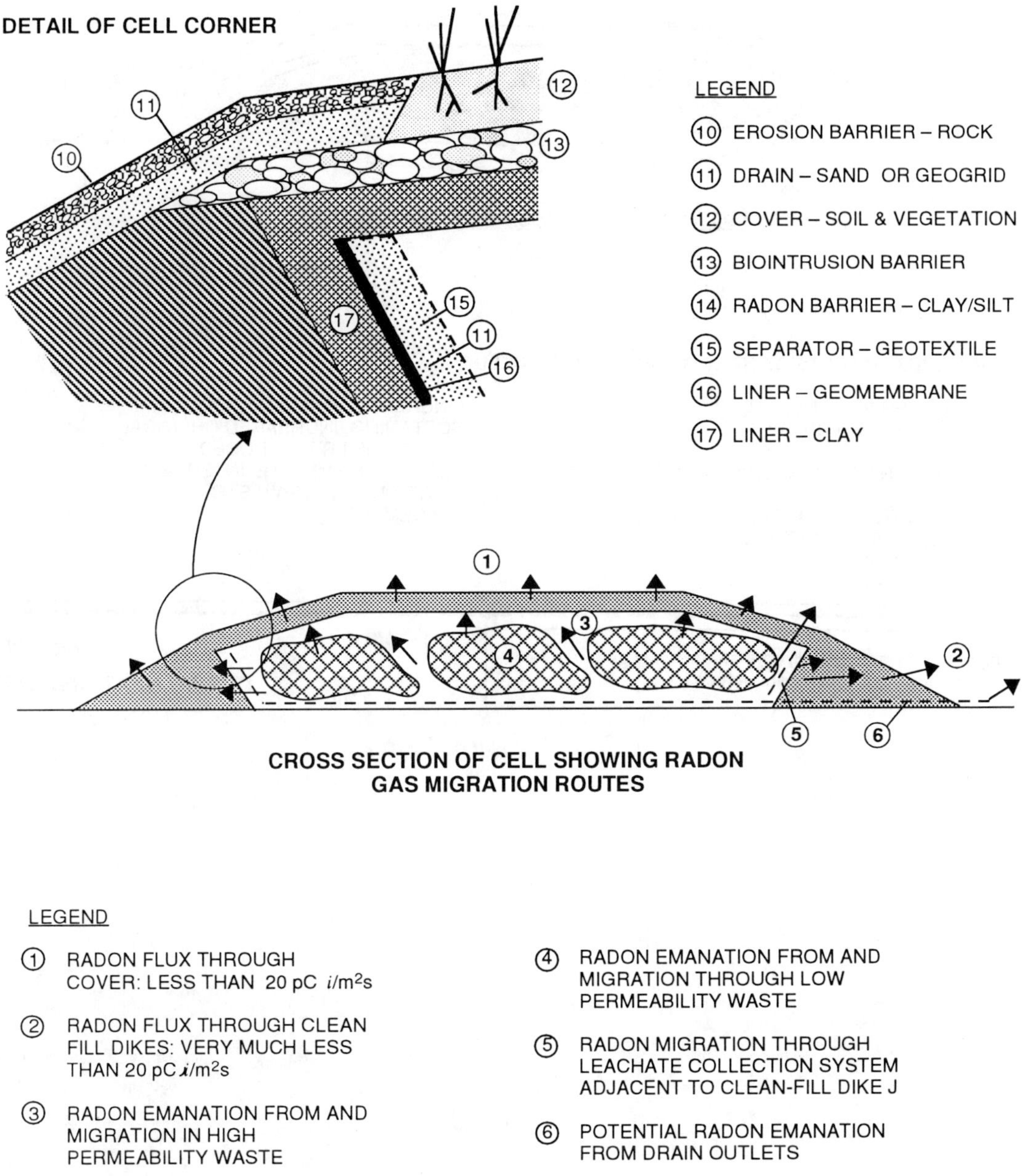

Radon gas emanates from the waste, migrates upward through the waste and cell drains, and exits the cell via and is attenuated by the cover radon barrier and clean-fill dike soils.

FIGURE 11.7

DISPOSAL CELL RADON FLUX MODEL

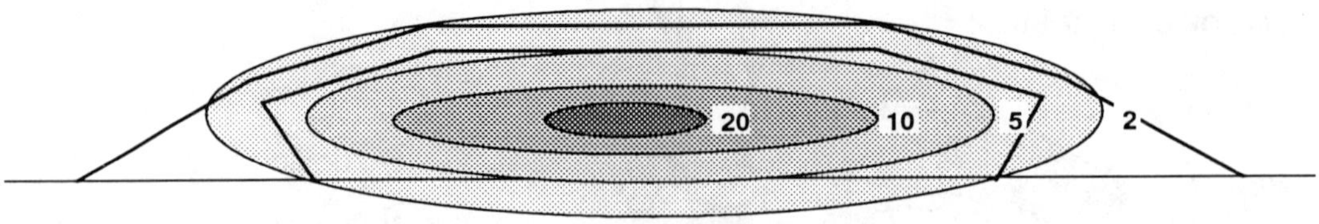

ISOTHERMS FOR CSS HEAT GENERATION

POSSIBLE EFFECTS OF HEAT GENERATION ON CELL

1 COVER VEGETATION: INCREASED GROWTH

2 RADON BARRIER: DRYING AND CRACKING

3 INFILTRATION BARRIER: PARTIAL SATURATION WITH LESS INFILTRATION

4 LINERS–GEOMEMBRANES: SOFTENING, INCREASED CHEMICAL SUSCEPTIBILITY, REDUCED CLAY/GEOMEMBRANE INTERFACE FRICTION DURING WASTE PLACEMENT

Setting of the Cement Stabilized materials may generate heat that could affect cell component behavior.

FIGURE 11.8

CURING HEAT GENERATION

Chapter 12

THE HYDROLOGICAL MODEL: BASAL LINERS, DRAINS, AND THE FOUNDATION

CELL BASAL LAYERS

Drains

Figure 12.1 shows the basal liner and drain system water balance. Seepage from the waste at the base of the disposal cell will be primarily diverted and removed by the drains; multiple liners impede seepage to the foundation. In the short-term, little or no seepage will pass through the liner that separates the upper drain and the lower backup drain (leak detection system). In the long-term, some seepage may pass through to the lower drain and flow laterally to the collection system. A very small amount of seepage may occur through the lowermost liner to the foundation of the cell.

Seepage rates through the basal drains and liners may vary over a relatively wide range and with time. The flow quantity from the cell varies in proportion to cover infiltration and also depends on the number and location of the drain outlets. The major factor affecting the variation of seepage with time is the deterioration of the geomembrane. Figure 12.2 shows conditions after cell abandonment, and possible failure of geomembranes and leachate collection systems. It shows that cover infiltration may flow to the groundwater or the surface, depending on cell layout features.

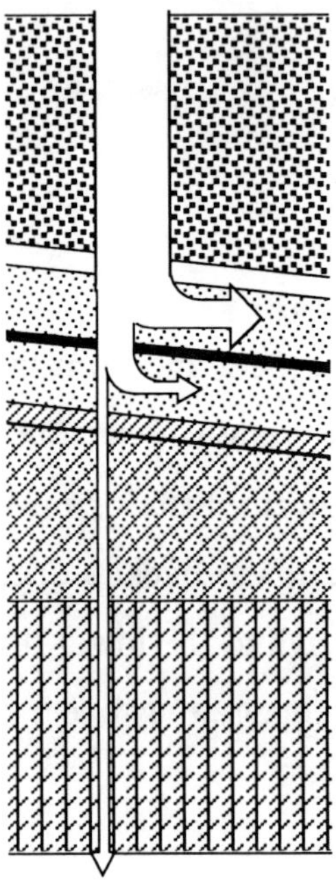

COMPONENT	FLUX (cm/s) (at base of layer)
WASTE	10^{-7} to 10^{-6}
DRAIN: SAND	
LINER: GEOMEMBRANE	10^{-10} to 10^{-6}
LEAK DETECTION: SAND	
LINER: BENTONITE MAT	10^{-10} to 10^{-8}
LINER: CLAY/SILT	
FOUNDATION: SILT/CLAY	10^{-10} to 10^{-8}

Seepage from the waste at the base of the disposal cell will be primarily diverted and removed by the drain; multiple liners impede seepage to the foundation.

FIGURE 12.

BASE LINER FLUX RATE

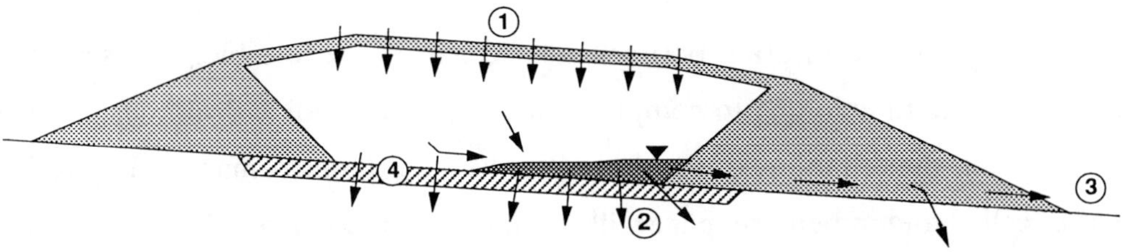

LONG-TERM THROUGH-FLOW MODEL I
COVER INFILTRATION EXCEEDS BASAL LINER SEEPAGE. OUTLET DRAINS BLOCKED, LINER SYSTEM STILL INTACT. SATURATED ZONE DEVELOPS AT CELL LOW POINT. INCREASED HYDRAULIC HEAD INCREASES SEEPAGE THROUGH FOUNDATION AND DIKE. SATURATED ZONE DEPTH AT EQUILIBRIUM IN X $_1$m.

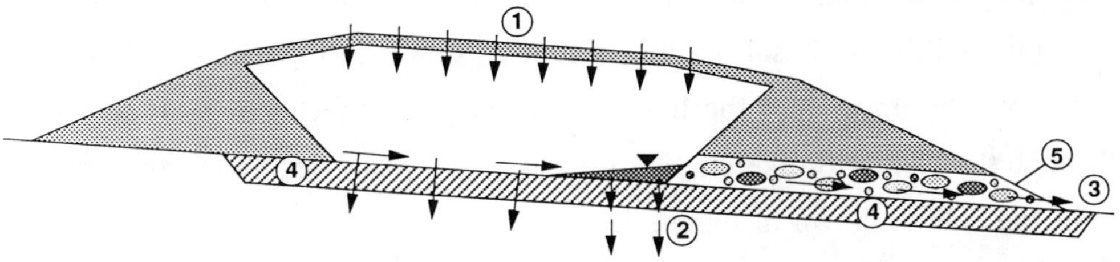

LONG-TERM THROUGH-FLOW MODEL II
COVER INFILTRATION EXCEEDS BASAL LINER SEEPAGE. A SMALL SATURATED ZONE DEVELOPS AT CELL LOW POINT. SEEPAGE OCCURS PRIMARILY THROUGH HIGH PERMEABILITY MATERIALS PLACED IN LOWER PART OF DIKE. SEEPAGE EMERGES PRIMARILY AS SURFACE FLOW.

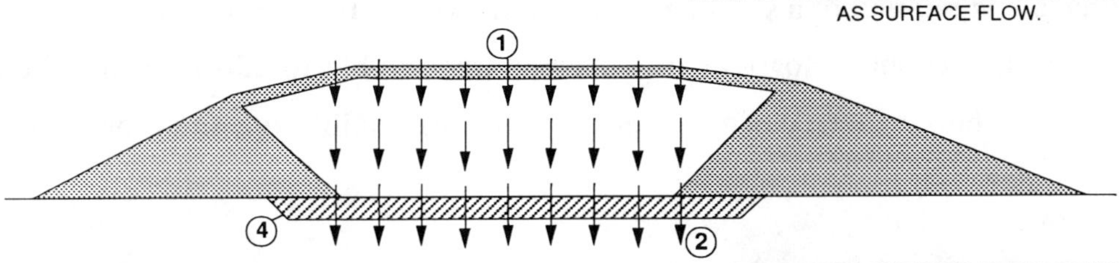

LONG-TERM THROUGH-FLOW MODEL III
THE POTENTIAL SEEPAGE THROUGH THE BASAL LINER (DEGRADED GEOMEMBRANES) EQUALS OR EXCEEDS COVER INFILTRATION OR SEEPAGE FROM THE WASTE. ACCORDINGLY FLOW IS ESSENTIALLY VERTICALLY DOWNWARD THROUGH THE CELL.

LEGEND

(1) COVER INFILTRATION
(2) BASAL SEEPAGE THROUGH LINERS
(3) SURFACE SEEPAGE FROM CELL
(4) GEOCHEMICAL BARRIER
(5) HIGHLY POROUS ZONE AT BASE OF DIKE.

After cell abandonment, and possible failure of geomembranes and leachate collection system, the cover infiltration may flow to the groundwater of the surface, depending on cell layout features.

FIGURE 12.2

LONG-TERM THROUGH-FLOW MODELS

A variety of alternative long-term drain performance scenarios are feasible. These range from continued, complete functioning to complete blocking. Alternative drain designs will have varying impacts on the reliability of the drain, and hence cell performance. Figure 7.3 suggests that after cell abandonment, seepage will continue to emerge from the cell, and will flow overland to distant water bodies, be evapotranspirated, or seep into the surrounding soils.

In the worst case, long-term scenario of blocked drains, the level of saturation in the waste would increase until equilibrium basal outflow occurs, as shown in Figure 12.3. The basal drains (either or both the upper and the lower drain) may fail by clogging. If the single outlet drain through the clean-fill dike clogs, a saturated zone may develop in the waste at the lower part of the cell as shown in Figure 12.2.

Gravel and cobbles may be incorporated into a base of a clean-fill dike. This zone performs as a broad, continuous high permeability layer through which seepage may continue out of the cell, thus precluding buildup of a saturated zone in the waste. In the extreme event that this zone of gravels and cobbles clogs (or it is deemed undesirable to allow uncontrolled long-term seepage to the northern perimeter of the cell), the equilibrium situation shown in Figure 12.2 will develop.

Liners

Figure 12.1 shows the basal liner and drain system water balance. Seepage from the waste at the base of the disposal cell will be primarily diverted and removed by the drains while multiple liners impede seepage to the foundation. In the short-term, little or no seepage will pass through the liner that separates the upper drain and the lower backup drain (leak detection system). In the long-term, some seepage may pass through to the lower drain, and

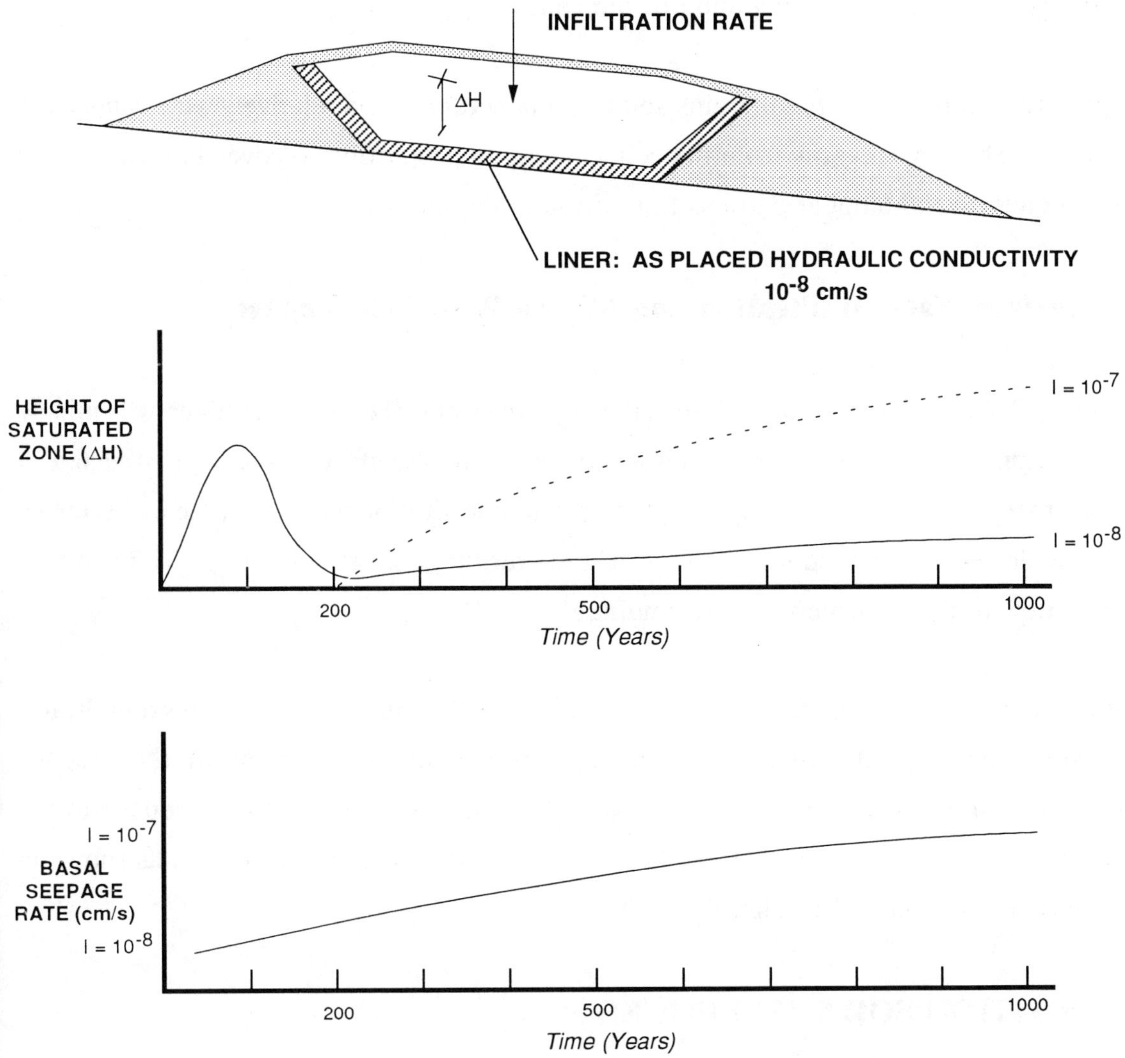

In the worst-case, long-term scenario of blocked drains, the level of saturation in the waste would increase until equilibrium basal outflow occurs.

FIGURE 12.3

DISPOSAL CELL PLUGGED DRAIN PERFORMANCE

flow laterally to the collection system. A very small amount of seepage may occur through the lowermost liner to the foundation of the cell.

Seepage rates through the basal drains and liners may vary over a relatively wide range, and with time. The flow quantity from the cell will vary in proportion to cover infiltration and also depends on the number and location of the drain outlets.

Site Surface Water Infiltration and Mixing With Cell Seepage

Figure 10.7 shows that the run-off from the top cover and the clean-fill dikes will mingle with precipitation on the perimeter buffer and potentially with cell seepage collected in sumps or trenches near the cell. During the period of institutional controls, the cell seepage will be collected and treated for discharge; thereafter the seepage will mingle with cell run-off and flow to the environment surrounding the cell.

During monitoring and operation, as shown in Figure 12.4, minor seepage to groundwater is feasible (but unlikely) through the basal liner and at drain outlets at the seepage collection facility. After cessation of cell operation and monitoring, minor seepage to the groundwater is feasible. The route and quantity depend on cell layout details, as shown in the alternatives described in Figure 7.4.

FOUNDATION SOILS AND ROCKS

Foundation Fill

If a foundation fill is placed beneath the cell, the fill will be partially saturated when placed. It may potentially become saturated as a result of consolidation by the waste. Ultimately,

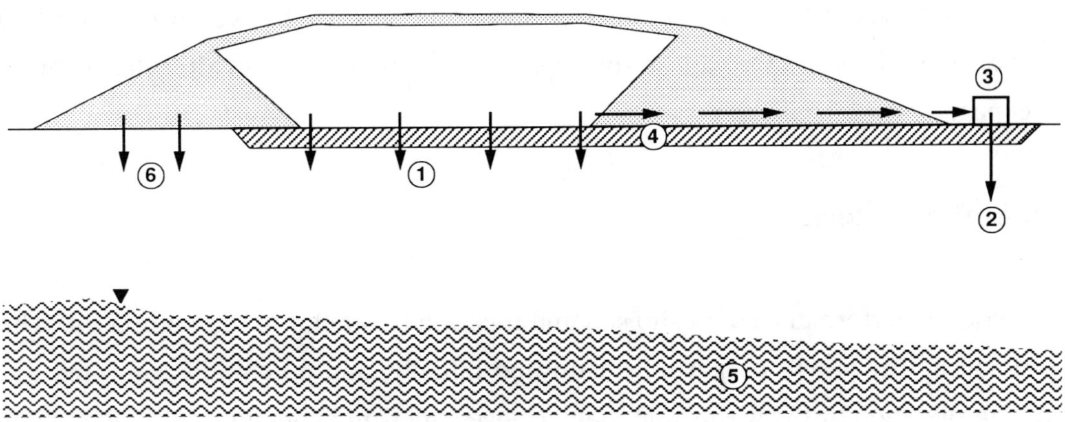

MODEL OF SEEPAGE FROM CELL TO GROUNDWATER DURING OPERATION AND MONITORING PHASE

LEGEND

(1) SEEPAGE THROUGH LINERS AND GEOCHEMICAL BARRIER ($<<10^{-8}$ $cm^3/cm^2/s$)
(2) SEEPAGE AT CELL PERIMETER FROM INTERCEPTION AND COLLECTION SYSTEM
(3) DRAIN INTERCEPTION AND COLLECTION SYSTEM
(4) LINERS AND FOUNDATION: GEOCHEMICAL BARRIERS
(5) GROUNDWATER
(6) CLEAN SEEPAGE THROUGH DIKES

Minor seepage to groundwater is feasible (but unlikely) through the basal liner and at drain outlets at seepage collection facilities.

FIGURE 12.4

SEEPAGE TO GROUNDWATER DURING MONITORING AND OPERATION

the foundation soil will reach a moisture content consistent with the equilibrium seepage from the liner.

If at any time during the operating life of the disposal cell seepage passes through the basal liner to the foundations and the vadose zone, geochemical processes will alter the chemistry of the seepage.

In-Situ Soil and Rock

The hydrological performance of in-situ soil and rock will be similar to foundation fill. That is, in-situ soil and rock will be partially saturated initially, and potentially saturated as a result of consolidation by the waste, and then achieve a moisture content consistent with the equilibrium seepage from the liner. Figure 12.5 describes the changes in density and moisture content which are expected over time in in-situ soils. In the long-term, seepage quantities from the cell to the vadose zone are a small part of the total flow system.

Figures 12.6, 12.7, and 12.8 describe changes to and determinants of vadose zone moisture content resulting from construction of the cell. Generally, building the cell reduces soil permeability, causing the vadose zone to dry out.

THE GROUNDWATER SUBSYSTEM

A comprehensive assessment of the performance of a disposal cell will include a detailed examination of the effect of cell seepage on the groundwater at the site. The literature on modeling groundwater flow and assessing the impact of influent contaminants on seepage water quality is vast, and we neither repeat nor attempt to add to the topic here. We note only that as shown in Figure 12.9 the quality of the groundwater may be protected in many diverse ways.

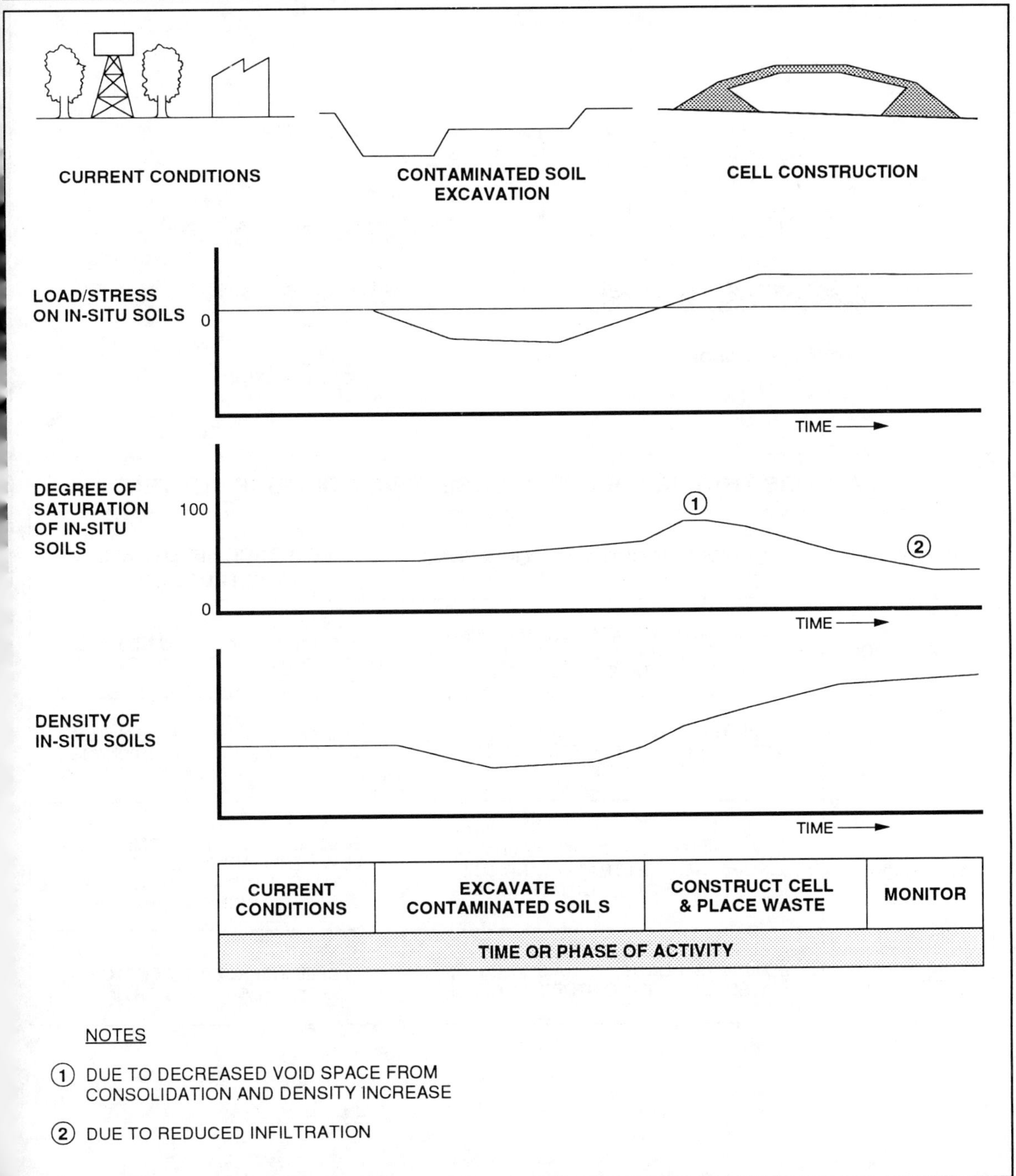

Changes in the load on the vadose zone soils cause the density and moisture content to vary with time.

FIGURE 12.5

IN-SITU SOILS RESPONSE TO CELL CONSTRUCTION

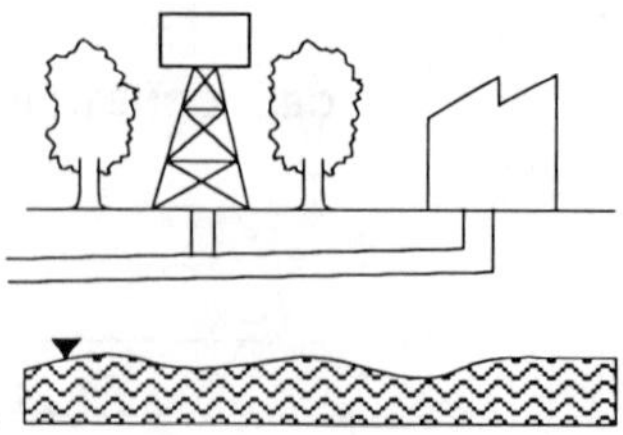

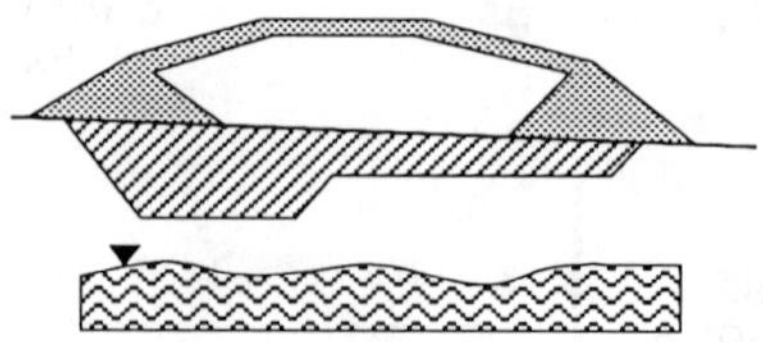

FACTORS THAT INFLUENCE VADOSE ZONE MOISTURE CONTENT		
FACTOR	**INFLUENCE FOR CURRENT CONDITIONS**	**INFLUENCE AFTER CELL CONSTRUCTION**
INFILTRATION	SIGNIFICANT INFILTRATION MAINTAINS SOIL MOISTURE CONTENT (STEADY STATE CONDITIONS)	REDUCED OR ZERO DUE TO CELL, REDUCED SOIL MOISTURE CONTENT
PIPELINES	DISTINCT CONDUITS FOR WATER MOVEMENT INCREASE SOIL MOISTURE CONTENT	REMOVAL OF PIPES ELIMINATES SOURCE, HENCE SOIL DRYING IS LIKELY
BACKFILLED TRENCHES	LOW-DENSITY, HIGH-PERMEABILITY FILL INCREASES INFILTRATION AND SOIL MOISTURE CONTENT	SELECTED FILL, WELL-COMPACTED, LOW-PERMEABILITY FILL INHIBITS INFILTRATION AND MOISTURE CONTENT INCREASE
FACILITIES	BUILDINGS, PONDS, ETC. A SOURCE OF WATER AND SOIL MOISTURE INCREASE	CELL INTERCEPTS SEEPAGE AND LEADS TO SOIL DRYING

Removing building, etc., and building the cell reduces soil permeability and infiltration, hence potentially causing the vadose zone to dry out.

FIGURE 12.6

VADOSE ZONE MOISTURE CONTENT DETERMINANTS

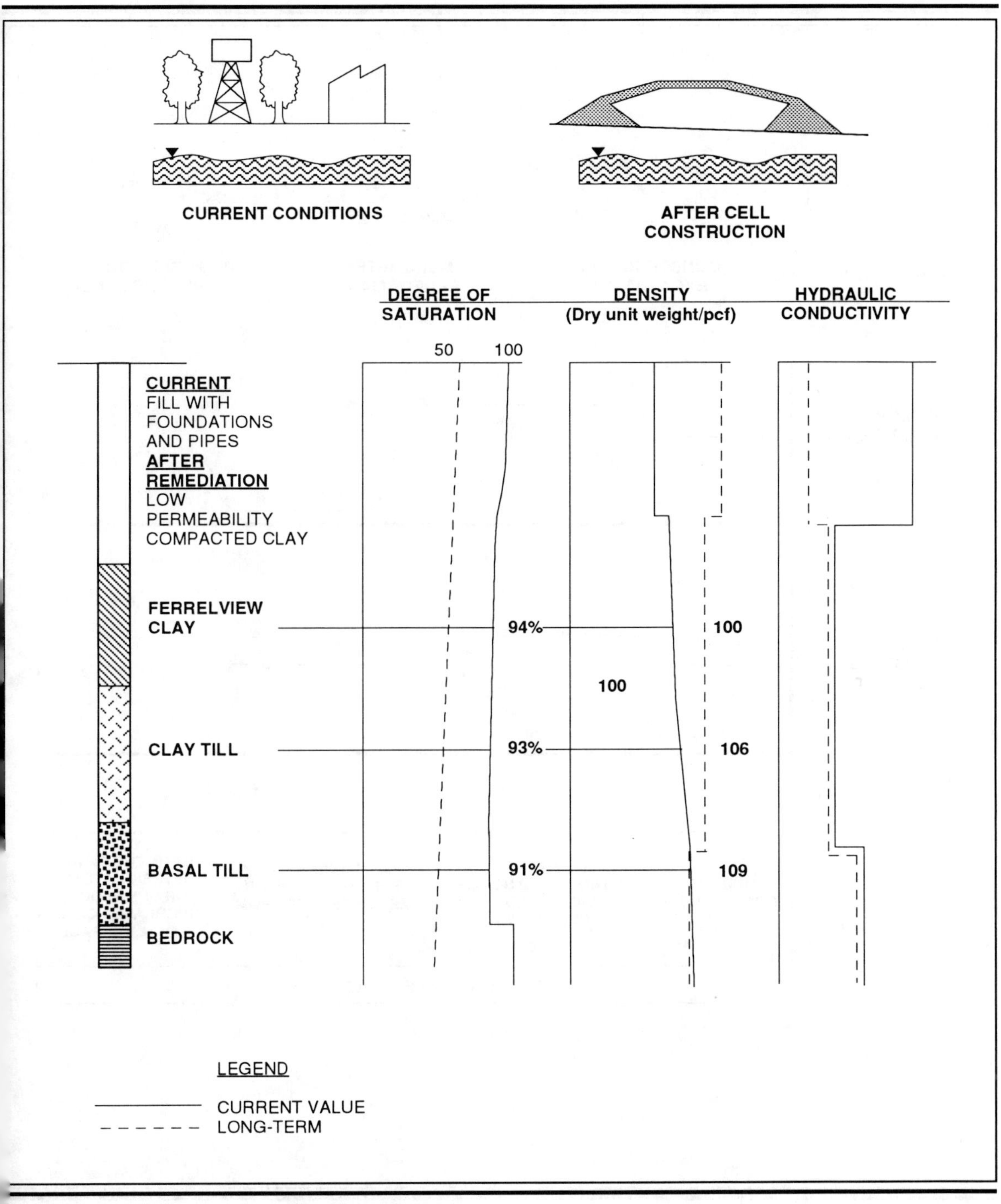

The density and moisture content of the vadose zone will change due to the presence of the disposal cell.

FIGURE 12.7

IN-SITU SOIL CONDITIONS DUE TO CELL CONSTRUCTION

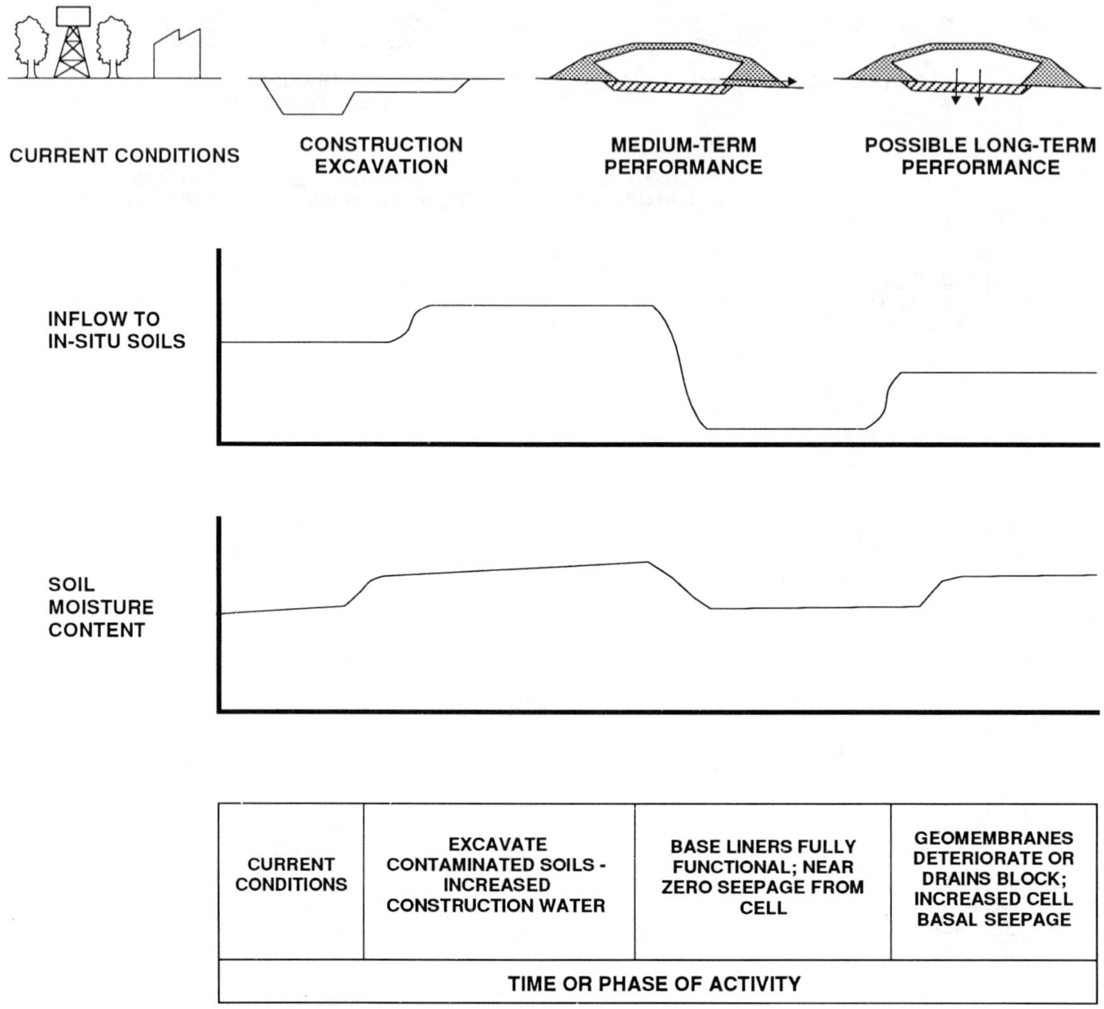

The infiltration to the vadose zone changes as the cell is built and its performance varies; this results in a commensurate change of the soil moisture condition.

FIGURE 12.8

VADOSE ZONE INFILTRATION

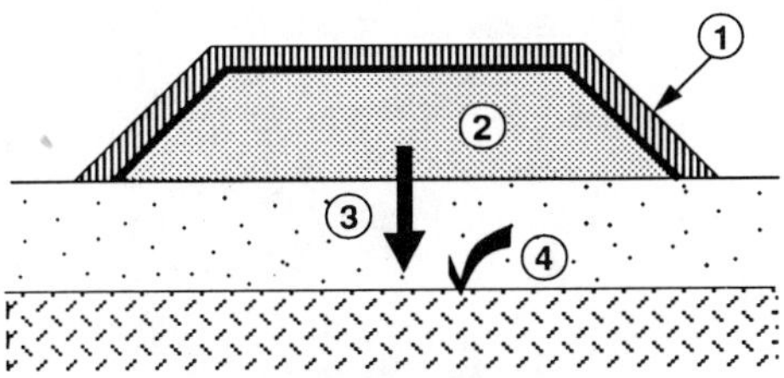

Leachate Quantity Reduced By:

- Cover Liner
- Water Moisture Content Reduction

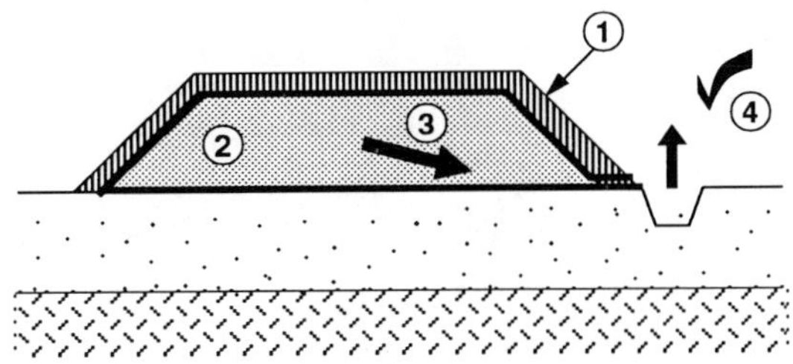

Leachate Quantity Reduced By:

- Cover Barriers and
- Interception by Liner and Leachate Collection

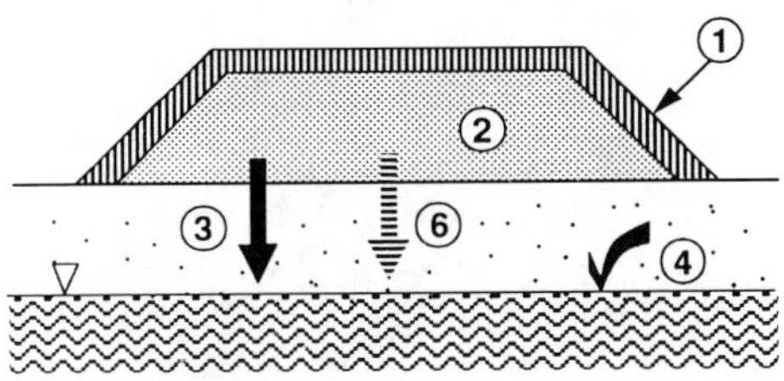

Leachate Quality At Phreatic: Surface Modified By

- Waste Stabilization
- Vadose Zone Geochemical Attenuation

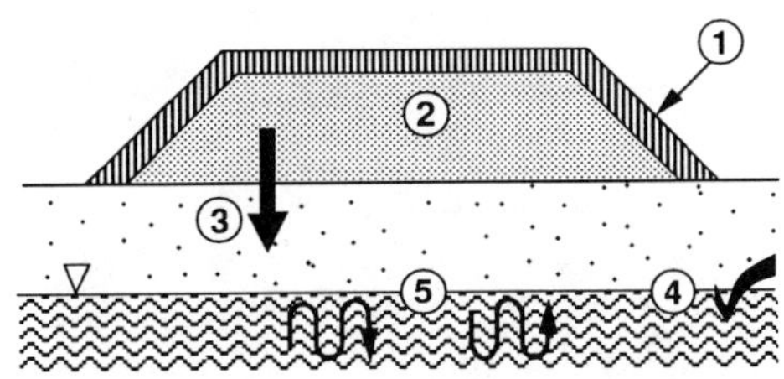

Groundwater Quality At Point Of Exposure Moderated By:

- Aquifer Dilution

LEGEND

1. COVER
2. WASTE
3. LEACHATE: CONTAMINATED
4. ACCEPTABLE LEACHATE QUALITY OR CONTROL
5. DILUTION BY MIXING
6. ATTENUATION BY GEOCHEMICAL PROCESSES

✓ ACCEPTABLE WATER QUALITY

The quality of groundwater may be protected by controlling the nature and quantity of the leachate from the cell, to and into the aquifer.

FIGURE 12.9

GROUNDWATER PROTECTION BY LEACHATE CONTROL

Chapter 13

THE GEOTECHNICAL MODEL: DEFORMATION AND CRACKING

GENERAL

This and subsequent chapters describe the performance of the cell as it is affected by the following forces: gravity, earthquakes, erosion from run-off, and site geomorphological change. In particular the section describes the performance of the cell as affected by consolidation of the foundation and the wastes, potential slope stability and the gradual long-term process of soil loss, and the change of shape of the dike and the cover. Figure 13.1 shows the main subsystems that are modeled and analyzed in a geotechnical performance assessment of a disposal cell.

FOUNDATION AND WASTE DEFORMATION

The load imposed on the foundation by the wastes may cause the soils of the site to consolidate and settle. The wastes that are placed in the cell may also settle as additional load is imposed on them by successive waste and cover placement. This consolidation of the foundation soils and wastes may result in deformation of the cover as shown in Figure 13.1. As shown in Figure 4.6, the total settlement may be significantly reduced by excavating compressible soils from the site prior to construction of the encapsulation system.

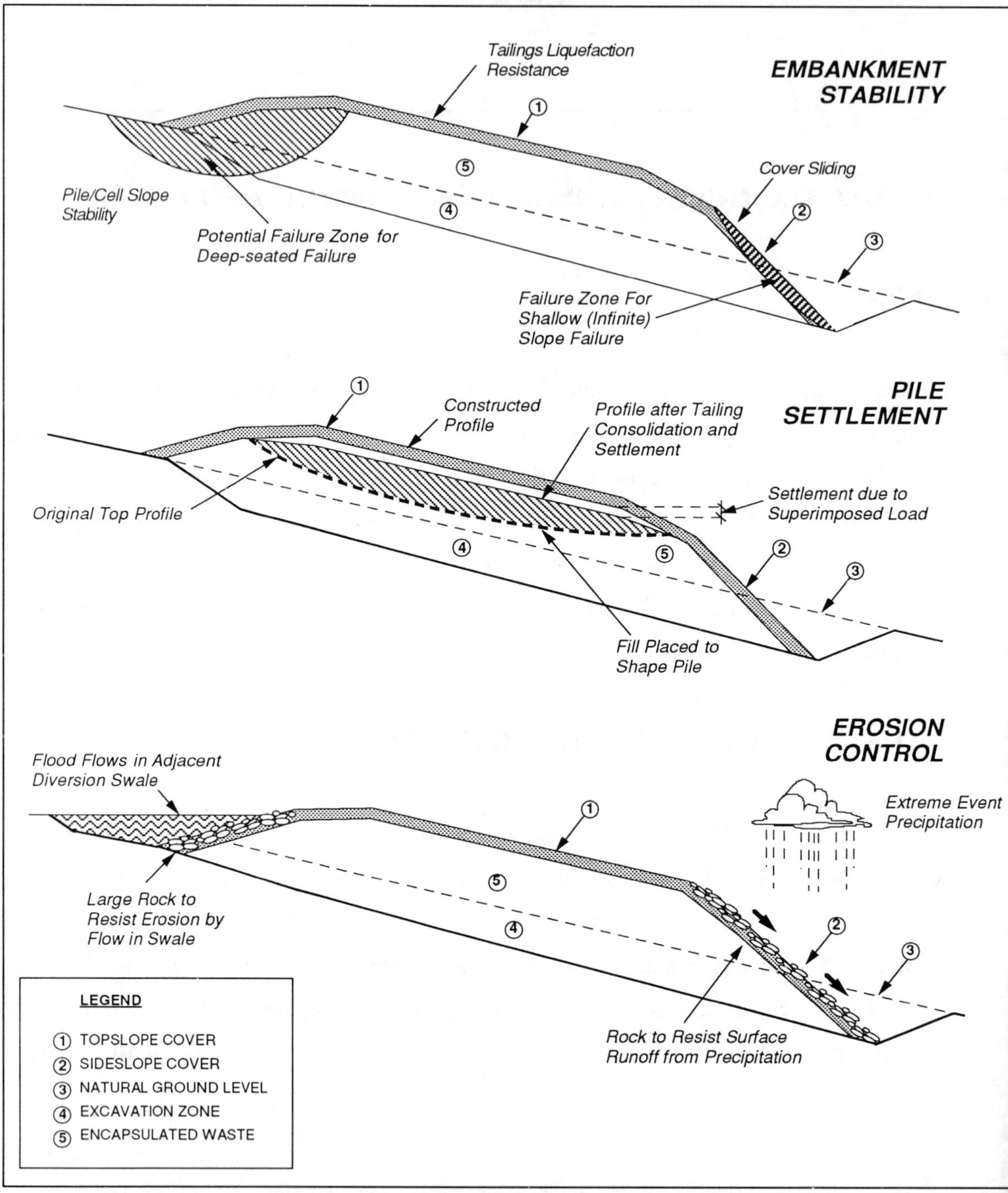

FIGURE 13.1

EMBANKMENT STABILITY, PILE SETTLEMENT, EROSION CONTROL

Settlement of the foundation and the wastes may negatively impact cell performance. Consequences may include: cover deformation, barrier cracking, liner regrading, dike deformation, flow concentration, and potential increased rates of radon and water infiltration flux and geomorphological change.

Figure 13.2 details the potential effects of uneven settlement of the cell foundation and the possibility that waste deformation might result in cracking of the radon barrier. Figure 13.3 indicates that cracking of the radon barrier may be prevented by limiting deformation-induced stresses to less than overburden-induced stresses. In addition, as shown in Figure 13.4, cracking of the radon barrier may be precluded by limiting the concave deformation and horizontal stresses on cover components. These various failure scenarios are precluded by the following design features: dense or compacted waste, avoiding discontinuities in waste type, use of thick cover components, and increased overburden stress on radon barrier.

COVER DEFORMATION AND CRACKING

COVER CRACKING

General

Cover cracking is of concern primarily if the radon or infiltration barrier cracks. Cracking may result from either soil desiccation or excess cover deformation.

FEATURES THAT PRECLUDE FAILURE

- Dense or compacted wastes
- Avoid discontinuities in waste type
- Thick cover components
- Increased overburden stress on radon barrier precludes crack opening
- Soil with higher plasticity in radon barrier

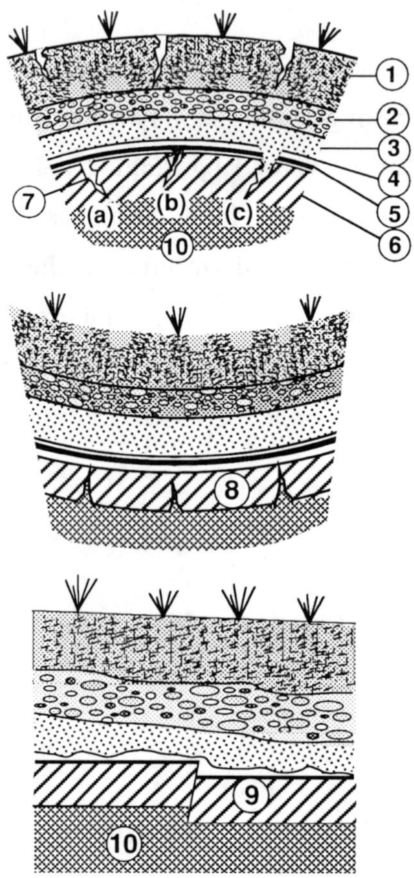

FAILURE SCENARIO I:
CONVEX DEFORMATION

Convex deformation induces cracking of radon barrier, and:
(a) Geomembrane and Geomat bridge crack
(b) Bentonite falls into crack and seals it
(c) Sand falls into crack through breached geomembrane and geomat.
Sands and cobbles do not crack.
Random soil cracks by heat and by soil movement process.

FAILURE SCENARIO II:
CONCAVE DEFORMATION

Concave deformation cracks bottom of radon barrier. Other soils placed into compression and do not crack. Increased radon flux through crack is stopped by bentonite layer.

FAILURE SCENARIO III:
SHEAR DEFORMATION

Shear crack develops. Bentonite is disrupted. Geomembrane stretches. Sands and cobbles deform. Minor potential additional radon and water flux through deformation plane.

LEGEND

(1) RANDOM SOIL
(2) COBBLE BIOINTRUSION
(3) SAND DRAIN
(4) GEOMEMBRANE
(5) GEOMAT: BENTONITE AND GEOTEXTILES
(6) RADON BARRIER SILT/CLAY
(7) CONVEX DEFORMATION CRACK
(8) CONCAVE DEFORMATION CRACK
(9) SHEAR DEFORMATION CRACK
(10) WASTE

Uneven settlement of the cell foundation and waste could deform and crack the radon barrier.

FIGURE 13.

DEFORMATION-INDUCED CRACKIN
OF THE RADON BARRIE

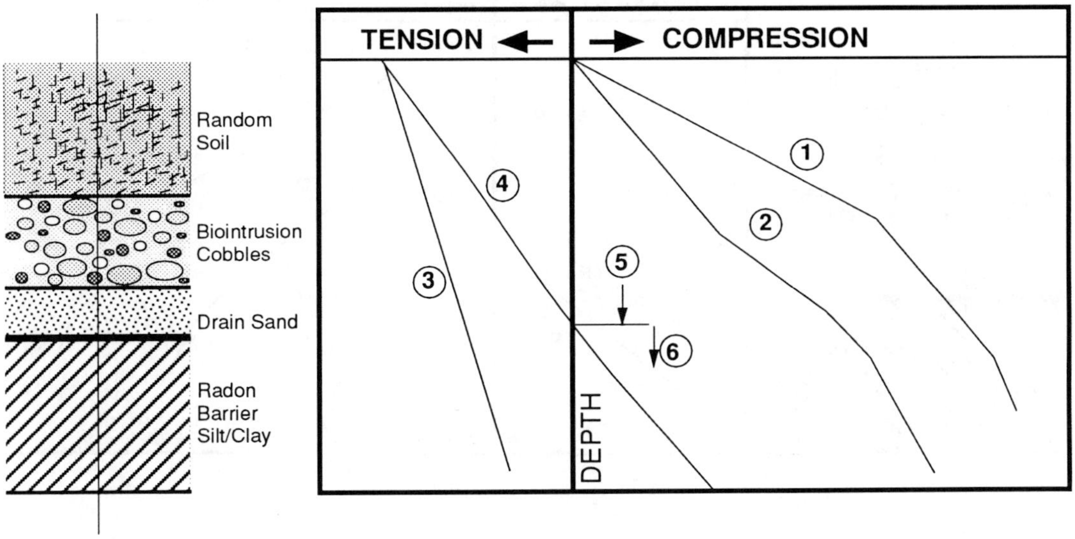

1. VERTICAL STRESS DUE TO OVERBURDEN (σ_V)
2. HORIZONTAL STRESS DUE TO OVERBURDEN ($\sigma_H = K_O\ \sigma_V$)
3. TENSILE STRESS INDUCED BY CONVEX DEFORMATION (σ_D)
4. RESULTANT HORIZONTAL STRESS ON COVER COMPONENTS ($\sigma_{RH} = \sigma_H - \sigma_D$)
5. DEPTH OF TENSILE ZONE: POTENTIAL TENSILE CRACKING
6. COMPRESSION ZONE: NO TENSILE CRACKING

Cracking of the radon barrier may be prevented by limiting deformation-induced stresses to less than overburden-induced stresses.

FIGURE 13.3

RADON BARRIER CONVEX CRACK AVOIDANCE

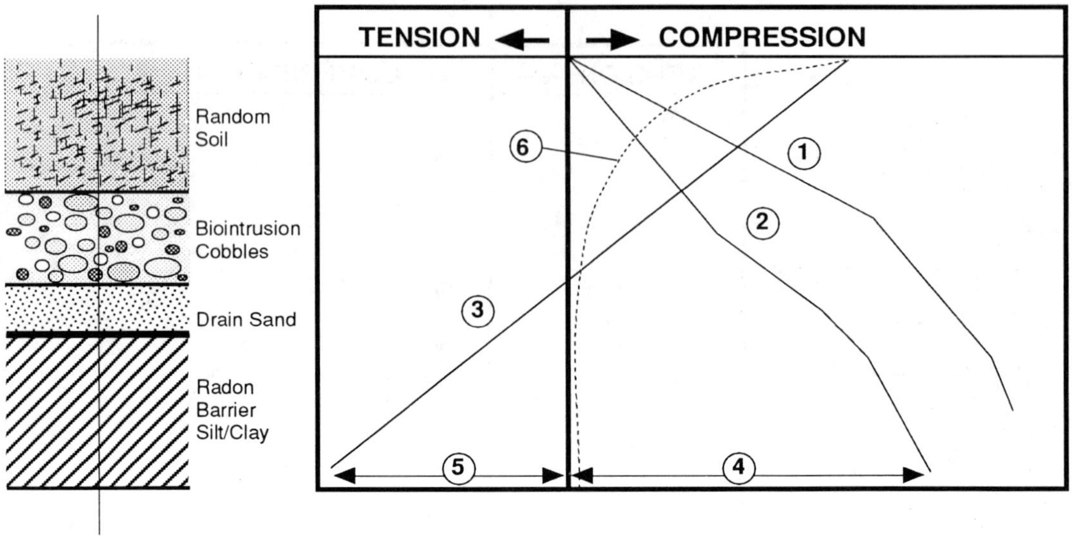

① VERTICAL STRESS DUE TO OVERBURDEN (σ_V)

② HORIZONTAL STRESS DUE TO OVERBURDEN ($\sigma_H = K_O \sigma_V$)

③ HORIZONTAL STRESS INDUCED BY CONCAVE DEFORMATION (σ_D)

④ MAXIMUM COMPRESSIONAL STRESS AT BASE OF RADON BARRIER (σ_4)

⑤ MAXIMUM STRESS AT BASE OF RADON BARRIER TO PRECLUDE TENSION CRACKING ($\sigma_4 = \sigma_5$)

⑥ RESULTANT HORIZONTAL STRESS ON COVER COMPONENTS TO AVOID TENSILE CRACKING (σ_{RH})

Cracking of the radon barrier may be precluded by limiting the concave deformation and horizontal stresses in cover components.

FIGURE 13.

RADON BARRIER CONCAV
CRACK AVOIDANC

Desiccation Cracking

Desiccation-induced cracking may be prevented by locating the susceptible element deep in the cover below the depth of significant seasonal drying (see Figure 13.5). Alternately, soils that have a low cracking potential may be used to construct the radon and infiltration barrier. Detailed description of the desiccation-induced cracking phenomenon is provided by the DOE (1991).

Desiccation occurs when fine-grained soil is exposed to the evaporative effects of the atmosphere. Desiccation cracking can occur as a result of either exposure to evaporation or because of evapotranspiration resulting from root penetration. As pore water is removed from the soil, negative pore water pressure (tension or suction) is created. The negative pressure in the water increases the effective stress in the soil, with a resulting decrease in volume. This decrease in volume overcomes the cohesion between the soil grains, and cracks occur.

While locating the susceptible element deep in the cover below the depth of significant seasonal drying can prevent long-term exposure to evaporation, short-term exposure can result from poor construction practices or unforeseen delays, as shown in Figure 13.6. On the UMTRA Project site in Tuba City, Arizona, a three-month exposure during summer resulted in cracks up to 460 mm deep. Figure 13.7 shows the relationship between the site moisture content, shrinkage limit, and depth for one of the test pits excavated into the cracked radon barrier. Significant cracking generally extended to about 12 inches; incipient cracks extended to the depth at which the moisture content was equal to the soil shrinkage limit (i.e., the soil moisture content at which the soil ceases to change volume with decreasing moisture content.)

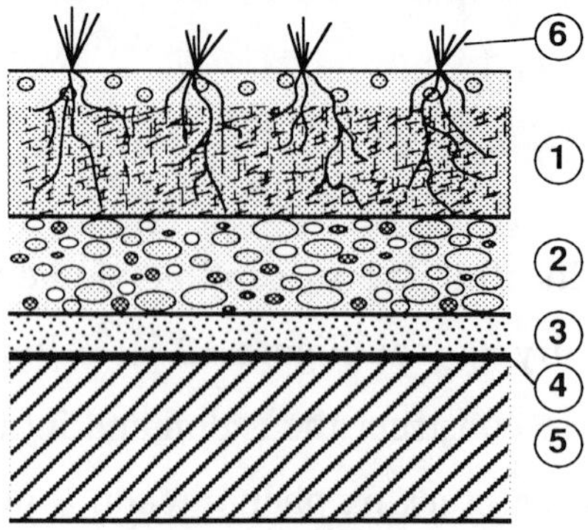

NORMAL SITUATION:

Vegetation grows on cover in soils at reasonable moisture content.

FEATURES THAT PRECLUDE FAILURE

- Hardy vegetation
- Thick soils
- Noncohesive soils: cobbles and sands are noncracking
- Geomat with bentonite

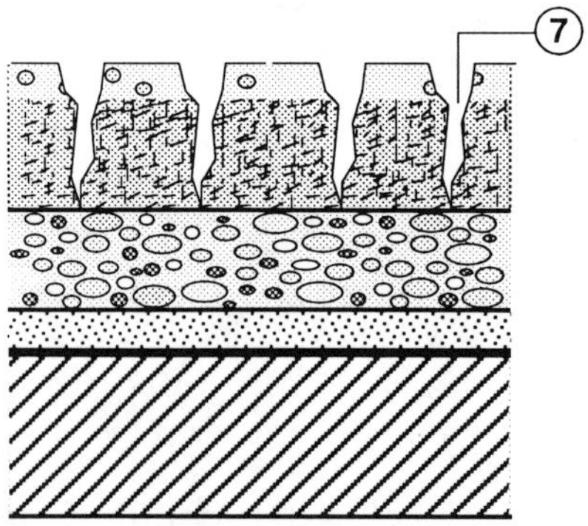

FAILURE SCENARIO I:

Drought kills vegetation and causes dessication cracking of upper soils. Bentonite layer (in geomat) precludes dessication of radon barrier.

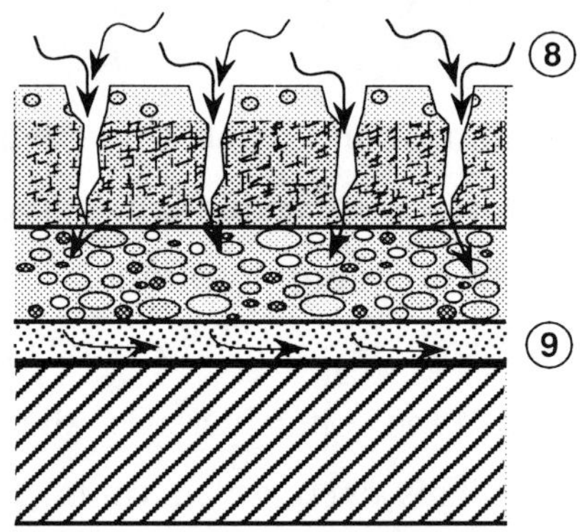

FAILURE SCENARIO II:

Intense precipitation at end of drought causes increased flow in cracks to drain. Increased infiltration to waste precluded by bentonite layer (in geomat) and intact radon barrier soils.

LEGEND

(1) RANDOM SOIL
(2) COBBLE BIOINTRUSION
(3) SAND DRAIN
(4) BENTONITE GEOMAT & GEOMEMBRANE
(5) SILT/CLAY RADON BARRIER
(6) NORMAL VEGETATION
(7) DESSICATION CRACK
(8) STORM INFLOW
(9) INCREASED DRAIN FLOW

Prolonged dry weather could kill cover vegetation, crack soil, and create preferential flow paths.

FIGURE 13.

DROUGHT-INDUCED COVER FAILURE SCENARIOS

NORMAL SITUATION:

Radon/Infiltrative Barrier Clay/Silt compacted into place over the waste. Soils kept moist and timeously covered.

PROCEDURES TO PRECLUDE FAILURE:

- QA/AC control
- Spray moistening of soil
- Timely placement of Geomat and Geomembrane
- Long-term performance of Bentonite

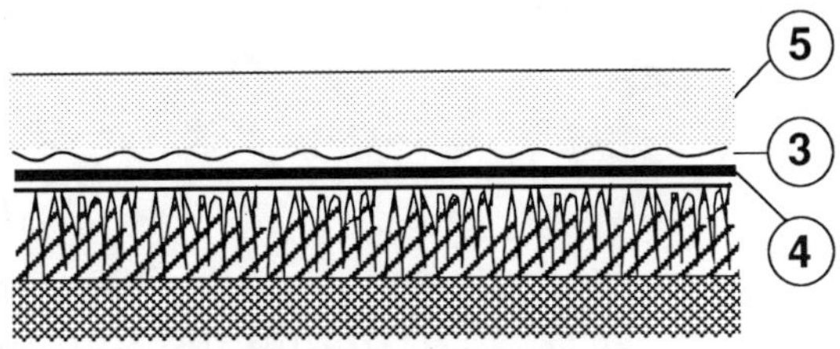

FAILURE SCENARIO II:

Cracked Clay/Silt covered by Geomat (Bentonite) and Geomembrane.

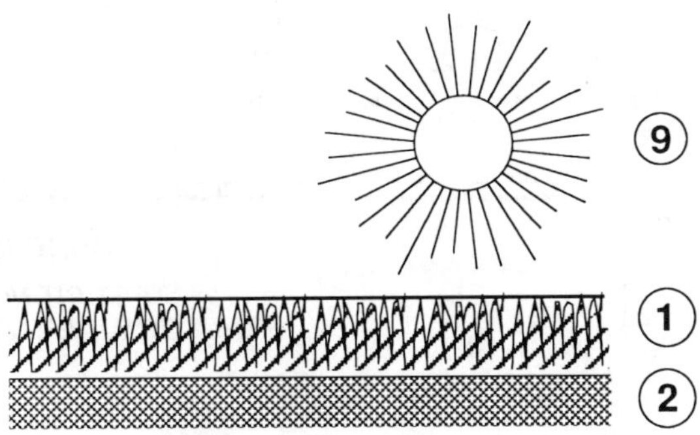

FAILURE SCENARIO I:

Sun drying of unprotected clay/silt induces dessication cracks.

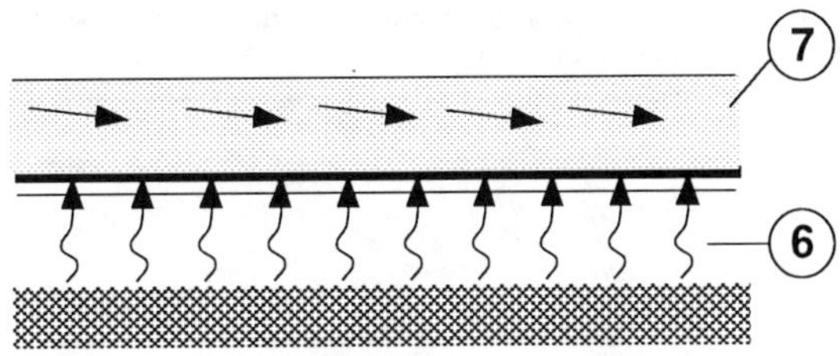

LONG-TERM PERFORMANCE SCENARIO:

Radon Flux through cracks is contained and attenuated by Bentonite layer. Potential infiltration is controlled by Bentonite.

LEGEND

① CLAY/SILT OF RADON BARRIER
② ENCAPSULATED WASTE
③ GEOMEMBRANE
④ BENTONITE LAYER IN GEOMAT
⑤ DRAIN SAND
⑥ RADON FLUX IN CRACKS
⑦ WATER IN DRAIN
⑧ COMPACTOR
⑨ SUN

Dessication cracking of the radon barrier during construction may impair its function. The use of a Bentonite layer (Geomat) can mitigate potential negative consequences.

FIGURE 13.6

SOIL CRACKING FAILURE SCENARIO

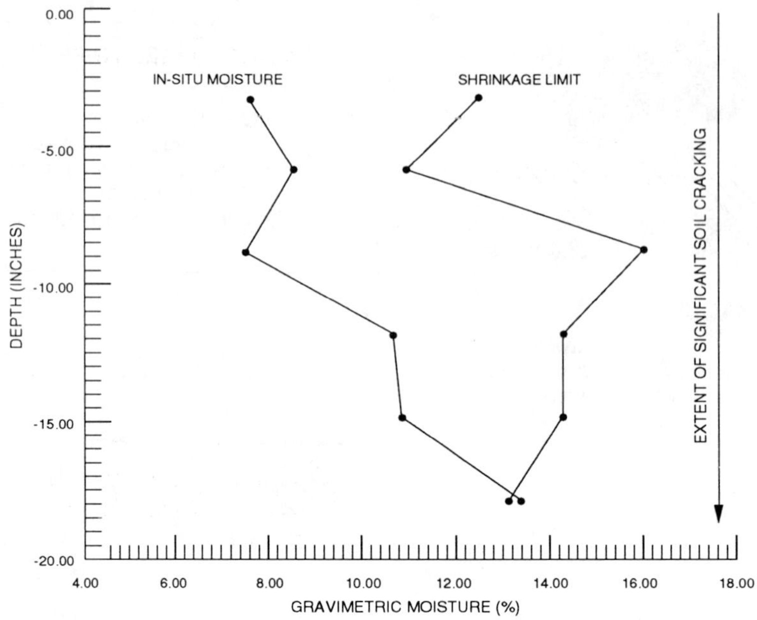

We excavated a test pit, took samples, and noted that the cracking extended to about 20 inches. These are the lab results.

FIGURE 13.7

TEST PIT NO.1
TUBA CITY, ARIZONA

The relationship between soil suction (negative water pressure) and depth is such that soil suction decreases with depth. The significance of this is that, as explained above, the moisture content associated with a soil suction of minus 15 bars pressure (15,000 cm) is the default design moisture content for radon flux calculation. If the soil has cracked due to desiccation at the 15-bar moisture content, then this is an undesirable and unrealistic design default parameter. If this happens, the radon flux at minus 15-bar soil suction will be controlled by the cracks, not the soil moisture content. The design implications are twofold:

- The cover design should preclude drying of the radon barrier to the extent that desiccation cracking is likely.

- The default moisture content for radon flux calculations should be higher than that associated with desiccation cracking.

Desiccation cracking during construction must be prevented by not leaving critical materials exposed. Unavoidably exposed surfaces must be kept moist by spraying and wetting.

The increase in hydraulic conductivity caused by desiccation cracks will not be eliminated when the soil is rewetted. Cracks frequently fill with coarser grained particles which prevent the closing of the cracks, and the soils, when rewetted, experience volumetric expansion vertically as well as horizontally so the cracks do not close. With sufficiently high confining pressures, the cracks may close and the hydraulic conductivity may decrease.

The relationships between desiccation and the shrinkage limit of the soil and the shear strength of the soil have not been fully explored in the literature available, nor has the literature examined the theories of fracture mechanics to explain the depth of cracking. The relationships between desiccation cracking and compactive moisture have been studied, and indicate that higher compactive moisture makes soil more susceptible to desiccation cracking. This introduces a problem because reducing compactive moisture also makes the soil less plastic and more susceptible to deformation cracking. It is therefore important that the cover design carefully examine the relative possibilitics of thc various causes of cover cracking.

Deformation Cracking

Uneven or differential settlement of the disposal cell contents, and hence of the cover, may cause tensile or shear strains in the soils of the radon or infiltration barrier. If the deformation-induced strains exceed the soil strength, the soil will crack. In essence, the technical approach to evaluate this problem involves:

- Define potential cover deformation
- Calculate resulting cover tensile and shear stresses and strains
- Establish soil yield (or cracking) stresses and strains. This may be done by direct measurement or correlate soil type and soil properties to yield stresses
- Compare predicted to actual stresses and strains.

If this process predicts unacceptable cover cracking, one or more of the following design changes may be made:

- Select a soil that has a higher cracking resistance (generally a more plastic soil)
- Reduce cover deformation by placing disposal cell contents in a dense, less deformable state

- Change the stress distribution in the soil. This may be done either by increasing the load on the cover component (add overlying layers), or shaping the cover to reduce the rate of change of deformation (round out the cover at the intersection of top- and sideslopes).

The permissible tensile failure strain (i.e., the point at which the soil cracks) is increased by increasing the Plasticity Index (PI) of the soil. The equation (DOE, 1989, Technical Approach Document) which relates strain at failure to the plasticity index of the soil is as follows:

$$e_f = 0.05 + 0.003PI$$

where,

e_f = strain at failure.

PI = plasticity index.

Generally, the PI of a soil is an inherent function of the soil type. To alter (increase) the PI of a soil, bentonite may be added. Even moderate amounts of bentonite effect a significant increase in the soil's PI, hence crack resistance. Accordingly, the most appropriate procedure for reducing or avoiding radon and infiltration barrier cracking is the addition of bentonite to the soil in the critical zone. In that the change of PI with addition of bentonite is soil-specific, actual tests on proposed soils must be performed to establish design parameters.

An incidental extra benefit of adding bentonite to the infiltration barrier soil may be a reduction of soil hydraulic conductivity. This potential extra benefit may also justify the increased cost of bentonite addition.

Cover Deformation Cracking Analyses

Analytical procedures to evaluate the stresses and strains induced in the cover due to deformation of the cell and its contents involve standard procedures in soil and materials mechanics. Accordingly, they are not discussed in detail here. Suffice it to say that two-dimensional, finite element grids may be compiled and used with standard, readily available computer codes to calculate a detailed pattern of cover stresses and strains. These same finite element grids may be used to study the effect of changing the cover geometry (rounding corners) or increasing the load on the radon barrier on stresses and soil cracking.

Deformation of the cover may induce tensile, compressive, and shear stresses and strains in the radon infiltration barrier. Compression will not crack the soil. Tensile stresses may cause open cracks in the soil; two adjacent blocks of intact soil may, in effect, separate. Shear will cause two adjacent, intact blocks of soil to slide past one another, but the plane of failure is unlikely to be an open zone or void.

The only comprehensive technical approach to determine the stress distribution in the radon/infiltration barrier is use of a suitable computer code that incorporates finite element analyses procedures and accounts for the soil-specific, stress-strain relationships. A number of such programs are available although use of them is neither simple nor inexpensive.

In order to avoid the cost and difficulty of the computer codes, procedures to assess cracking have been developed. The simplest procedure in essence views the radon barrier as a deforming beam. Stresses are calculated from the deformed shape of the beam. The crack potential evaluation implies that the radon barrier soil will act more or less as a concrete or steel beam, with tensile stresses at the top and compressive forces at the bottom. In practice, a high-plasticity clay placed as an extensive flat sheet will behave more as a flexible membrane than as a rigid beam. The beam analogy analysis may present an inappropriate picture of cover cracking potential; it is, however, generally conservative and simple, hence often used in the event of tensile strain. It may be assumed that the lower third of a cracked radon barrier is in compression, thus uncracked. If the actual thickness of the radon barrier is at least three times the theoretically required thickness, even a cracked radon barrier will still function to reduce radon flux to below standards and prevent excessive infiltration.

It is not within the scope of this book to detail the complex theories of soil deformation behavior and response or the mathematical techniques that may be used to calculate the stresses in liners, membranes, or beams subject to deformation. Suffice it to say that such procedures and codes are available and should be used where appropriate.

Evaluation of Alternative Covers

Cover cracking is not considered to be a significant factor in the comparison of alternative covers. To the extent that cover cracking may be a factor, the simple rule of thumb is that the thickness of the material over the radon barrier to lower the crack potential is adequate for comparison of covers.

The following details should be defined and established at a particular site before a comprehensive evaluation of cover cracking is practical:

- The potential settlement pattern of the cell foundation or contents
- The soils that may be used to construct the radon and infiltration barrier
- The precise geometry of the cover.

Design Guidelines

In preparing the final design of a radon/infiltration barrier, the materials shall be selected and the barrier should be designed and placed to resist both desiccation and deformation-induced cracking by:

- Selecting suitable construction soil
- If appropriate, modifying construction soils by adding bentonite
- Placing an adequate thickness of overlying layers
- Increasing the thickness of the radon/infiltration barrier
- Minimizing deformation by suitable placement of disposal cell contents
- Removing deformation susceptible foundation strata
- Shaping the cover to reduce excessive rates of change of deformation.

The best way to evaluate the stress distribution in the cover is with a finite element model. If radon/infiltration barrier, deformation-induced cracking becomes an issue, the use of a standard finite element stress analysis code is recommended.

Chapter 14

THE GEOTECHNICAL MODEL: EROSION

TOP COVER SYSTEM AND MODEL

Erosion of the cover may be caused by short-term precipitation run-off and the long-term loss associated with geomorphological change. As shown on Figure 14.1, with time, the upper surface of the cover may be affected by bioaccumulation, erosion, or development of a deflated rock layer. This may be quantified as shown in Figure 14.2 which illustrates that with time, sheet erosion may reduce the thickness of the soil layer.

DIKE SYSTEMS AND MODELS

Figure 14.3 shows alternative models that may be quantified to assess dike erosion and face retreat due to sheet erosion. As shown in Figure 6.6, even significant gully formation by erosion of the clean-fill dike will not expose the encapsulated wastes. Figure 14.4 shows how vegetated dikes may be protected from erosion by extreme precipitation run-off by flattening the slope and varying flow concentrations (as discussed below).

TECHNICAL AND ANALYTICAL PROCEDURES

The Universal Soil Loss Equation (USLE) may be used to calculate the amount of soil lost from a vegetated cover over 1,000 years or any selected design period. This USLE approach is used to quantify models such as those shown in Figures 14.1 and 14.2. The USLE does

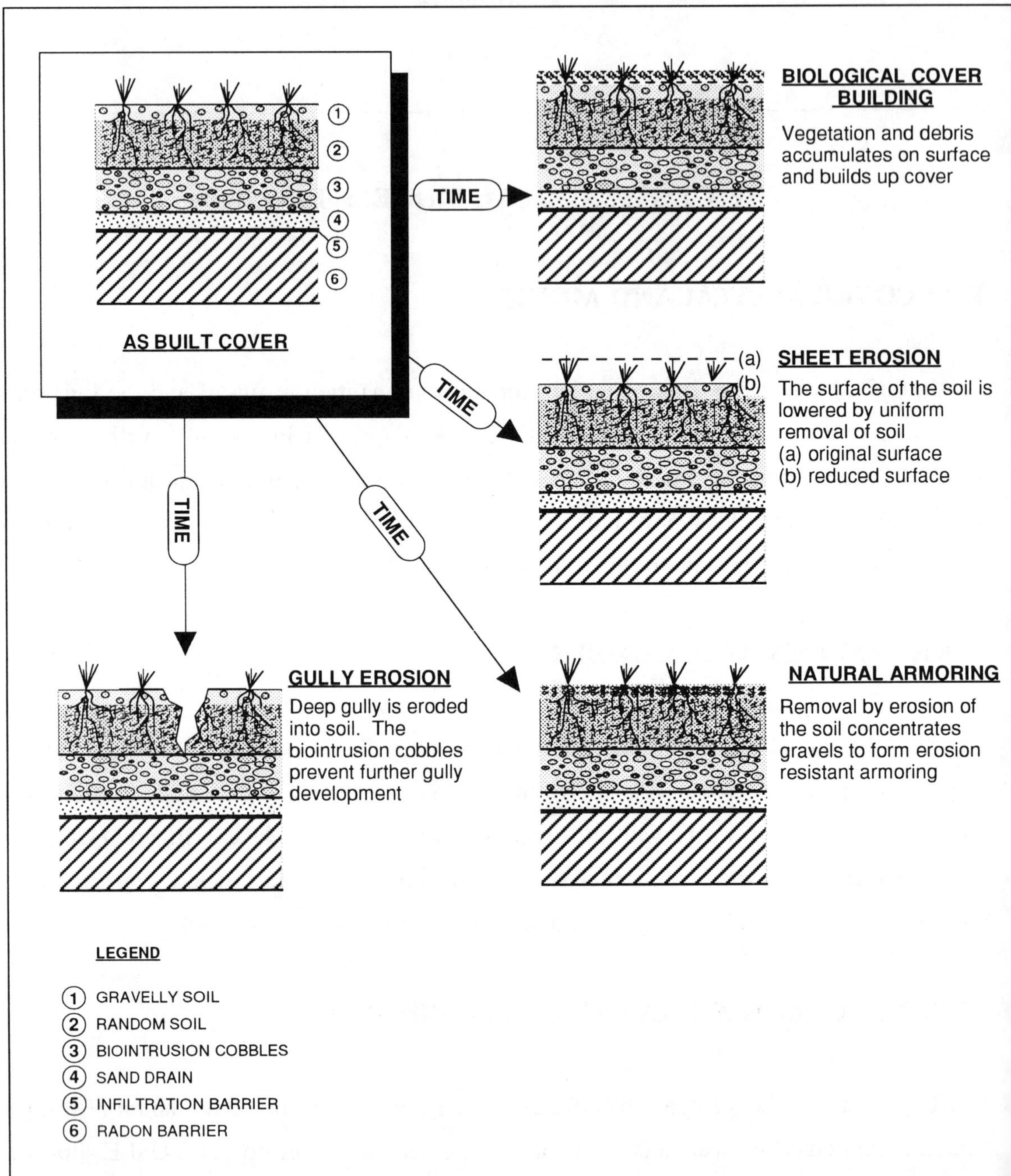

With time the upper surface of the cover may be affected by biological cover building, erosion, or development of a rock armor.

FIGURE 14
TOP COVER EROSION - ALTERNATIV
PERFORMANCE SCENARIO

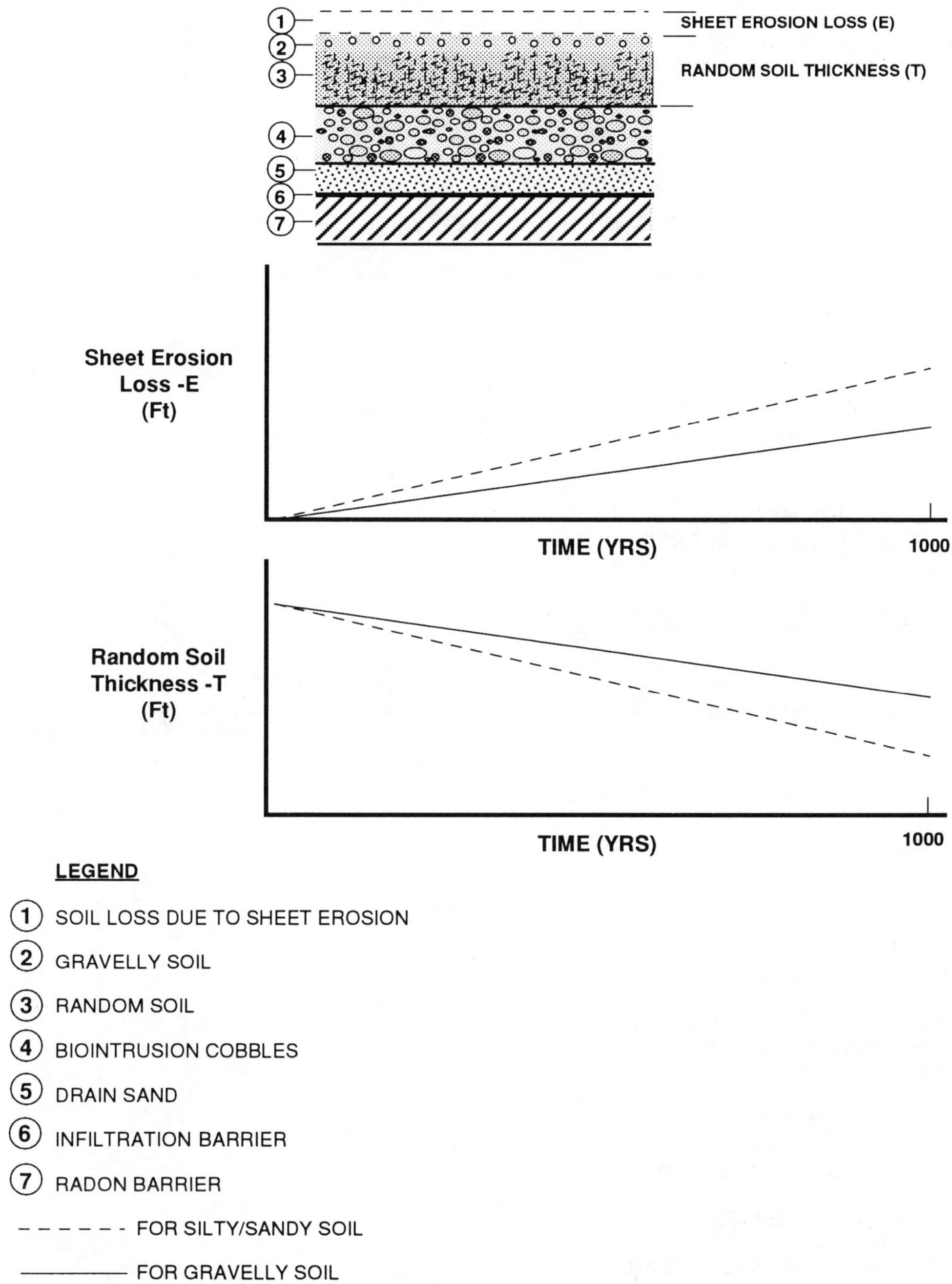

With time sheet erosion may reduce the thickness of the soil layer.

FIGURE 14.2
COVER SHEET EROSION RATES

SOIL SLOPE EROSION MODEL 1:
PARALLEL SLICE EROSION

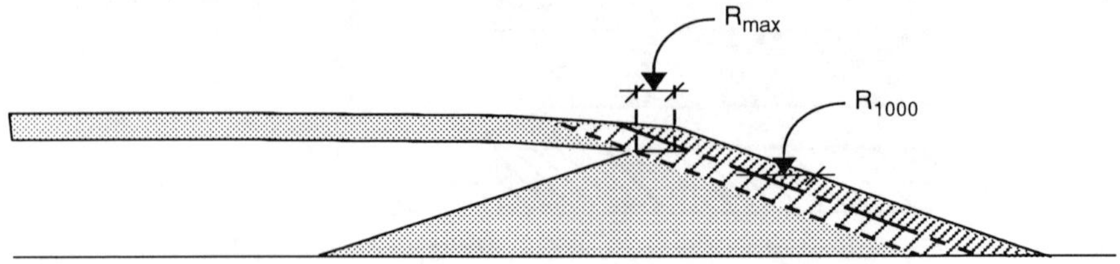

SOIL SLOPE EROSION MODEL 2:
SLOPE FLATTENING MASS BALANCE

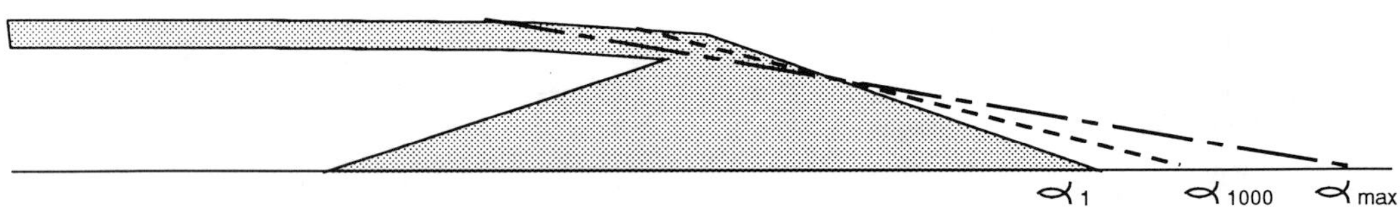

LEGEND

R_{1000} – SLOPE FACE RETREAT IN 1,000 YEARS

R_{max} – MAXIMUM FACE RETREAT FOR WASTE EXPOSURE: TIME OF RETREAT >1,000 YEARS

α_1 – INITIAL SLOPE ANGLE

α_{1000} – SLOPE ANGLE AT 1,000 YEARS

α_{max} – MAXIMUM SLOPE ANGLE FLATTENING FOR WASTE EXPOSURE: TIME >1,000 YEARS

Face retreat in slope flattening of soil dike may occur depending on surrounding area geomorphology change.

FIGURE 14.
SOIL CLEAN DIKE EROSION MODEL

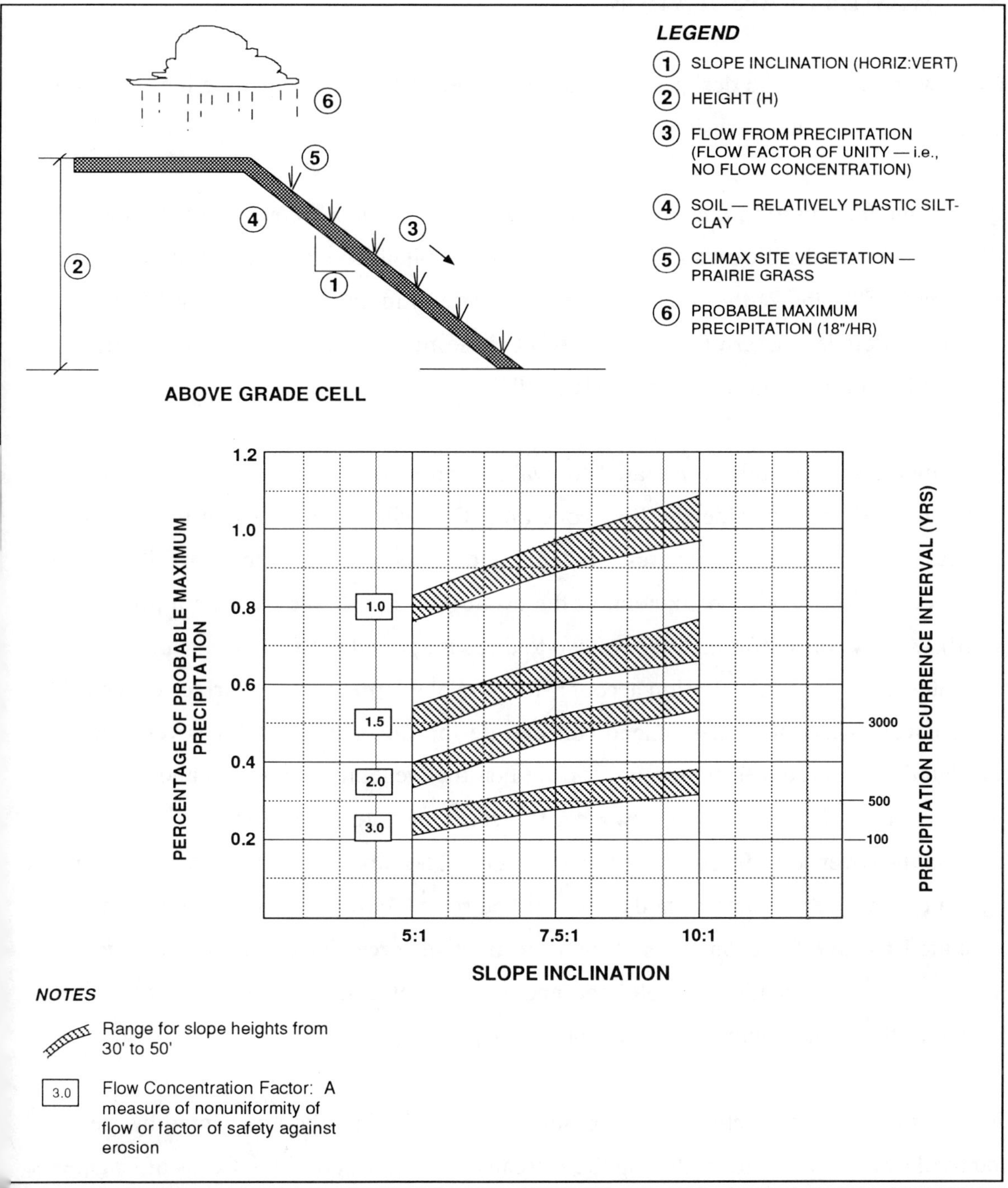

The erosion resistance of a vegetated slope increases as the slope is flattened.

FIGURE 14.4

SOIL EROSION VS SLOPE

not, however, enable the designer to study, evaluate, or design to preclude the initiation of soil or rock rill or gully erosion.

The two methods used to analyze the erosional resistance of soil covers to the initiation of rills or gullies, or significant soil loss, are: 1) the permissible velocity method, documented by Chow (1959); and 2) the tractive force (shear stress) method discussed by Temple et al. (1987). A detailed description of the use of these methods is given in the DOE (1989) Technical Approach Document and NRC (1990).

These methods, originally formulated to evaluate erosion in channels, have subsequently been modified to determine the flow conditions that will initiate erosion of a soil cover. The design engineer first calculates the depth and velocity of the water that will flow over the cover, or down the embankment, in the event of the probable maximum precipitation (PMP). A flow concentration factor of two to three is applied to the theoretically calculated flow rates; this flow concentration accounts for possible depressions or surface irregularities in the cover surface that may indeed concentrate the flow. If the cover is vegetated, the calculated flow is reduced to account for the fact that vegetation impedes and slows water flow over the soil. For a given flow rate, the engineer determines the tractive (or shear) force on the cover soil caused by the flowing water. The capacity or ability of a soil to resist a given tractive force is determined or established from laboratory tests. Comparison of the calculated tractive force on the soil, and the tractive force that is known to initiate soil erosion, or gully formation, enables the engineer to determine if erosion of the cover is likely and the design changes, if any, that are necessary.

For designing rock-protected slopes, conservative, standard methods accepted by the NRC and used by many agencies including the Bureau of Public Roads, U.S. Corps of Engineers, Bureau of Land Management, and DOE, are utilized. The Safety Factors Method is used on the flatter (2 to 5%) topslopes; and the Stephenson's Method on the steeper (>10%) sideslopes. Both methods, in essence, involve calculating the flow conditions (velocity and depth) that will just initiate movement of the individual rocks (gravels or boulders) that make up the rock cover (DOE, 1989, Technical Approach Document).

If the upper cover components are soil over a rock layer, the designer may accept that gullies can occur in the soil, and size the underlying rock to resist further erosion by flow in the gully.

As noted by the NRC (1990), design details to control erosion effectively may be in conflict with design details that limit infiltration. For example, an uppermost layer of clean, coarse rock is unquestionably the best and most assured way to prevent erosion. Such a rock layer, however, precludes the development of a dense stand of vegetation which is the most effective way to reduce infiltration. Indeed, such a rock layer further promotes the growth of deep rooting species that may physically disrupt the infiltration barrier leading to high infiltration and exhume radon gas fluxes.

Case History 4: Evaluation of Alternative Covers

Only the uppermost erosion protection layer(s) is required for analysis of the erosion resistance of a cover. Any one of these five erosion protection systems could be used on the five cover alternatives discussed in detail in this book:

- Rocks
- Rock overlain by soil, vegetated with sod-forming grass
- Gravelly soil with sod-forming grass
- Gravelly soil with clump-forming bunch grass
- Vegetated soil.

Table 14.1 shows the rock size requirements for these five alternatives based on a probable maximum precipitation design storm (18.4 inches in the midwest where the cover is assumed to be located) event for the top- and sideslopes of a standard rock-protected cover. The same rock size is also applicable if soil is added to fill the voids in the rock. If this is done, care must be taken not to increase the volume of the void space in the rock and also not to leave a significant depth of soil covering the rock which would lead to gullying.

TABLE 14.1

EROSION PROTECTION REQUIREMENTS FOR TOP- AND SIDESLOPES OF STANDARD ROCK-PROTECTED COVER UNDER PMP CONDITIONS

COVER CONFIGURATION	ROCK SIZE (d_{50}), INCHES (SLOPE LENGTH-FEET)
Standard Rock Cover: Topslope	
2% Topslope	1.5(65/250/450)
5% Topslope	1.5(65) 1.5(250) 2.0(450)
Standard Rock Cover: Sideslope	
10% Sideslopes with 2% Topslopes	1.7(465) 1.9(650)
10% Sideslopes with 5% Topslopes	1.9(465) 2.0(650)
20% Sideslopes with 2% Topslopes	3.5(465) 3.6(650)
20% Sideslopes with 5% Topslopes	3.7(465) 4.0(650)

Table 14.2 shows the rock size requirements for vegetated soil overlying choked rock. The vegetation was assumed to be a sod-forming grass. It was anticipated that the prairie-type grasses would not grow well due to the lack of adequate rooting zone depth.

A number of soil covers were examined for use on both the topslope and sideslope using PMP conditions, which are shown in Table 14.3.

Tables 14.4 and 14.5 show the results of additional analysis for sideslopes using vegetated gravelly soil. The analysis was performed for both a prairie-type "bunch" grass and for a more lawn-type "sod" grass. Two methods were used to analyze sod-type grass, both the tractive force and permissible velocity. The permissible velocity method is not considered applicable for clump-type grasses on slopes steeper than 10 percent.

TABLE 14.2

ROCK SIZES FOR VEGETATED SOIL OVER CHOKED ROCK

		ROCK SIZE (d_{50}), INCHES				
SLOPE TYPE	**SLOPE CONFIGURATION**	**PMP**	**1/2 PMP**	**500-YR**	**300-YR**	**200-YR**
Topslope 2%	65 ft	4.7	3.1	2.1	2.1	2.0
	250 ft	9.3	6.8	4.4	4.3	4.2
	450 ft	10.8	8.3	5.5	5.4	5.2
Sideslope 20%	Top(2%) = 65 ft Side(20%) = 400 ft	13.2	9.4	6.0	5.8	5.7
	Top(2%) = 250 ft Side(20%) = 450 ft	18	14.2	8.8	8.5	8.3
	Top(2%) = 450 ft Side(20%) = 200 ft	>18	16.3	9.9	9.8	9.3

TABLE 14.3

RESULTS OF EROSION RESISTANCE OF BARE OR VEGETATED SOIL COVERS UNDER PMP CONDITIONS

		TOPSLOPE			SIDESLOPE	
		2%	5%		10%	20%
Bare Soil or Gravel Mulch	Topslope = 65 ft. Sideslope = 400 ft.	Unstable	Unstable	Topslope = 2%	Unstable	Unstable
				Topslope = 5%	Unstable	Unstable
	Topslope = 250 ft. Sideslope = 450 ft.	Unstable	Unstable	Topslope = 2%	Unstable	Unstable
				Topslope = 5%	Unstable	Unstable
	Topslope = 450 ft. Sideslope = 200 ft.	Unstable	Unstable	Topslope = 2%	Unstable	Unstable
				Topslope = 5%	Unstable	Unstable
Vegetated Soil	Topslope = 65 ft. Sideslope = 400 ft.	Stable	Stable	Topslope = 2%	Unstable	Unstable
				Topslope = 5%	Unstable	Unstable
	Topslope = 250 ft. Sideslope = 450 ft.	Stable	Unstable	Topslope = 2%	Unstable	Unstable
				Topslope = 5%	Unstable	Unstable
	Topslope = 450 ft. Sideslope = 200 ft.	Unstable	Unstable	Topslope = 2%	Unstable	Unstable
				Topslope = 5%	Unstable	Unstable
Vegetated Soil and Gravel	Topslope = 65 ft. Sideslope = 400 ft.	Stable	Stable	Topslope = 2%	Unstable	Unstable
				Topslope = 5%	Unstable	Unstable
	Topslope = 250 ft. Sideslope = 450 ft.	Stable	Unstable	Topslope = 2%	Unstable	Unstable
				Topslope = 5%	Unstable	Unstable
	Topslope = 450 ft. Sideslope = 200 ft.	Unstable	Unstable	Topslope = 2%	Unstable	Unstable
				Topslope = 5%	Unstable	Unstable
Sacrificial Slope Requirements	Topslope = 65 ft. Sideslope = 400 ft.	Not feasible	Not feasible	Topslope = 2%	Not feasible	Not feasible
				Topslope = 5%	Not feasible	Not feasible
	Topslope = 250 ft. Sideslope = 450 ft.	Not feasible	Not feasible	Topslope = 2%	Not feasible	Not feasible
				Topslope = 5%	Not feasible	Not feasible
	Topslope = 450 ft. Sideslope = 200 ft.	Not feasible	Not feasible	Topslope = 2%	Not feasible	Not feasible
				Topslope = 5%	Not feasible	Not feasible

TABLE 14.4

RECURRENCE INTERVALS OF PRECIPITATION THAT CAUSE GULLIES ON 20% SIDESLOPES (TRACTIVE FORCE METHOD)

COVER	FCF	SLOPE CONFIGURATION		
		Top = 65 ft. Side = 400 ft.	Top = 250 ft. Side = 450 ft.	Top = 450 ft. Side = 200 ft.
Sod	2	1/2 PMP < DS < PMP	1/2 PMP < DS < PMP	1/2 PMP < DS < PMP
Clump	2	DS~ 1/2 PMP	500-Yr < DS < 1/2 PMP	DS~ 1/2 PMP
Sod	3	1/2 PMP < DS < PMP	500 yr < DS < 1/2 PMP	1/2 PMP < DS < PMP
Clump	3	500-Yr < DS < 1/2 PMP	500-Yr < DS < 1/2 PMP	500-Yr < DS < 1/2 PMP

TABLE 14.5

RECURRENCE INTERVALS OF PRECIPITATION THAT CAUSE GULLIES ON 20% SIDESLOPES (MAXIMUM PERMISSIBLE VELOCITY METHOD*)

COVER	FCF	SLOPE CONFIGURATION		
		Top = 65 ft. Side = 400 ft.	Top = 250 ft. Side = 450 ft.	Top = 450 ft. Side = 200 ft.
Sod	2	500-Yr < DS < 1/2 PMP	DS ~ 500-Yr	500-Yr < DS < 1/2 PMP
Sod	3	DS ~ 500-Yr	DS < 200-Yr	DS < 200-Yr

annot use this method for analyzing covers with clump grasses if slopes are greater than 10%.

F - Flow Concentration Factor
\- Design Storm
P - Probable Maximum Precipitation

The assumptions made for this analysis were that the soil would have a mean diameter (d_{50}) of 1 inch. The "bunch" grass was assumed to be the prairie-type climax vegetation. The "sod" grass was assumed to be similar to typical lawn grasses, such as Kentucky Bluegrass. The height of the bunch grass was assumed to be greater than that of the sod grass. Sod grasses were assumed to be in good condition. The bunch grasses were assumed to be in fair condition because maintenance would have been less frequent and less intensive.

Table 14.4 shows the results of the tractive force method. The design storm which will cause gullying for each type of vegetation has been identified. Because the PMP does not have a time interval associated with it, frequency intervals are not given. It can be seen that the sod-forming grass will withstand a greater storm event than the bunch-forming grass. It can also be seen in this table that the flow concentration factor which is used has a large effect on the design storm which can be withstood. These flow concentration factors are ones typically used in design, but are not well defined by empirical or theoretical means for vegetated slopes.

Table 14.5 shows the results of the erosive forces using the permissible velocity method. This method could not be used for bunch grasses. Extrapolating the results of the sod grasses based on the previous results would indicate that bunch grasses would fare even worse than the sod grasses.

An examination of the results shows that these vegetated slopes will not withstand the PMP without gullying. The use of a lesser design storm may still satisfy regulatory intent, in which case a vegetated sideslope could be used. If a lesser design storm is accepted, the additional surveillance and maintenance required must also be authorized to avoid failure.

Guidelines for Final Design

Sideslopes

To achieve satisfactory erosion resistance of disposal facility sideslopes, the following alternatives should be considered:

- Clean-fill dikes
- Rock cover with a perpetual vegetation control program
- Vegetation with a perpetual maintenance program, or a lesser design storm or acceptance of an increased risk of erosion.

Clean-fill dikes have the significant disadvantage that they reduce the available cell volume and require considerable volumes of construction material. Conversely, clean-fill dikes protected from erosion with a layer of durable rock can be left unattended. Deep-rooted vegetation that establishes through the rock will not detrimentally affect disposal short-term cell performance. Thus, the increased cost of clean-fill dikes may offset the long-term costs of maintenance.

Topslopes

In order to minimize infiltration, which is considered a desirable objective, it is recommended that the uppermost component of the cover be a soil on which as robust a stand of vegetation as the locale will support is established.

To provide satisfactory resistance to erosion, the topslope inclination should be a nominal two percent. Of course, it is difficult to construct so flat a slope; in practice, the slope will be more or less flat. To preclude development of a reverse slope, the disposal cell contents should be placed in such a manner that differential settlement is less than that which would reduce a two percent topslope to less than zero (or inverse slope).

If it is not possible to place cell contents to reduce differential settlement to the above-specified extent, then steeper topslopes consistent with predicted cell content's differential settlement will have to be adopted.

If slopes steeper than two percent are mandated by consideration of disposal cell content differential settlement, then it may be necessary to:

- Use an upper layer cover component soil with an increased gravel content
- Accept a reduced erosion resistance factor of safety, or
- Reduce the cover design life to 200 years, as a 1,000-year design life is not reasonably achievable.

Chapter 15

THE GEOTECHNICAL MODEL: STABILITY

GENERAL

This section discusses three processes by which the functional integrity or stability of the cover may be affected. These are:

- Gravity- or earthquake-induced slope instability
- Desiccation-induced radon and infiltration barrier cracking
- Freezing- and thawing-induced damage of the radon and infiltrative barrier.

There is no fundamental reason for grouping these three topics together in one section, except perhaps the fact that all of them, if they occurred, would open cracks in the cover, and in particular the radon and infiltration barriers. These openings or cracks would increase both the water and gas permeability of the cover, thus increasing infiltration and radon emulations.

TOPCOVER STABILITY

The very flat inclination of the topcover precludes instability by gravity-induced forces. It is not standard practice to calculate the static factor of safety of the topslope cover.

If bentonite or a geomembrane is included in the topcover, a sufficiently large earthquake could cause sliding of soil and rock layers overlying the bentonite or geomembrane. The factor of safety against such sliding is readily calculated using conventional pseudostatic or dynamic stability analysis.

Although the bentonite in the geomat is a potential failure plane in the event of an extreme earthquake, Figure 15.1 indicates that cell features including flat slope, use of noncrack susceptible sands and cobbles, and application of deformable random soil will minimize the long-term effects of seismic activity.

DIKE STABILITY

Figure 4.5 shows the two primary dike slope stability failure modes for static conditions. These are:

- Shallow or surficial sliding. The potential failure surface is close to and parallel to the outer edge of the cover.

- Mass instability of the slope by failure along a plane or an arc substantially within the cell and passing through the dike or the waste.

Figure 15.2 shows the primary slope stability failure modes for dynamic or earthquake loading. Standard slope stability analysis procedures are used to quantify the potential for such encapsulation system performance.

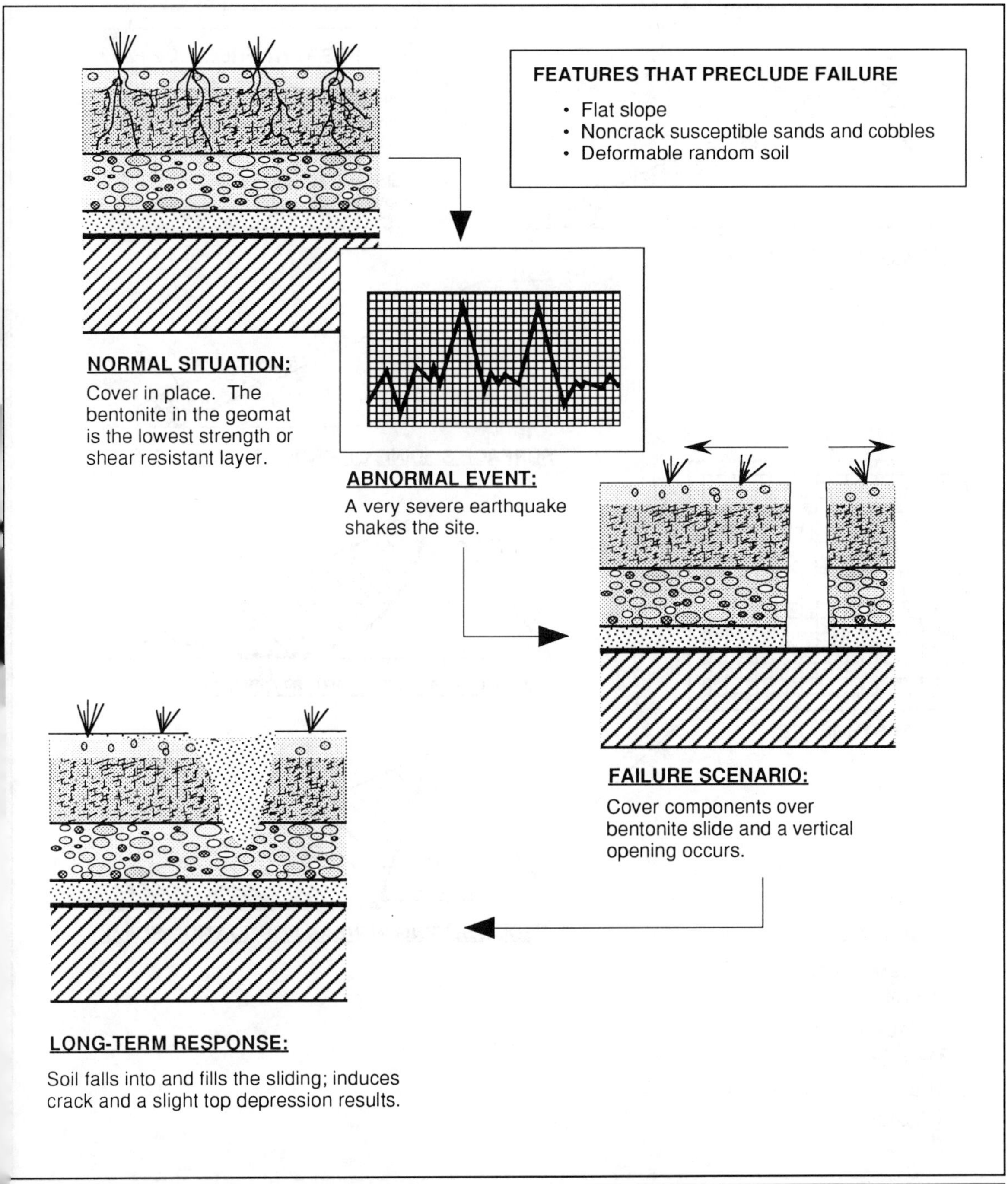

The bentonite in the geomat is a potential failure plane in the event of an extreme earthquake.

FIGURE 15.1

EARTHQUAKE-INDUCED COVER SLIDING

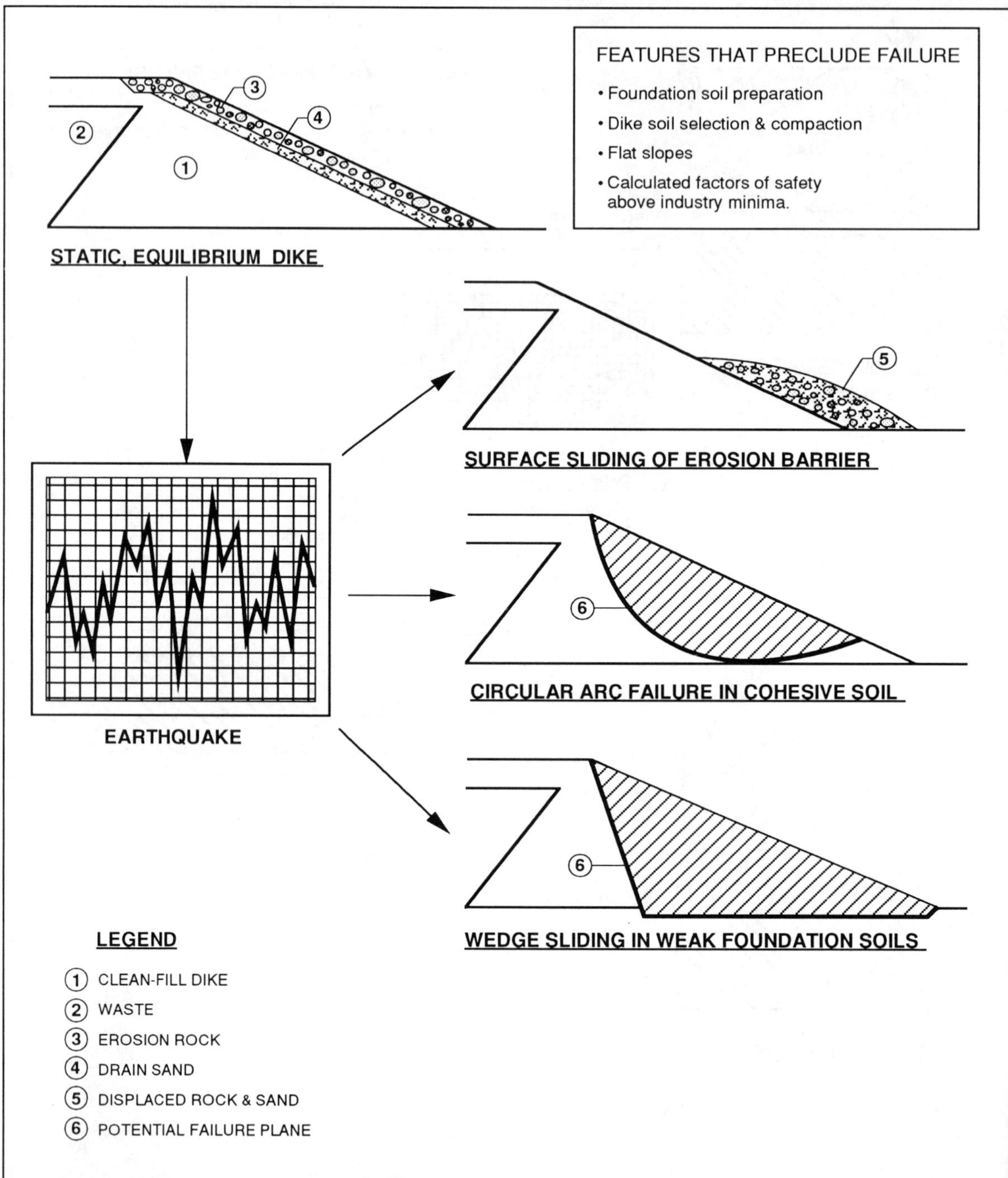

An earthquake could induce slope instability and potentially cause dike soils to displace.

FIGURE 15.2

DIKE POTENTIAL SLOPE FAILURE MODES

Case History 5: Assessment of Alternative Covers

Sideslopes and Topslopes of Natural Materials

Table 15.1 lists reasonable strength values, taken from the literature, for various materials that may constitute the various components of the five alternative covers described earlier in this book.

Table 15.2 lists the factors of safety for 20:1 and 2:1 slopes in both static and dynamic, end-of-construction and long-term, conditions for all infinite slope failure surfaces likely to occur through each of the possible materials of the alternative covers. As shown in this table, all factors of safety exceed the minimum required by standard practice.

Sideslopes with Geosynthetic Materials

In order to reduce infiltration through the cover and to meet RCRA requirements, a geomembrane may be required for both the top- and sideslopes. An example of a geosynthetic includes very low density polyethylene (VLDPE), a layer of bentonite sandwiched between two geotextiles. The sliding resistance at the interface between the geosynthetic and the adjacent soil is the parameter that controls the sliding stability of covers that incorporate geosynthetics. For the topslope of two percent, to obtain a factor of safety of 1.5, the angle of shearing resistance at the interface must be at least 1.7°. For a 20 percent slope, the angle must be at least 17°.

Table 15.3 lists the interface angle of sliding resistance of some common geosynthetic/soil systems. By using the textured VLDPE, it is evident that a geomembrane can be used on the sideslopes of the disposal cell if careful consideration is given to the material used as a subbase.

For topslopes, stability is predicted in that the factor of safety exceeds that conveniently considered sufficient. Thus, with careful design, such low-permeability elements may be included in top- and sideslope covers if consideration of longevity does not otherwise preclude their use.

TABLE 15.1

DESIGN PARAMETERS FOR SLOPE STABILITY ANALYSIS

Material	Moist Unit Weight (pcf)	Saturated Unit Weight (pcf)	Cohesion (psf)	Friction Angle (degrees)	Effective Cohesion (psf)	Effective Friction Angle (degrees)
Riprap	135	155	0	35	0	35
Bedding Filter/Drain	125	135	0	35	0	35
Biointrusion	130	150	0	38	0	38
Frost Protection	110	120	500	20	150	30
Radon Barrier	120	130	990	0	250	20
Growing Medium	105	115	300	15	100	25

TABLE 15.2

FACTORS OF SAFETY FOR INFINITE SLOPE FAILURE SURFACE

	20:1				5:1			
Material	Short-Term Static	Short-Term Dynamic	Long-Term Static	Long-Term Dynamic	Short-Term Static	Short-Term Dynamic	Long-Term Static	Long-Term Dynamic
Riprap	35.0	2.5	35.0	2.5	3.5	1.4	3.5	1.4
Bedding Filter/Drain	35.0	2.5	35.0	2.5	3.5	1.4	3.5	1.4
Biointrusion	39.1	2.8	39.1	2.8	3.9	1.6	3.9	1.6
Frost Protection	83.1	18.2	48.3	6.3	8.4	2.5	4.9	1.7
Radon Barrier	56.2	19.5	39.0	6.7	5.6	2.0	3.9	1.3
Growing Medium	156.3	38.1	70.9	14.0	15.9	4.3	7.2	2.2

TABLE 15.3

ANGLES OF INTERNAL FRICTION FOR GEOSYNTHETIC/ SOIL SYSTEMS AND VARIOUS SOILS

SLIDING SURFACE	SMOOTH HDPE	TEXTURED HDPE
Clay	16	24
Ottawa Sand	17	26
Concrete Sand	23	29
Geotextile (Mirfi 140N)	11	29

FROST PROTECTION

Technical Approach

Repeated freezing and thawing of a soil, such as is conventionally used for radon and infiltration barriers, alters the structure of the soil and may cause an increase in both the permeability and radon flux rates through the soil.

In order to prevent freezing and thawing of the radon and infiltration barrier, soil layers whose functions will not be impeded by frost action must be placed over the radon and infiltration barrier. The thickness of these overlying layers should equal or exceed the design frost penetration depth.

The site-specific frost penetration depth is a function of:

- Climate (the number of days in which the temperature falls below 0°C)
- Temperature (the average and extreme cold temperatures)
- The soil properties of the overlying cover layers.

The DOE (1989) Technical Approach Document provides a detailed description of the technical approach to calculate the depth of freezing as a specific calculation. The following is a summary of the technical approach.

The calculation of the frost penetration depth uses a computer program developed by the U.S. Army Corps of Engineers Cold Regions Research and Engineering Laboratory (CRREL). The CRREL program converts the air freezing index. The air freezing index is obtained from temperature data (which is provided by the National Climatic Data Center in Asheville, North Carolina).

The soil parameters required as input into the design calculation, in addition to layer geometry, include dry density and moisture content. While these values are based on conditions at the time of placement for the analysis, sensitivity analyses should be performed to allow for material variation, as might occur during the design life of the cell. The computer program calculates the depth of freezing using standard heat/cold flow equations.

Case History 6: Evaluation of Alternative Covers at Midwestern Site

Table 15.4 lists the parameters used to calculate the 200-year recurrence interval frost penetration depth at a Midwestern site. Table 15.5 provides the depth of frost penetration calculated for the five cover alternatives.

TABLE 15.4

MATERIAL PROPERTIES OF COVER COMPONENTS

MATERIAL TYPE	MOISTURE CONTENT*	DRY DENSITY (pcf)
1) Random Soil	11	115
2) Frost Protection	13	108
3) Radon/Infiltration Barrier	14	105
4) Filter/Drain/Bedding	13	110
5) Erosion Protection	2**	130

* All moisture contents were calculated to 65% of saturation.
** Moisture content of erosion protection material was estimated at 2% - gravimetric.

TABLE 15.5

FROST DEPTH PENETRATION

ALTERNATIVE COVER	FROST DEPTH PENETRATION (inches)
No. 1 Simple Vegetated/Rock Cover	37.8
No. 2 RCRA Cover	38.7
No. 3 Double-Drain Cover	36.2
No. 4 Multicomponent Cover	41.6
No. 5 Erosion Resister	42.5

A 200-year recurrence interval is adopted, as this is considered to be the limit of rational extrapolation of the limited period of available and pertinent data. Moreover, this probably yields an extreme, or conservative, prediction of frost depth penetration. To witness, consider that the Uniform Building Code for the site area requires a frost cover depth of 40 inches. The Uniform Building Code is known to be conservative, and the depth it gives closely matches the 42 inches determined from calculations.

Chapter 16

THE GEOTECHNICAL MODEL: THE DRAINS

DRAIN DESIGN

Technical Approach

The cover drain, or bedding layer, is conventionally placed over the radon or infiltration layer. The functions or purposes of this layer include:

- Protecting the fine-grained soils of the underlying layers from pitting or "punching" damage by the overlying coarse-grained gravel or cobbles. Such pitting could result in depressions in which water could pond to seep subsequently into the infiltration barrier.

- Preventing erosion of the fine-grained soils of the radon and infiltration barrier by water flowing through the drain and being shed off the cover.

- Expediting the passage or flow of water off the cover. This phenomenon is significant in reducing the source of water available to seep into the infiltration barrier, and hence in reducing total water flux into the disposal cell.

Selection of a suitable soil for the drain or bedding layer accordingly involves careful balancing of the potentially conflicting demands of:

- Coarse, permeable materials to increase flow or shedding rates, thus lowering total infiltration

- Fine soils to most effectively eliminate erosion of the underlying fine-grained soils.

A comprehensive evaluation of the performance of drain or bedding materials is provided in Abt et al. (1989). The following is a summary description of the conclusions of that study.

The design of the drainage or bedding layer is dependent upon cover slope, median grain size, overtopping discharge, sediment removal, interstitial velocity, and drainage rate. A bedding layer is required to protect the radon barrier from erosion. The average interstitial velocity can be calculated as a function of the slope and the median size of the bedding material. Using standard geotechnical filter criteria results in adequate bedding for the riprap and protects the radon barrier from erosion, but retains residual water and does not drain well. A median bedding layer grain size of 8 mm is recommended; finer bedding materials are highly susceptible to entrainment into overtopping flows. The bedding drainage time can be estimated using an empirically developed equation, but this estimate does not include the time to drain residual water retained at or near the radon barrier/bedding interface. The transition area from top- to sideslopes is highly vulnerable to flow turbulence and increases localized velocity due to water surface drawdown. Additional erosion protection at and near the transition is warranted.

To calculate the shedding potential or flow rate through the drain, conventional technical procedures for assessing flow through porous media may be employed. In particular, the maximum flow quantity per unit width of drain is given by:

$$q = k * S * t$$

where,

k = soil hydraulic conductivity.

S = inclination or slope of the drain.

t = drain thickness.

To preclude ponding of water in the drain over the infiltration barrier, the most effective approach is to select drain soils with as high a permeability as is consistent with erosion control. Practical experience (at the Shiprock, New Mexico, UMTRA Project site) indicates that if too fine-grained a drain material is used, not only is water shedding inhibited, but the continual presence of moisture is conducive to plant growth. For both the so-called rock (no vegetation) and vegetation-supporting covers, avoidance of roots in the drain layer is desirable because roots in the drain increase the potential for root penetration and degradation of the radon and infiltration barrier.

As shown in Flume Studies described by Abt et al., clean, fine, coarse-grained sands are the preferred material for drains in covers. Such sands are sufficiently fine to control erosion, but sufficiently permeable to shed water rapidly and preclude the permanent saturation that is conducive to root establishment and sustenance.

Chapter 17

THE ECOLOGICAL MODEL: VEGETATION

VEGETATION AND ENCAPSULATION

Properly used, plants can be vital contributors to the performance of a disposal cell. They can stabilize the soil of a cover's upper surface against damage by rainsplash, sheetwash, and incipient gully formation. Their roots can bind soil and rock together, even stitching up damage by minor erosion. They admirably serve the water balance of a cell, piping moisture from considerable depth back to the atmosphere and thereby preventing water from percolating, dissolving, and mobilizing the encapsulated waste. They render the cover green, "environmentally safe" to neighboring citizens, blending the offending cell into the native landscape. They grow, reproduce, and respond aptly to changes in weather and climate.

Plants also die, sometimes in catastrophic concert. They burn, topple, or find themselves churning in the rumen of a beast who cares nothing about hazardous wastes, cell-performance, or even the EPA. Untamed, they can drive their roots downward, penetrating a cell's encapsulating layers and intruding upon the underlying waste. During the cataclysmic storm that takes a containment system to its envelope, they may be slurried in rivers of mud cascading down the sideslopes, breaching the cell and exposing its contents to the environment from which they had been isolated.

The truth is that a properly selected and established plant community can contribute in many ways to the performance and longevity of a disposal cell, not to mention its appearance and potential public acceptability. However, the unreliability of vegetation -- the risk of absence during the moment of greatest need -- is a liability that must be accommodated in the design of a vegetated cover.

That having been said, this chapter promotes the treatise that, with a few rare exceptions, disposal cells for all types of wastes in all kinds of environments should be crowned with vegetated covers. A vegetated cover, in the mind of this author, is a relatively precisely defined thing, certainly distinct from the rock covers and soil covers described in Chapter 8. A vegetated cover is a cover in which vegetation is incorporated as an integral feature. The choice of soil and its placement is dictated at least in part by the anticipated needs of the vegetation, as well as the intended role of the vegetation in maintaining water balance. The vegetation itself is selected based on a thoughtful ecological analysis of the environs of a cell, as well as the intended performance and longevity of that cell. During the design process, any soil cover can be rendered a vegetated cover simply by including an ecologist and groundwater hydrologist on the team, and by giving them the charter and license to optimize the cover's suitability for a well-adapted plant community.

The chapter begins by revisiting the various attributes of vegetation in a vegetated cover, and provides further guidance on how to design to their advantage. It also addresses liabilities and failure scenarios, along with design strategies that exploit vegetative virtues while minimizing both the likelihood and consequences of these failure scenarios. The ecological model upon which this and the following chapter are based is illustrated in Figure 17.1. A vegetated cover is ecologically functional, with biological cycles confined to the upper layers. Favorable processes such as transpiration and organic cover building enhance the performance and longevity of the cell. In most cases, a rock cover is ecologically dysfunctional. Organisms may initially be excluded by the hostile outer layer, but when invasion invariably occurs, the biological influences will degrade the cell. These processes, known as biointrusion, progressively damage the encapsulating system and accelerate the release of contaminants to the environment.

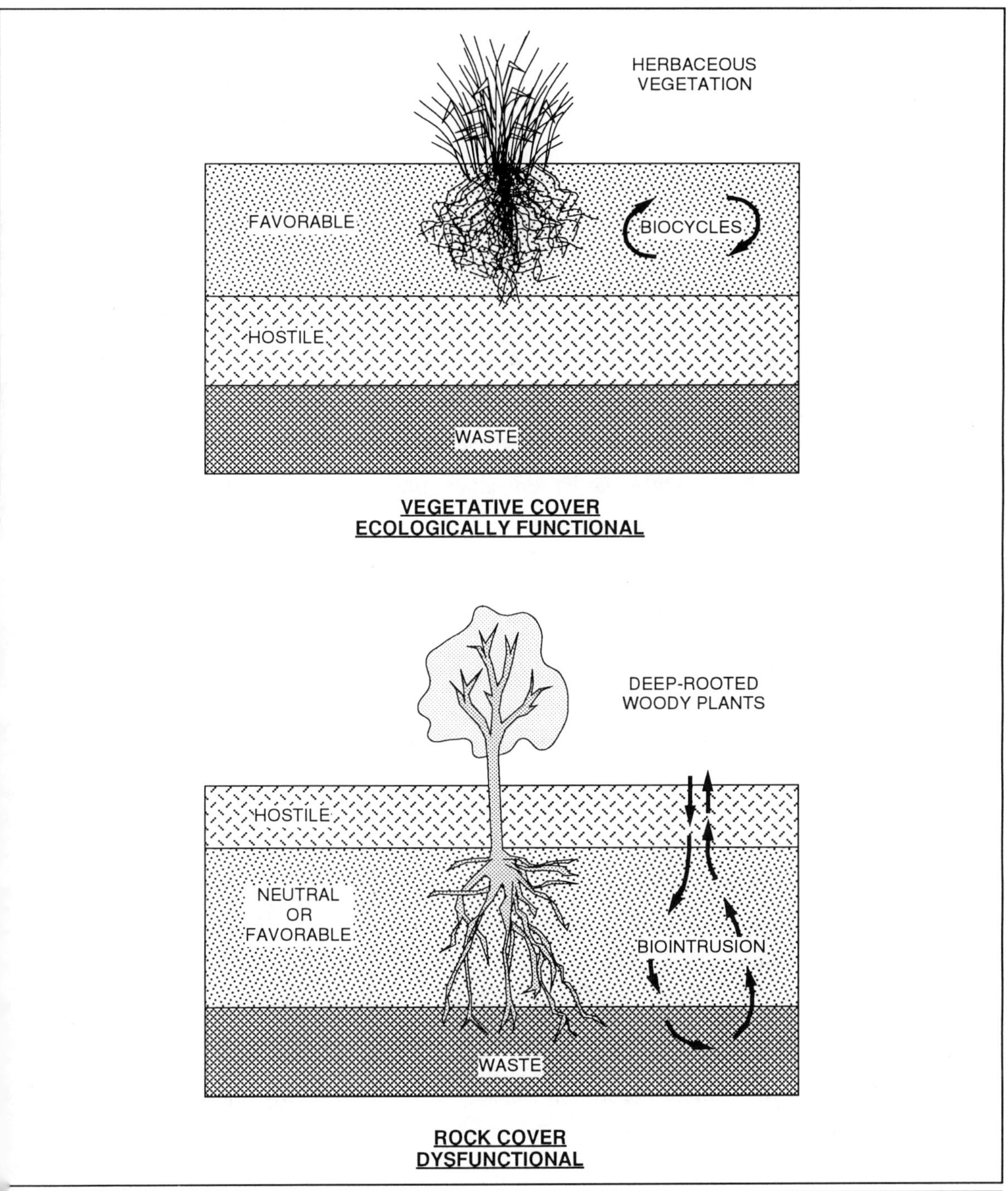

FIGURE 17.1

ECOLOGICAL MODELS FOR FUNCTIONAL AND DYSFUNCTIONAL COVER DESIGNS

BIOINTRUSION REVISITED

Perhaps the most compelling reason to incorporate vegetation as an integral feature of a cover has to do with biointrusion which, as mentioned above is the process whereby plants and animals damage disposal cells. Biointrusion is addressed at length in Chapter 18, and the details need not be dwelled upon. However, it is worth emphasizing that, except for rare circumstances, vegetal colonization of a cover is inevitable. Plants are remarkably aggressive at invading even the thinnest cracks with the scarcest supply of soil. Designers of rock covers have underestimated this aggressiveness, and have been surprised at the number of plants that have established themselves among the rock interstices. As their roots advance in the underlying soil, the plants modify its permeability, provide access for animals and potentially transporting contaminants. The engineered soil that was presumed protected beneath the rock armor is relentlessly transformed by roots and their faunal associates into a biological soil.

This pedogenic process -- the transformation of a simple abiotic medium to a biologically complex one -- is the predictable consequence of ecological succession, the encompassing biological principle that describes how organisms perpetually modify their environment so as to be ever-more suitable for subsequent generations of their own kind, or of organisms of a higher sere (successional stage). Roots, microbes, and burrowing animals loosen and agitate the soil, adding organic matter and encouraging the movement of nutrients and water. In doing so, they tend to increase its permeability to water and contaminants, both liquid and gaseous. Whatever placement criteria and performance objectives were intended for the originally installed engineered soil, these will be irreversibly changed as the result of the biological processes.

The solution to this is to abandon any strategy that depends upon the exclusion of organisms from the cover's upper layers. Instead, the cover should be designed to accommodate organisms and exploit their beneficial influences. Soil should be selected and placed in such a way that facilitates plant establishment (on the designer's terms) and encourages their self-perpetuation. Chapter 18 describes how layers of sand, rock, asphalt, and geosynthetics may be buried beneath the favorable rooting medium to discourage the advancement of roots and subterranean organisms into the cell's deeper layers. These buried layers will provide variable degrees of resistance to root penetration, depending upon the amount of physiological stress that drives roots and animals downward against this boundary. The key to resisting biointrusion, and the take-home message of all the ecological chapters of this book, is to provide such a favorable biological environment in the cover's uppermost layers that plants and animals are not compelled to advance downward. The soil must be placed, and bounded underneath, in such a way that biological cycles are, except in the rarest of circumstances, confined to these upper layers. In other words, if water and nutrient cycles remain in the upper layers, and if the underlying layers are hostile in terms of water and nutrient availability, then the great majority of biological activity will remain in the favorable zone. Intrusion into underlying zones, and biological modification of those zones toward increased bio-hospitability, will be minimal.

An important attribute, then, of vegetated covers is that biointrusion and its concomitant damage is minimized. If the soil/vegetation system is an integral component, the designer and constructor assert control over the biological conditions in and atop the cover. These conditions can be established and fine-tuned with reasonable assurance that the biotic components and processes of the cover, and their confinement to the upper layers, will persist well into the cell's future. Conversely, if the cover is designed to exclude plants, then the constructor can only wait to see if plant colonization is indeed resisted; if plants do indeed invade, then a perpetual program of plucking, cutting, burning, or poisoning will be needed to keep the cover in its original, unvegetated state.

WATER BALANCE

Another attribute of vegetation for use in cover design is transpiration, the process whereby plants extract water from the soil, transport that water up through the vascular system to the leaves, and exhale it as a metabolic byproduct. Evaporation and transpiration are generally the best fate for meteoric water on a disposal cell; if water is evapotranspired, then it neither runs off, potentially eroding soil, nor percolates to depth, contacting and mobilizing the encapsulated waste. By installing a carefully conceived plant community, as well as the appropriate medium to support it, the engineer is emplacing a self-renewing system that can be relied upon for the long-haul to intercept water from the root zone and transpire it back to the atmosphere.

This is not to say that the cover should be without, in its deeper layers, the appropriate combinations of liners and drains, as described in Chapter 16, to impede percolation. As mentioned at the outset, plants die, sometimes catastrophically because of fire, drought, or overgrazing, and their favorable influence is interrupted. But the contribution of these engineered systems is finite and ever-diminishing due to the clogging of drains and leaking of liners. Unlike the overlying plant community, these systems neither grow, reproduce, nor mend themselves.

Plant transpiration is a simple process, essentially powered by the sun. Roots absorb water from the surrounding soil and pipe it to the crown, incidentally carrying nutrients in the rising stream. The nutrient laden water is drawn by capillary action up to the leaves, where it participates in metabolic processes, including that most blessed phenomenon known as photosynthesis, without which we would have no salads (much less vegetated covers). Water is a byproduct of, as well as a feedstock for, plant metabolism, and it is vented when necessary as vapor through tiny leaf apertures, known as stomata. These eye-shaped openings wink shut during periods of drought, metabolic inactivity, or sustained dormancy, for instance, during winter. When moisture is available in the soil, and when solar energy is available to drive the metabolic powerplants, water will be sucked from the root zone, elevated to the leaves, and eventually vented to the atmosphere.

Because roots extend deep into soil and probe into subsurface pockets of moisture, transpiration is a much more powerful process than evaporation for removing subsurface water. The depth of the evaporative zone differs significantly from one soil profile to another, principally based on the soil's ability to conduct heat downward and moisture upward. For instance, because of the large size of sand particles and the interstices among them, a layer of sand either on the surface or below will impede the capillary rise of moisture and greatly reduce evaporation. A layer of rocks or gravel, if clean of interstitial fines, will do the same; if on the surface, rocks will insulate the soil and greatly reduce evaporation, essentially serving as a one-way valve for water movement into the soil. In such situations, transpiration is really the only way to raise water from the depths to the atmosphere.

Interestingly, a veneer of rocks is highly desirable for vegetated covers in arid environments. This is partly because of the supplemental erosion control provided by the rock, as discussed later in this chapter. But it is also because the rocks will reduce evaporation and thereby increase the supply of moisture for plants. Although the inhibition of evaporation may seem contrary to the broader goal of reducing percolation into the cell, the truth is that plant productivity (hence transpirational capacity) will be much higher on a "rock-mulched" soil than on a soil that is fully exposed to insolation, heating, and evaporative stress. An unmulched soil in an arid environment may yield so much water to evaporation that no moisture remains for plants to germinate, grow, or survive, resulting in the barren, often unstable landscapes that are popularly regarded (by the biologically naive) as deserts. In fact, deserts can be biologically both diverse and productive when moisture is blocked from evaporation and otherwise harvested into soil fines by rocks, both on and below the surface. The rock-rich surface layer that results from the removal by deflation of soil fines is known as desert pavement, a condition that promotes plant productivity and that should be emulated by designers of cells in arid environments. Given the right combination of rock, soil, and drought-hardy plants, a vegetated cover can be established even where precipitation is relatively meager and infrequent.

Of course, too much rock on the surface will prevent the establishment of a robust, desirable plant community, and will instead promote the invasion of deep-rooting species that do nothing but damage the cover. Unfortunately, there is a paucity of empirical evidence to consult when determining just how much rock should be applied to the surface to discourage evaporation and erosion without unduly restricting plant productivity. As a conservative guideline, I recommend that at least 70 percent of the surface area of a cover be composed of soil fines, at least for the initial seeding and establishment of the plant community. (The percentage of rock may gradually increase due to deflation.)

Rock mulches are useful in arid environments for maximizing vegetal productivity, which in turn provides maximum transpirational capacity to maintain the desired water balance. A design imperative for all environments, if the objective is to maximize vegetal productivity, is to select the best suited soil and to place it at the optimal density and thickness. A good soil mothers the plant community, receiving and accumulating water and nutrients, then supplying them generously when needed by the overlying vegetation.

With respect to moisture, the best soil will behave like a sponge, readily absorbing moisture from precipitation, storing it, and then releasing it to plants when they are transpiring. A soil with too much clay is slow to take up moisture, and once the moisture has been absorbed, is very recalcitrant in releasing it to plants. Plants rooted in a clay-rich soil will wilt even when that soil holds more than 7 percent of its own weight in water due to the high tension with which the water is held. Conversely, a sandy soil readily accepts the infiltration of water, but holds very little for availability to plants; rain falling on a sand-dune will infiltrate and drain below the root zone. Hence, loams (soils with a relatively even mix of sand, silt, and clay particles) are generally best for plants because there is adequate sand to encourage infiltration, while there is enough silt and clay to provide storage capacity.

Of course, there is plenty of room for creative textural layering in cover design. For instance, at the Jackpile uranium mine in west-central New Mexico, we capped uriniferous sandstone ore with a layer of clay-rich spoil derived from Mancos shale overlain by a sandy, droughty soil created by crushing overburden from Tres Hermanos sandstone. Our intent was that water would readily infiltrate to the loamy transition between the shale and sandstone layers, there to remain perched and highly available to plants (Reith et al., 1990). Water was essentially bounded in the root zone, blocked from percolating by the underlying clay and protected from evaporation by the mulching effect of the rock and noncapillary sand.

The thickness and placement density of soil is another important consideration in optimizing the performance of a vegetated cover relative to water balance. The principle is simple: The designer must assure that enough soil is in the upper cover to accept and store all of the moisture that could enter that cover before transpiration is able to remove it. In other words, storage capacity must be provided for the total accumulation of water that occurs during episodes of plant dormancy. If all the soil's capillary surfaces are occupied by water and still more water infiltrates, then the consequence will be breakthrough -- the drainage of water from the bottom of the rooting medium into the underlying cover layers.

Winters generally pose the greatest challenge because most plants are dormant, protecting themselves against frost damage, and moisture often accumulates as a thickening snow mat. This snow may melt from below and charge the soil water before most plants are exposed to the atmosphere, much less transpirationally active. Thus, the accumulation of moisture in soil during winter dormancy generally determines the minimum acceptable storage capacity for a soil layer over a landfill.

Instrumented test plots are often used to determine optimum soil thicknesses for different regions where disposal cells are beingconstructed; for instance, 1.4 meters was found to be optimum for landfill covers at the Idaho National Engineering Laboratories. Weighing lysimeters were used to establish a thickness of 1.5 meters for silt loam to be placed in protective "barriers" over waste too hazardous to excavate at the Hanford site in Richland, Washington. Studies determined that 70 cm of soil were needed for landfills at Los Alamos National Laboratories, which are perched high on the shoulder of the Jemez Mountains in New Mexico and which therefore receive a hefty accumulation of snow. Besides optimizing soil placement, experiments are in progress with the use of pine trees on cell covers because of their ability to transpire, albeit slowly, even in the dead of winter if the soil water is liquid and the atmospheric conditions are reasonably hospitable.

The problem with instrumented test plots is that they may not, during their relatively short period of operation and observation, experience the full range of conditions to which a vegetated cover may be exposed during its design life. Experiments with supplemental moisture, temporary defoliation, and other anomalous conditions may partially alleviate this problem. Water-balance models are also needed for optimizing cover design and predicting long-term performance under a wide range of conditions. These models were first applied to agriculture with the intent of optimizing water usage by crops in environments where water was scarce or expensive. The models most widely used for cover design are CREAMS, HELP, and UNSAT. The models have been at least partially validated by comparing their predictions to the observed behaviors of water in instrumented test plots such as mentioned above.

There are a number of ways that water balance models can be used to optimize soil placement. First is to determine optimal soil thickness with a computer model, such as HELP, and representative site climatic data, the thickness of soil that prevents moisture breakthrough may be established.

Another parameter to optimize is placement density. This may be done by modeling the responses in the behavior of water to differences in the hydraulic conductivity of the soil in the root zone. If the soil is compacted in excess of the optimum value, the hydraulic conductivity is too low to allow for full infiltration, hence water runs off the cell; if the soil is insufficiently compacted, then water moves gravitationally through the root zone before it can be intercepted by roots and transpired to the atmosphere, the result being increased percolation and lateral drainage.

Once soil thickness and placement density have been optimized, the model can then be used to evaluate performance under historic, simulated, and anticipated long-term conditions.

Historic data can be mated with a stochastic climate simulator to evaluate cover performance over much longer time frames which include the extreme conditions (drought or deluge) that might be expected over such a time frame. Furthermore, models can be used to test cover performance in response to climatic change that might plausibly occur during a cell's design life. In designing barriers for the Hanford site, where the objective is to provide up to 10,000 years of protection to the underlying waste, precipitation regimes with double the current annual average rainfall are being used. This increased precipitation is distributed in the same monthly pattern, only double, as occurs currently.

Water balance models such as CREAMS, HELP, and UNSAT can be a great advantage to the cell designer, and can provide a basis for optimizing soil placement when instrumented test plots are impractical. As with all models, the user must perform sensitivity analyses that bound the range of anticipated conditions, and must respect the uncertainty and imprecision that is invariably associated with any model.

Given that models and test plots are useful tools for designing vegetated covers, common sense must always be foremost in the process and, when the others are unavailable, can be used to provide assurance that the cover will protect the underlying waste from moisture.

The basic principle to exploit is that, where vegetation flourishes, water is being efficiently used.

EROSION RESISTANCE

The third attribute of vegetation in a vegetated cover is its ability to protect the cover's upper layers from the ravages of erosion. Plants are wonderfully effective at preventing erosion, and our world's terrestrial flora deserves full credit for slowing the rate at which continents are washed to the sea. The stabilizing influence of vegetation is widely appreciated by landscape architects, highway engineers, foresters, farmers, and reclamationists, all of whom rely upon plants to protect sloped or otherwise vulnerable soils from removal by sheetwash, gullies, and deflation. There seems to be a fundamental distrust of vegetation by many who design disposal cells. The designers lack confidence that any plant community is sufficiently robust to prevent gullies from developing and expanding during the cataclysmic delivery and migration of water of the PMP. The PMP is a conservative and hypothetical value derived from an evaluation of a site's climatic and topographic setting. The intensity of rainfall and magnitude of overland flow during a PMP substantially exceeds that which would be encountered during a hundred-year storm, which is the design storm for many civil design projects.

As is the case with many hypothetical standards, the PMP is strictly a calculated value that has never been, nor likely will be, empirically validated. If we are to design a disposal cell for perpetuity, then we need to establish some upper limit for the meteorological circumstances that must be weathered. The PMP represents just such a limit and, because of its conservatism, is an important reason why cell designers distrust vegetation.

Another reason why plants are distrusted, as mentioned at the outset, is the possibility of their absence during the moment of greatest need. If a major storm occurs when vegetation is for some reason absent -- for instance, because of drought, fire, or herbivory -- then catastrophic erosion may occur, the cover will cleave, and the cell will fail as the once-encapsulated wastes are vented to the environment. One problem with this outlook is a lack of appreciation by overly cautious engineers of the favorable influence of plant residues that stabilize soil even when the plants themselves are dead or dormant. Nonetheless, the reality is that cells must be designed for longevity, and static systems are needed that will be immune to such natural phenomena as fire, herbivory, and drought. For this reason, vegetal resistance to erosion is an attribute to be fully exploited, but generally as a supplement to other engineered resisters of erosion such as proper cell shaping, soil selection, and the use of erosion-resistant materials such as rock.

The mechanisms whereby plants protect soil from erosion are:

Interception: foliage and plant residues intercept rainfall energy and prevent soil compaction from raindrops

Restraint: root system physically binds or restrains soil particles while aboveground residues filter sediment out of run-off

Retardation: aboveground residues increase surface roughness and slow velocity of run-off

Infiltration: roots and plant residues help maintain soil porosity and permeability

Transpiration: depletion of soil moisture by plants delays onset of saturation and run-off.

Absent from this list is the role that plants play in resisting deflation, both in their direct blanketing of the soil and in their provision of windbreaks that interrupt the laminar movement of wind across the earth's surface. It is worth noting that, except for transpiration, all of the above stabilizing attributes of vegetation will be provided even when the plants are dormant or dead. The favorable role of a plant community in stabilizing against erosion will attenuate only gradually after plants die, as decomposition and weathering slowly disassemble the interwoven organic molecules that bind the soil.

The role of vegetation, among other factors, in determining erosion rates is quantified in the universal soil loss equation (USLE) and modified universal soil loss equation (MUSLE). Both are based on regression models and are empirical formulas for predicting soil losses due to sheet or rill erosion. The stabilizing influence of plants is represented in the USLE as a cover or crop management factor and the MUSLE as a factor that encompasses not only vegetation, but mulching and chemical emulsification as well.

Deflation and sheetwash erosion are less serious concerns to the cover designer than is gully formation. During intense storms, gullies will develop rapidly on unprotected surfaces, will concentrate overland flow, and will expand in a positive-feedback loop until flow attenuates or an underlying resistant layer is encountered. Once formed, the gully remains to concentrate flow during subsequent storms and thereby facilitate its own expansion. One attribute of vegetated covers is the ability of plants to stitch up rills and encourage sediment deposition that will eventually eliminate the concentration of flow. Gullies are rills that have exceeded the threshold beyond which such self-repair will not occur. Only human intervention (i.e., maintenance) can repair a cover that has been lacerated with gullies.

One of the most effective agents against gullies on a cell is a layer of rock, generally consisting of alluvial cobbles or crushed strata sized to resist mobilization during a PMP and tested for resistance to weathering. If the outermost layer of a cell is rock, then the cell is maximally protected against gully erosion, but may be susceptible to biointrusion damage and may have water-balance problems. The best strategy is to underlay a vegetated cover with a layer of rock that not only serves as a deep root barrier, but also provides backup protection against gullies that incise the soil during periods of vulnerability. This approach is reflected in designs such as the "multicomponent cover." The problem that must be addressed any time soil is layered over rock is that, the thicker the soil, the greater the velocity of water that will move through a hypothetical gully in that soil. This means that very large rock will be required to resist the flow concentrations that might form in the generous layer of soil needed to provide biointrusion protection and water balance. As is often the case in engineering, the challenge is to reconcile these opposing requirements.

The UMTRA Project has evolved through a long and complicated series of perspectives on the merits of vegetated covers relative to rock covers. UMTRA Project engineers now fully recognize the advantages of vegetated covers relative to biointrusion, but they find it difficult to elicit regulatory acceptance of any kind of soil cover (vegetated covers included) due to their perceived susceptibility to gully formation. The NRC has responded to the increasing number or requests on the part of UMTRA engineers for regulatory approval of vegetated cover designs by issuing the following list of elements to be addressed in a solicitation for approval:

- A detailed geomorphological study of the site and its surrounding area to establish the relationship between the watershed, run-off rates, vegetation, slopes, and gully development.

- A mathematical analysis to demonstrate that a gully on a given slope will not, as a result of the average annual precipitation, intrude into the tailings.

- A mathematical calculation to determine the slope length required, for a given slope inclination, to develop a gully.

- Calculation of the potential soil loss from vegetated and unvegetated slopes due to sheetwash erosion.

- An empirical evaluation of the potential for the formation of deep gullies on soil sideslopes and soil perimeter dikes, with the objective of showing that even in the event of gully development the gully will not break through to the tailings.

- Placement of rock beneath the soil to control erosion in the event of the development of a gully in the soil.

- A rigorous cost-benefit comparison between the use of rock and soil covers in enhancing erosion protection.

The NRC does not require that all of the above be done. They note that one or more approaches may be needed to evaluate and confirm the anticipated performance of a proposed cover design. It is left to the judgment of the designer of the disposal cell to choose the combination of appropriate technical analytical methods to be used a particular site.

Practical experience with the NRC on UMTRA has shown that the regulators insist on highly conservative, worst case analyses before a soil cover will be approved. As a consequence, UMTRA designers seem to be moving toward a cell design where a relatively flat topslope is perched above rock-armored clean-fill dikes. The elevated rock "parapets" enforce a high base level that reduces gully potential. The steep rocky sideslopes are also gully resistant and, though susceptible to invasion by deep-rooted plants, are designed such that roots will not penetrate buried protective layers.

Chapter 18

THE ECOLOGICAL MODEL: BIOINTRUSION

VEGETATION AND ENCAPSULATION

Biointrusion refers to the damaging effects of organisms on covers that are intended to contain waste. The primary concern is intrusion by higher plants, especially woody species such as trees or shrubs, and by burrowing mammals, ranging from rodents such as mice and ground squirrels up to larger predators such as badgers and foxes. Insects and microbes can also damage covers, although their presence and activity is often closely associated with the roots of large plants or burrows of animals that have successfully invaded a cover. Intrusion by humans is a special case that involves sociology as much as design.

This chapter first addresses the hazards of biointrusion in terms of cover degradation and environmental consequences. The chapter then describes approaches that may be used to resist biointrusion by plants and animals. I briefly discuss the prevention of human impacts on covers, although there is no practical way I can think of what would deter a determined, well-equipped team of human intruders.

HAZARDS

Covers are designed to encapsulate waste. Organisms have evolved to use resources. If either the cover or the waste constitutes a resource, then organisms will attempt to use that resource for energy or habitation.

PLANT EFFECTS ON SOIL

Earthen materials such as rock and soil are important ingredients for many waste covers because of their reliability for long-term performance. Yet these substances are highly attractive to organisms. Soil is, of course, a rooting medium for vegetation, which relies on the soil for mechanical support, nutrients, and moisture. Once a plant manages to germinate in a soil, it will extend its roots downward and laterally to harvest nutrients and water, and to enhance its attachment to the substrate. Figure 18.1 illustrates typical vertical root distribution for mature plants in clay soil. As roots advance through soil, they introduce organic matter and excrete acids to dissolve nutrients. Upon their death, roots leave conduits of decaying wood and eventually voids through which liquids and gases may pass readily. Plant roots also introduce a host of organisms to the soil, including bacteria, fungi, insects, and other small multicellular animals.

For many covers, soil is placed with a specific purpose in mind; this may involve radon attenuation, frost protection, or infiltration control. The ability of the soil to perform its intended function may depend largely on its degree of compaction, yet unwanted plant roots and their biological hitchhikers will most assuredly loosen the soil and modify its character.

It is also important to realize that the effects of roots on a soil far outlast their tenure as living appendages. The conduits of increased permeability left by decaying roots may serve as avenues for waste emanation, or for unimpeded water movement into the waste, generating leachate far in excess of predicted quantities. The empty or humus-filled root channels will be tunnels of biological activity, with microbes, ants, termites, earthworms and other organisms busily engineering the soil to better suit their own existence. The tunnels will also serve as the first corridors of access for rodents, soon to be followed by predatory weasels and badgers. The message here is that if an engineered soil is to remain an engineered soil, it must be isolated.

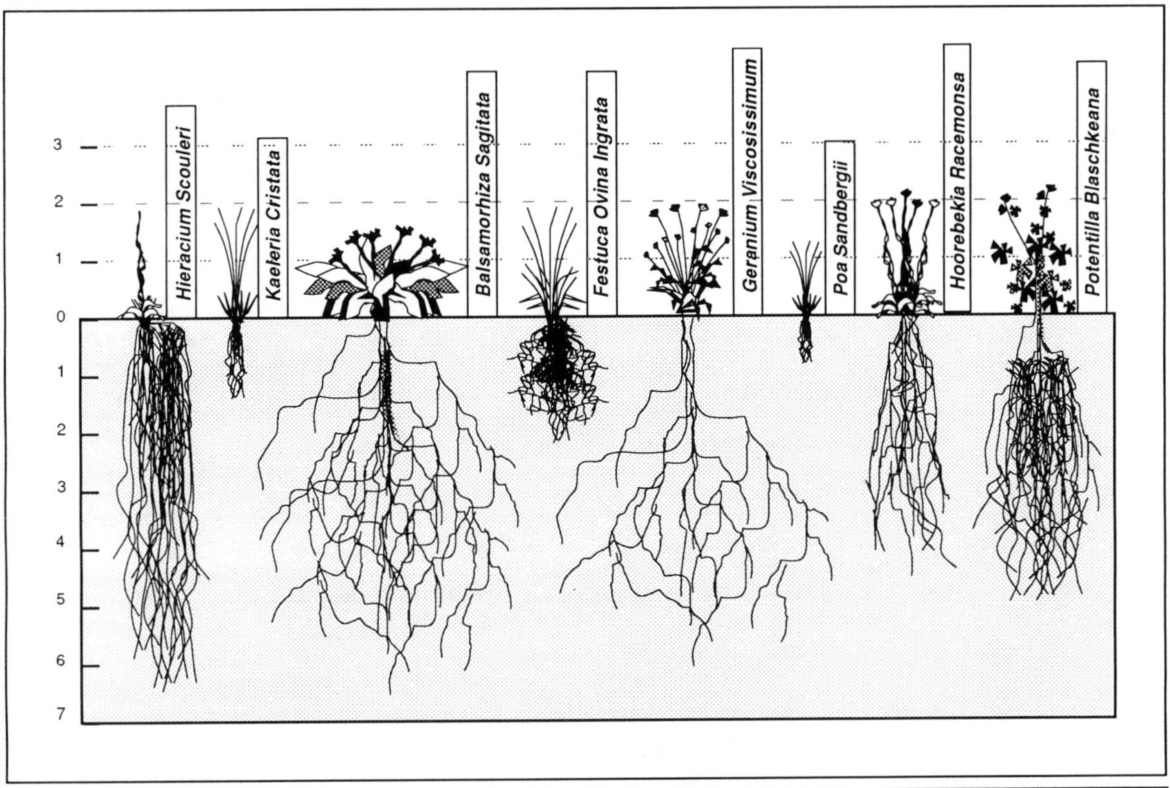

FIGURE 18.1
ROOT AND STEM RELATIONSHIPS OF PRAIRIE PLANTS

This leaves the cover designer with two choices; either assume that roots will be modifying the soil from its initial compacted, homogeneous abiotic condition to a more permeable, heterogeneous biological condition; or install components that will protect buried soil layers as long as possible from the advance and influence of roots.

Must we always assume that root presence in a cover is negative, worthy of the epithetical term "biointrusion?" Absolutely not! Plants and their roots can be vital contributors to a cover's long-term stability and performance. Roots must be confined to those portions of a cover where their role is beneficial. They must be prevented from penetrating the cover, contacting the waste, and essentially breaching the containment. Depending on the waste form, roots in contact with the waste, or even in its proximity, may transport dissolved or gaseous contaminants to the surface.

CONTAINMENT TRANSPORT BY PLANTS

Various studies have shown that plant roots can transport contaminants into the aboveground portion of the plants. Plant uptake of contaminants by plants rooted in uranium mill tailings from Mexican Hat, Utah; Naturita, Colorado; and Ambrosia Lake, New Mexico shows that plants growing on these tailings readily accumulated molybdenum (Mo) and selenium (Se) at levels toxic to animals. In addition, four-wing saltbush was found to be highly enriched with uranium (U). Plant uptake of contaminants near an active uranium mill in Colorado shows that soil and plants that came into contact with tailings pond seepage had very high levels of Mo and U. Plants grown on tailings had U levels an order of magnitude higher than plants grown on noncontaminated soil, while levels of radium-228 (Ra-228) were two orders of magnitude higher in plants grown on tailings than the controls. Plants and soil in uranium mine dewatering areas were found to have elevated levels of U-238, thorium-230, (Th-230), Ra-226, and polonium-210 (Po-210).

Radon gas is another contaminant that is readily taken up by plants and then emanated into the air. Radon is very soluble in water and is transported into the plant as the roots pick up the radon-laden water. Once in the plants, radon is transported via mass flow of water through the plant and then out the leaves into the air. Radon emanation out of plants takes place during the day and the night, and there is little relationship between the amount of water transpired and the levels of radon exiting the plant. Radon is probably transported through the leaf cuticle and epicuticular wax. If the tailings are covered with soil and then vegetated, the release of radon from plants may be much greater than from the soil cover alone.

Morris and Fraley (1989) studied the release of radon from tailings covered with 20 inches of soil and overburden and 20 inches of soil and overburden plus a 12-inch clay cap. Radon release from plants grown on these two study plots was highest during later summer and early fall and lowest in the winter. Release rates from plants grown over clay-capped tailings were an order of magnitude higher than background, ranging from approximately 25 to 30 pCi/m^2s during the peak months of July through October. These levels exceed the EPA standard of 20 pCi/m^2 for the disposal cells constructed for uranium mill tailings in the Uranium Mill Tailings Radiation Control Act (UMTRCA). Radon release rates from plants grown in soil over tailings ranged from approximately five to 12 pCi/m^2s for the same time period. Morris and Fraley (1989) indicate that clay with its low permeability and high moisture content "may serve as a radon 'trap', slowing diffusion sufficiently for the concentration of radon to increase." Furthermore, only six percent of the root material was in the upper portion of the clay layer (the remainder was in the soil/overburden layer), indicating that a relatively small amount of root biomass in the clay layer will result in heightened radon release from vegetation.

The point is that roots can transport contaminants surfaceward even if the layers overlying the waste are not fully penetrated by those roots. The phenomenon described for radon may conceivably apply to any contaminant moving from the waste into an overlying soil layer, although the release of contaminants to the environment (and subsequent exposures to nearby animals) should be less of a problem for a dissolved substance such as a metal or metalloid than for a gaseous element such as radon. A number of investigations are underway on this issue of radon transport by plants rooted in soil over tailings; the outcome of these investigations may indicate that failure to isolate the radon barrier from plant intrusion essentially defeats the whole purpose of the barrier.

PLANT INVASIONS OF ROCK COVERS

It is easy to foresee the invasion by plants and their roots of unprotected soil layers in a cover, and to predict the process whereby roots will modify the soil's ability to isolate waste. Engineers have been surprised at the aggressiveness with which plants colonize covers armored by an outer layer of rock. The Uranium Mill Tailings Remedial Action (UMTRA) Project provides a good example of how both designers and regulators have underestimated the ability of plants to invade rock covers and extend their roots into the underlying layers of soil.

The objective of the UMTRA Project has been to design, construct, and license disposal cells for uranium mill tailings at 24 locations around the United States. Before the first cell was designed, a multi-disciplinary program of investigations was conducted to address a broad range of environmental factors. Included in the first wave of investigations were studies suggesting that the placement of rock armor over soil would cause moisture accumulation that would render the underlying soil highly favorable for plant growth. The rocks would serve as a one-way valve for water movement, encouraging infiltration while resisting evaporation. Studies were also conducted on precipitation and erosion potential at the various UMTRA Project sites, and it was concluded that the erosion hazard could best be addressed by covering most cells with layers of rock sized to resist movement during the probable maximum precipitation or flood at each site. When the results of the various investigations were compiled into a summary technical document, the issue of plant colonization was either disregarded or relegated to secondary status behind erosion prevention. The technical document asserted that:

> In practice...a substantial cover thickness of rock riprap on the top and sides of the pile will satisfy requirements for the control of animal burrowing and root penetration.

Accordingly, a number of cells were constructed with the basic cover configuration shown in Figure 18.2. Unfortunately, no quantitative guidance was provided on what constitutes "a substantial thickness of rock," with the result that several of the covers were dramatically invaded by volunteer plants. In 1991 unanticipated vegetation was growing in abundance on the covers of the cells.

The plant invaders at Shiprock and Burrell ranged from annual forbes and grasses up to large, potentially deep-rooting trees and shrubs. A number of plants at Shiprock and Burrell were carefully excavated to understand root penetration and damage associated therewith. In most cases, the majority of roots occupied the bedding layer immediately beneath the rock layer or the uppermost portion of the clay-rich radon barriers. Some root extension downward into the deeper reaches of the barrier was encountered. Figure 18.3 includes representative sketches from these excavations. The presence of the roots in the radon barrier (directly overlying the tailings) raised the specter of radon transport as described above. Studies may someday be conducted on UMTRA to determine if plant-mediated radon transport indeed occurs in accordance with Morris and Fraley's (1989) observations.

The experiences on the UMTRA Project as related to plant biointrusion are summarized in a series of documents issued by the DOE. As a result of the biointrusion studies, a number of rock covers will either be modified to deter future plant invasion -- for instance,

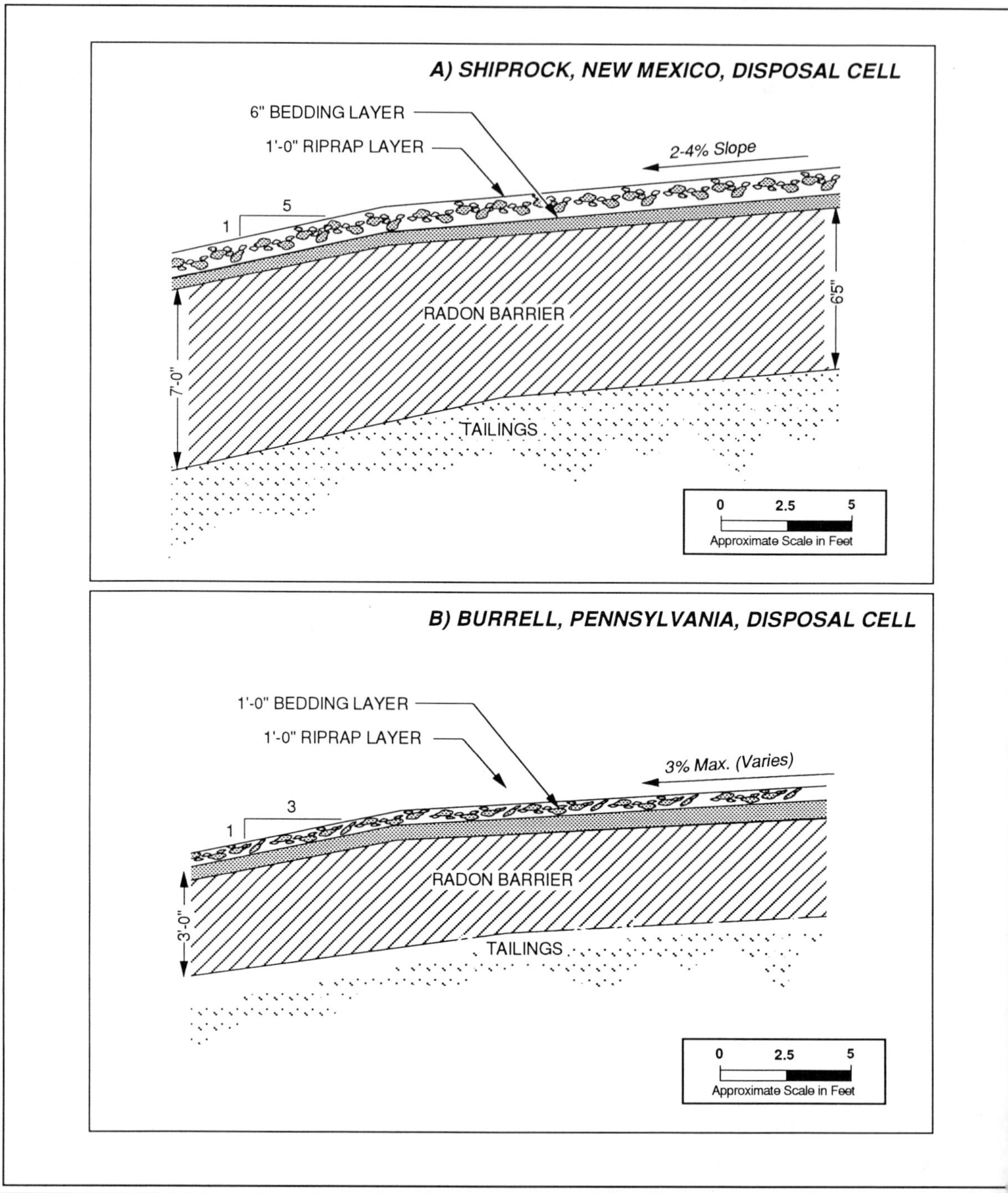

FIGURE 18.2

TYPICAL CROSS SECTIONS OF ROCK-COVERED CELLS

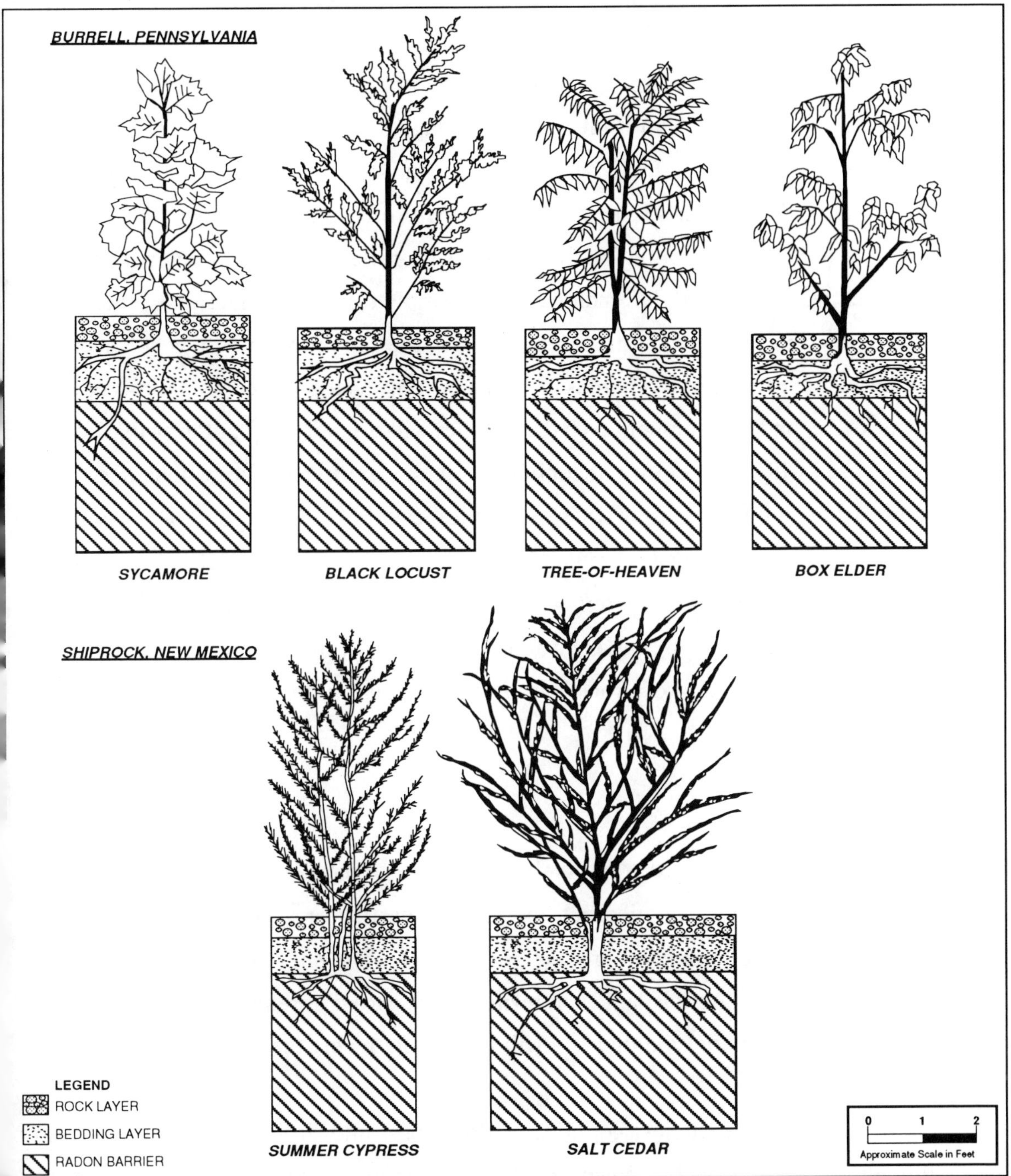

FIGURE 18.3
ROOTING PATTERNS OF TREE SPECIES GROWING ON THE BURRELL, PENNSYLVANIA AND THE SHIPROCK, NEW MEXICO DISPOSAL CELL

by adding additional thickness of rock -- or subjected to perpetual and frequent vegetation removal, depending upon the disposition of DOE managers. Other rock covers, for instance in Clive, Utah, and Green River, Utah, seem to be resistant to plant invasion, at least during their first few years of existence. It is of course difficult to determine whether the absence of plants from a rock cover, or a region of a rock cover, reflects some characteristic of the rock layer which truly excludes plant colonizers, or simply reflects the fact that seeds have simply not yet reached the unvegetated areas. This is the big problem with rock covers in terms of biointrusion prevention. There is great uncertainty as to what combinations of rock thickness, rock size, and environmental characteristics will discourage plant growth. If rock armor is to be used with any measure of confidence, then extensive studies will be necessary to determine what combinations of these variables can be relied upon to resist plant colonization. Even if effective combinations are found, the retardation of vegetal colonization will probably only be temporary. This is because some rate of particle deposition will probably occur in any environment, causing even the thickest and cleanest rock layer to be transformed into a rock/soil matrix that will ultimately be colonizeable.

ANIMAL INTRUSION

As mentioned earlier, the hazard of animal intrusion is generally closely associated with plant intrusion. Roots introduce microbes, insects, and earthworms, which devour the root tissue and deposit degraded organic products in the soil. The small size of insects belies their ability to transport materials; ants and termites can excavate remarkable quantities of materials per year and, if unaccounted for, will eventually invert the layers of a landfill.

Burrowing animals such as moles, gophers, prairie dogs, and ground squirrels will also intrude into unprotected layers of a cover. Burrowing animals have intruded into waste sites at Hanford, Washington; Idaho National Engineering Laboratory near Idaho Falls; Oak Ridge, Tennessee; and Los Alamos, New Mexico. These organisms both transport waste to the surface and alter the hydrologic character of the overlying material. They may also render the surface of a cover more susceptible to erosion by creating channels of concentrated flow.

As mentioned above, initial colonization of a rock surface by plants generally precedes animal invasion. Windborne seeds of aggressive colonizers (weeds) will be deposited on a rock pile. The successful invaders will then serve as seed sources for subsequent generations. At some point during the process, animals will be attracted; their dispersal and impacts will be closely associated with the successive generations of plants.

Animals may be the vector that introduces plants. This is almost inevitable if a rock-covered cell is built in a habitat with nut-hoarding rodents. For instance, a rock cell was recently constructed near Lowman, Idaho, in a ponderosa pine forest populated with squirrels. It is expected that the rock pile will be an attractive environment for squirrels to cache their pine nuts which, if forgotten, will germinate into seedlings that will flourish in the soil layers and underlying waste. Squirrel activity and pine reproduction will probably increase steadily during the cell's first decade. The most serious breach of the cell, preventable through regular tree cutting, will occur if a large pine were to topple, fully exposing the underlying layers. The Lowman cell, which is situated beneath a long slope with potentially high erosion hazard, is a case where the biointrusion hazard was dismissed in deference to a single-minded design against erosion.

HUMAN INTRUSION

Human intrusion into wastes has generally resulted from the ill-advised desire by humans to use certain wastes for productive purposes. The classic example occurred in Grand Junction, Colorado, where local residents used uranium mill tailings as construction fill at more than 5,000 locations around a former mill site. These locations, known as vicinity properties, have required expensive and complicated cleanups to reduce radon hazards to homeowners and commercial building occupants. Unfortunately, there are many other examples where uranium tailings were used for construction purposes, evidently because of their ready availability and superior texture. In almost every case, though, these materials were taken before a serious attempt was made to encapsulate the tailings, and certainly before the hazards of radon were fully appreciated. Other intrusions into waste, for instance into commercial landfills, have probably been rather limited invasions for research purposes.

Now that toxic and radioactive hazards are more widely recognized, the naive and casual use of wastes such as occurred at Grand Junction is unlikely to recur. However, there are at last two possible scenarios for human intrusion that should be considered in cell designs. First is the possibility that a cell may blend so completely into the landscape that, if it is not otherwise posted, it may be unwittingly used as a site for construction or resource excavation. It is possible that a builder might begin sinking a foundation through a cover into the underlying waste, although the unnatural stratigraphy of the cover and unnatural condition of the waste would likely be noticed and the damage repaired without further incident. Perhaps a more likely possibility, with a more insidious hazard, is that a well might be sunk through the cover, waste, and liners in search of water, oil, or other buried resource. Such an event could be, depending on the waste form, immediately hazardous to the driller; it would also disrupt the containment, creating conduits for waste migration that conceivably could long go undetected. This situation is unlikely because of the unnatural cover/waste/liner stratigraphy that any reasonably alert driller would notice, but the probability of an undetected cell penetration can be further reduced if engineers wish to leave nothing to chance.

The other scenario is that the waste or the materials used for encapsulating the waste are sufficiently valuable to attract resource developers, legitimate or otherwise. Several of the rock-armored UMTRA cells discussed above were constructed with high-quality rock imported from some distance. Local builders and developers may be unable to resist the temptation to use these materials for their own projects. Possibly the Grand Junction mistakes may be repeated, but what the best designers can do is to conspicuously post the hazard and hope that good sense prevails over greed.

PREVENTION

There are two basic approaches to resisting biointrusion and concomitant cell damage. The traditional approach has been to attempt to design and construct a cover with an outer layer that is hostile to plants and animals. Based on our experiences with rock covers, this seems rather futile. Too much remains unknown about what combinations of rock size, rock thickness, and environmental characteristics are needed to assure even a temporary resistance to plant invasion. Further, the inevitability of particle deposition and gradual transformation of the rock surface to a rock/soil matrix means that all rock covers will eventually succumb to colonization and, if not accounted for, damage by plants and their faunal associates. Other materials for resisting plant invasion -- for instance, using concrete, asphalt, or synthetic membranes to cover a cell's outer surface -- are rarely considered seriously because of the unnatural characteristics of the material, the unsightly appearance of the cell, and the uncertain longevity with respect to all the characteristics required for effective encapsulation. Certainly those who maintain roads and sidewalks will attest to the aggressiveness with which plants colonize cracks and, given time, enlarge those cracks in concrete and asphalt. The same may be said for plastics used in mulching, although it must be granted that the thickness, quality, and durability of the concrete, asphalt, or synthetic membranes on a cell cover would likely exceed those characteristics of their counterpart materials used in roadbuilding, gardening, and other everyday applications. The message here is that plant exclusion from cell covers is an uncertain enterprise, best avoided by the designer who has alternatives.

The approach we recommend for preventing biointrusion and concomitant cell damage is to design a cover in which the outer layer is as hospitable as possible to a plant community dominated by shallow-rooted vegetation. A generous layer of soil should be provided that will host a carefully selected and established plant community. The best community is generally a native grassland that resists invasion by deep-rooted trees and shrubs. This would be a desert grassland in arid climes, a prairie in more humid areas, and a meadow in alpine or subalpine settings. The basic concept is to provide an uppermost layer of soil that is sufficiently favorable to both plants and animals to minimize the environmental stresses that would cause the organisms to advance downward into the cell's deeper layers. The soil should be a high-quality rooting medium, serving as a generous bank for the nutrients and water that plants require to thrive. The thickness and textural composition must be adequate to retain moisture even during periods of excess precipitation or plant dormancy, so that nearly all the water that falls on the cell is transpired by plants rather than lost by seepage to deeper layers. The idea is to initiate a vigorous and self-sustaining biological cycle, but to confine the cycle to the upper layers of the cover; this confinement is achieved not by force, but by creating such a favorable biological environment in the upper layers that the physiological stress which might otherwise encourage deep root penetration (or deep animal burrowing) is removed.

Thus our recommended approach to resisting biointrusion and root damage is to select the desired plant community, engineer the cell's upper layers to accommodate that community, and then utilize the plant community for its ability to achieve the cell's intended performance objectives. We use the term "vegetated cover" to describe a cover as such, where the soil and plants have been selected based on an ecological analysis of the environment surrounding the intended cell location. A vegetated cover can contribute to cell performance and longevity by: (1) resisting the invasion of unwanted deep-rooted plants; (2) resisting erosion by wind and water, albeit not as effectively as a layer of rock; and (3) transpiring water back to the atmosphere, which is usually the most desirable (least damaging) fate for water falling on a cell. Other attributes of a vegetated cover relate to its ability to self-heal in the sense that vegetation will replace itself after disturbance by drought, fire, or herbivory. Vegetation will also colonize and stabilize rills and small gullies, although there is certainly a threshold for gully development beyond which plants will not resist gully enlargement. The most important attribute of a vegetated cover is, in our thinking, that such a cover can truly be a key element toward the achievement of the long-held ideal of a maintenance-free cell for perpetual encapsulation. Only our ancestors will know for sure (and we oft wonder if they will care) whether it is indeed possible to design and construct a cell that lasts as long as the landscape in which it is sited. If such a cell is ever built, it will no doubt be covered with a soil/plant consortium that is fully adapted to the local environment, and that represents the designer's best attempt to emulate the conditions present in the naturally stable meadows, grasslands, or prairies found in the environment around the cell.

A comprehensive discussion of vegetated covers, their attributes and liabilities, and their design and construction appears in Chapter 17 of this book. Suffice it to say that the greatest liability of a vegetated cover is its vulnerability to erosion and breaching during major storms. Unless the cell is engineered to have surfaces which are barely inclined or, if steeper slopes are required, to have steep slopes that are very short, then a major precipitation event will cause flow concentrations that catastrophically mobilize and transport the cover's carefully selected soil. The sudden breech of containment that would result from such a storm would defeat the whole purpose of the cell. Chapter 17 describes cell configurations that may compensate for this vulnerability. Rock mulches are also discussed which, in combination with vegetation, can stabilize soil against erosion.

BURIED BIOBARRIERS

Our premise here is that an adequate thickness of favorable growth medium will eliminate the physiological stress that would otherwise motivate roots, burrowing mammals, and even subterranean insects such as ants and termites from venturing downward into the deeper layers of the cell. Another way of saying this is that roots and their microbial associates are not expected to advance from a favorable environment into an unfavorable one. Accordingly, an extra measure of protection against biointrusion may be achieved by underlaying the favorable medium with one that is hostile to life. The traditional approach to this has been to place a layer of loose cobbles beneath the rooting zone.

Root penetration of the barrier occurrs only where fines filter into the interstitial spaces between the rocks. Thus, if the interstitial voids can be kept free of capillary fines -- the sand, silt, and clay particles upon which capillary moisture may be stored -- then the rock layer will be devoid of the capillary moisture that fine root hairs need to survive and elongate. Thus for the loose-cobble biobarrier to work, a layer of choked rock or geotextile must be placed over the large cobbles to resist the migration of fines into the interstitial voids.

Tom Hakonson and his group at Los Alamos National Laboratory performed a series of important tests and demonstrations of this concept, among other concepts for cell design, relative to water balance and biointrusion control. Hakonson tested biointrusion layers in clear plexiglass tubes where root behavior could be directly observed and in larger-scale test plots with radio-labeled layers, such as in the Integrated Test Plot configuration illustrated in Figure 18.4. Hakonson's findings regarding the performance of buried cobble layers were mixed. Root penetration of the loose-cobble biobarrier occurred, although generally as a consequence of excess precipitation that saturated the overlying soil and caused moisture to percolate through the cobbles. Hakonson noted the importance of the soil/rock transition as a capillary barrier that will resist the downward movement of water except when the upper layer is saturated. Thus, the capillary barrier can be an additional factor that keeps the water cycle, and associated biological activities, in the upper portion of the cover, but only if the soil layer has the right combination of texture and thickness to prevent its saturation during periods of excess precipitation or plant dormancy, when transpiration is not operating to prevent moisture accumulation.

Our disposition toward loose-cobble biobarriers is that they can be relied upon as filters against root advancement, discouraging but not altogether impeding the progress of roots

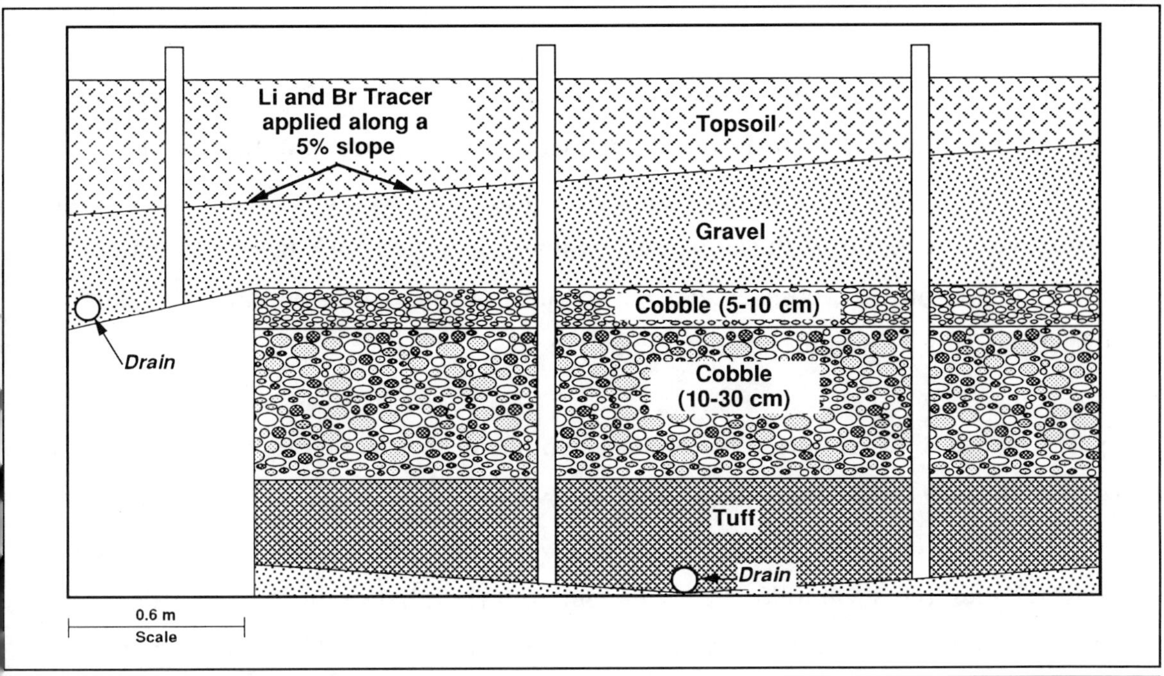

Improved trench cap design in the Integrated Test Plots.

FIGURE 18.4

INTEGRATED EXPERIMENT: IMPROVED PLOTS

into deeper layers. The effectiveness of such barriers will be strongly correlated to the thickness and quality of the overlying rooting medium. Of course, buried rock layers have additional value as deterrents to burrowing animals, as long as the rocks are large enough to be relatively immobile to those animals. Finally, a buried rock layer can temporarily impede a human intruder, possibly alerting a driller or foundation builder that additional site investigation is warranted prior to development.

The performance of other buried materials -- geomembranes, concrete, asphalt, etc. -- will also depend heavily on the characteristics of the overlying media, as well as the location and role of the buried layer relative to the cover's water balance. A good rule of thumb is: where water goes, roots will follow. So if the buried layer is below the zone of water cycling in the cover, or if it reliably resists water penetration, then it should remain unpenetrated by roots. Cracks that occur due to desiccation, freeze/thaw, or simply time will, if penetrated by water, be quickly penetrated by roots. The subsequent enlargement of those roots, with oft-demonstrated sidewalk-buckling strength, will widen the cracks and render the layer wholly ineffective as a barrier to the next wave of both plant and animal intruders.

Another option for retarding root advancement is the Typar Biobarrier, which uses a technology developed by Pacific Northwest Labs. This product consists of a geofabric into which are woven time-release capsules impregnated with the herbicide trifluralin. The herbicide repels roots or kills those roots that contact it, but does not kill the entire plant. The capsules release the herbicide at a rate which, given that the herbicide adsorbs onto soil particles, results in a two- to four-inch-thick wall of soil that will retard root advancement. The performance of life may exceed 100 years. The Typar Biobarrier is an innovative product that seems well suited to a broad range of applications; however, its cost of about two dollars per square foot, combined with its somewhat limited performance life, may limit its utility for disposal cell covers.

PREVENTION OF HUMAN INTRUSION

As mentioned earlier, the prevention of human intrusion is as much a sociological issue as it is a technical one. The topic is discussed briefly here for the sake of completeness, especially as it relates to biointrusion measures for plants and animals. At the heart of the issue is the recognizability of a disposal cell as an unnatural feature on the landscape. Advocates of rock covers point to the fact that these surfaces are unnatural and readily identifiable as something which should be investigated before building or drilling. A cell that is gently contoured to minimize erosion and then vegetated with a locally adapted plant community might be virtually unrecognizable as a structure that encapsulates a hazardous material.

A rock cover would therefore, because of its unnatural appearance, reduce the susceptibility of a cell toward accidental human intrusion relative to a vegetated cover. The rock may have value for some constructive purpose, thus inviting unscrupulous resource seekers to abscond with the cell's layer of armor. It will also signify the presence of a buried material that may invite unwanted investigations and excavations.

As discussed in Chapter 17, one promising cell configuration includes a vegetated cover on the topslope and steep rock covers on the sideslopes. This may have the benefit of discouraging inadvertent drilling and construction because of its unnatural appearance, yet having less vulnerability to "rock robbers." This reduced vulnerability is related to the fact that the steep rock-armored sideslopes would be underlain by clean-fill dikes, which has two advantages from the standpoint of cell vulnerability to human intrusion. First, rock removal would not cause catastrophic damage to the cell because the resultant erosion would, at least initially, only remove and disperse the underlying clean fill. Secondly, the rock slopes would rather quickly be invaded by deep-rooting trees and shrubs, which would not adversely affect cell performance because their roots would be in clean fill. The trees and shrubs would serve to obscure the rock, making the cell appear somewhat more natural except to those who investigated it closely.

For the moment, the best tools against human intrusion are durable and conspicuous markers identifying the contents of a cell and their potential hazards, as well as fencing and maintenance of records in the local property files. A buried loose-cobble layer is undoubtedly an effective deterrence to diggers or drillers should the monuments be removed and the institutional controls lost. Scientists have also contemplated burying durable plates with words and symbols to signify the presence of noxious materials below. The sophistication and expense of such measures will of course depend on the degree and longevity of the hazard. As with the entire issue of biointrusion, the response to potential human actions must be kept in perspective relative to the cell's many other performance objectives.

PRINCIPLES AND PRACTICALITY IN PREVENTING BIOINTRUSION

There are several notions that should be considered when attempting to design a cover that will, among its many performance objectives, resist biointrusion and damage therefrom.

- Draw upon the basic principles of soil ecology and root physiology to create a favorable environment in the uppermost layer of the cell. Be confident that if nutrients and water are effectively retained in the uppermost layers, then plants and animals will stay there too.

- Carefully study the natural environment of the cell to find natural plant communities dominated by relatively shallow-rooting grasses. Determine whether the absence of deep-rooting trees and shrubs is due to a natural ecological condition or to human intervention. If the native soil properties do indeed favor grasses over woody species, then emulate those properties in the rooting medium for the cover.

- Do not expect to achieve absolute resistance to biointrusion. Design a system where root or animal penetration to the deeper layer will be the exception and not the rule. Natural systems have variability, and a well-designed cover also will.

- Keep the issue of biointrusion in perspective relative to other cover design objectives, as well as cost. You may have to settle for suboptimal material -- for instance, an unconsolidated fill excavated from the footprint of the cell -- for part of your rooting medium. It may be sufficient to have only the upper 150 mm of the rooting medium comprised of a material that is highly favorable in terms of texture, organic content, and nutrients; this will support the germination and initial establishment of your selected plant community. You can then rely upon the roots of your first-generation plants to gradually transform the underlying material from its rather neutral or sterile condition into an ever-improving growth medium.

- View the use of a buried biointrusion layer, such as a loose-cobble biobarrier, as an extra measure of assurance against root and animal penetration beneath a well-engineered layer. If rock is readily available and will serve some other purpose (e.g., backup erosion protection or drainage), then go ahead and use it. But the cover should be engineered such that the rock layer is the secondary or redundant feature to resist biointrusion.

The implementation of these notions, and their integration with other considerations in cell design, is discussed in Chapter 17. Also presented in that chapter are specific examples, including cross-sections of covers designed and built with biointrusion resistance as genuine design objectives.

At the risk of sounding parochial, we believe that the best assurance of maximizing a cover's resistance to biointrusion relative to its many other performance objectives is to include a plant/soil ecologist as an integral member of the design team. With the above principles in mind, the ecologist will help the engineer to design with nature, not against it.

SECTION IV

MONITORING AND MAINTENANCE

Chapter 19

MONITORING PROGRAMS

GENERAL

Once the disposal cell or the encapsulation system construction is complete, it should be possible to walk away and leave the cell to start its long life in equilibrium with the forces of nature. In theory, if the cell is designed to last for 1,000 years and perform satisfactorily in the absence of human care or intervention, there is no need for the programs variously termed monitoring programs, or surveillance and maintenance programs.

Unfortunately, the logic of this proposition is not reflected in the laws and regulations that govern waste disposal cells and encapsulation systems. Multiple laws and, in particular, the regulators who enforce them, demand a full scale, and often very extensive post-construction monitoring program.

In spite of the logic that concludes that we should not monitor a cell after construction, we will, I know, continue to do so for as long as society is actively interested in a clean environment and in proper waste disposal. Justification for post-construction monitoring programs includes:

- The past record of waste facilities that monitoring confirms are not functioning properly

- A need and desire to gather scientific and engineering data about the performance of disposal cells

- The political and institutional need to gather data to confirm the safety of the system and establish protection of the environment, particularly groundwater

- The prudence of repairing failure or distress that does occur, and hence of prolonging the life of our disposal facilities.

This section contains three major chapters. The first discusses monitoring of the hydrologic performance of a disposal cell. The second discusses surveillance and maintenance systems to monitor the geotechnical performance of the cell. Ecological performance of encapsulation systems is described in the third section.

CHAPTER 20

HYDROLOGICAL PERFORMANCE MONITORING

General

This chapter describes and compares alternative systems that may be used to monitor the hydrological performance of a disposal cell. In particular, this report identifies technical approaches to monitoring infiltration through the cell's cover, seepage from the wastes that exit the cell via basal drains, and the migration of fluids (if any) through the cell's basal liners and foundation, to the vadose zone and the groundwater.

As shown in Figure 20.1, the chapter deals with:

- Cover infiltration monitoring
- Waste seepage monitoring
- Cell liner seepage monitoring
- Vadose zone monitoring
- Groundwater monitoring.

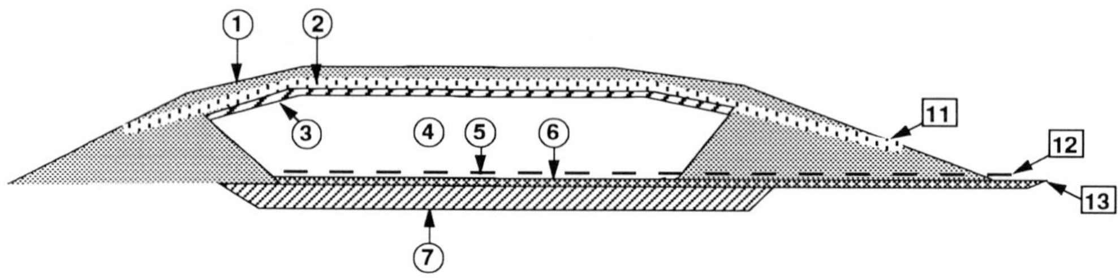

LEGEND

CELL COMPONENTS

1. COVER
2. DRAIN IN COVER
3. INFILTRATION BARRIER
4. WASTE
5. LEACHATE COLLECTION SYSTEM – — —
6. LEAK DETECTION SYSTEM
7. BASAL LINER AND FOUNDATION

CELL HYDROLOGIC PERFORMANCE MONITORING POINTS

11 COVER DRAIN FLOW RATE

12 LEACHATE COLLECTION FLOW RATE

13 LEAK DETECTION SYSTEM FLOW RATE

There are three primary points at which the water flow rates into and out of the cell may be measured.

FIGURE 20.1

STANDARD MONITORING POINTS FOR HYDROLOGIC PERFORMANCE

Approach

Figure 9.1 shows a schematic section through a disposal cell. The figure shows the three constituent parts of the disposal cell: namely, the cover, the vadose zone, and the groundwater. These constituent parts may be considered a system for purposes of analyzing the performance of the cell. Each of these constituent parts is comprised of a number of specific elements or components; for example, the cover may consist of an infiltration barrier, a drain, soil, and vegetation. The components function together as a system to achieve one or more objectives, such as, for the cover, impeding infiltration. In order to establish how the disposal cell performs, it is necessary to study the performance of the cell's constituent parts (the systems) and of the individual components that constitute the system. To quantify the performance of the cell and its constituent parts, one defines or establishes conceptual and analytical methods of the cell systems.

Using the performance model, the likely value (and feasible deviations) of selected parameters are calculated. For example, the quantity of water flowing in the cover drain, or the flow rate through the infiltration barrier, may be quantified.

Monitoring System Performance Requirements

The monitoring systems are used to measure, in the field, parameters that confirm that systems of the disposal cell are performing in accordance with design intentions and predictions. Alternatively, if the disposal cell and its constituent parts do not perform as designed and predicted, the monitoring systems discussed here should be able to detect deviations from design performance and provide ample opportunity for the implementation of corrective actions.

A disposal cell is designed to last up to 1,000 years and more. This gives rise to what may be termed the hydrological performance monitoring technical approach. There are three aspects of this technical approach:

- **Performance Models:** For the disposal cell and its constituent parts (systems), we define models that are used to quantify the performance of the system and hence the cell. The performance models establish a rational basis for selecting parameters to be measured in the field with monitoring instruments. Also, the performance models establish a framework for interpreting monitored parameters and assessing the adequacy of cell performance.

- **Access:** The cell must be designed so that now and in the future reasonable access is available to install hydrologic performance monitoring devices.

- **Instruments:** Those instruments that are currently available should be the basis of hydrologic performance monitoring systems. The selection of the instruments to be installed when the cell is constructed, and at any time in the future, should be based on the need to obtain data that when compared to the disposal cell performance model confirms anticipated cell behavior or indicates the need for corrective action, if any.

The subject of performance models is dealt with in Section III. This chapter concentrates on the issue of access, because consideration of and provision for access as part of the conceptual design of the cell may facilitate provision of suitable features in the constructed cell that will expedite instrument installation in both the short- and long-term. Instrument types are considered and discussed, but precise requirements for specific instruments are not considered. The technology of instruments is advancing rapidly. It is recognized that in the next fifty, and certainly the next 1,000 years, the instruments that may be used to monitor the hydrological performance of the disposal cell will change and advance. Accordingly, where appropriate, emphasis will be placed on access; it is better to provide access for future generations of engineers than to design disposal cell performance monitoring facilities to the needs and constraints of currently available instruments.

COVER PERFORMANCE MONITOR

Functional and Performance Requirements

While a small quantity of water may seep through the infiltration barrier into the cell, the quantity will be almost negligible in the short-term and may be difficult to quantify over the life of the cell. It is preferable not to monitor the cover performance, particularly with regard to infiltration. Rather it is better to rely on a sound design that is well constructed.

The cover controls infiltration to the cell and the wastes. A system may be installed to monitor the infiltration through the cover to the wastes. A cover infiltration monitoring system must function to measure parameters that enable the moisture flux through the cover to be estimated and quantified. The purpose of the monitoring is primarily to demonstrate that the cover is limiting the movement of moisture into the cell.

Figure 10.4 shows the performance model of the cover that establishes the basis of the functional requirement for the cover monitoring. This figure shows the water balance of the cover, and demonstrates that most incident precipitation runs off, is evapotranspirated, or shed through the cover drains. Only a small quantity seeps through the infiltration barrier into the cell. The cover infiltration monitoring facilities should therefore be able to measure the parameters that quantify the cover water balance, and also the flux of water through the infiltration barrier, as shown in Figures 10.9 and 10.13.

The performance requirements for the cover monitoring system are that the instruments must provide accurate and precise data to quantify the cover water balance and the infiltration through the infiltration barrier, the instruments must be replaceable if they fail, and the instruments must not damage the cover. Each technology must be simple to employ, use, and interpret.

The cover monitoring systems must provide for ready access to the various components of the cover to replace failed instruments or install new instruments that may be developed. Preferably, the cover infiltration performance monitoring system should be nonintrusive, i.e., it should not penetrate the cover. In particular, the cover monitoring system should be such that in the event of inadvertent or negligent abandonment of the cell, the disposal facility will continue to encapsulate the wastes and isolate them from the environment with no further human intervention.

Alternatives

Table 20.1 lists alternative cover infiltration monitoring systems. The alternative monitoring strategies incorporate fundamentally different approaches toward evaluating the movement of moisture through the cover.

In the first approach, the Darcy flux is inferred based on observed measurements of moisture content (θ) and matric potential (ψ) along profiles across the cell cover and, in particular, the infiltration barrier. To calculate the Darcy flux, the following relationships must be defined for the soils: the moisture characteristic curve, θ versus ψ, and the unsaturated hydraulic conductivity curves, K (θ) versus θ, and K (ψ) versus ψ.

The water budget approach incorporates meteorological data to estimate changes in precipitation (P), run-off (RO), evapotranspiration (ET), and drain flow (DF). The infiltration rate (R) through the infiltration barrier is calculated by subtracting RO, DF and ET from P, taking into account changes in moisture storage in the cover using matric potential measurements.

TABLE 20.1

ALTERNATIVE INFILTRATION BARRIER EVALUATION

EVALUATION CRITERIA	ALTERNATIVE	ADVANTAGES/ DISADVANTAGES
• Technical Suitability: Demonstrated technology, accuracy and precision	• Infer Darcy Flux (q) across the cover by measuring water content (θ), matric potential (ψ) and estimating θ versus ψ, K(θ) versus θ or K(ψ) versus ψ curves.	• Advantages - Can distinguish upward from downward movement - Can estimate spatial and temporal distribution of Darcy flux • Disadvantages - Requires redundant instrumentation in the cover - Requires abundant data/measurements/maintenance - Requires characterization lab/field data of the cover - Estimates of flux are prone to errors - Instrumentation can damage cover

TABLE 20.1

ALTERNATIVE INFILTRATION BARRIER EVALUATION

EVALUATION CRITERIA	ALTERNATIVE	ADVANTAGES/ DISADVANTAGES
• Longevity: Must operate during operational and maintenance phase, or be replaceable. If not, must plan for redundancy in instrumentation.	• Water Budget $R\downarrow = P - ET - RO \pm ds/dt$ where: R - recharge P - precipitation ET - evapotranspiration RO - run-off ds/dt - change in storage	• Advantages - Minimal instrumentation/damage of cover - Most of the measurements are collected away from the cell • Disadvantages - Requires abundant monitors and data to estimate ET - Will require ψ data to estimate ds/dt for cover - Measurements prone to errors - may not be sensitive to small changes - Difficult to demonstrate accuracy and precision
• Minimize damage to cover	• Suction lysimeters (moisture only)	• Advantages - Simple, cheap, do not require abundant instrumentation - Minimizes collection of data • Disadvantages - Cannot use redundant instrumentation - longevity is problem - May induce moisture movement - Used to measure velocity of a pollutant/tracer; is not a direct, quantitative measurement of flux

TABLE 20.1

ALTERNATIVE INFILTRATION BARRIER EVALUATION

EVALUATION CRITERIA	ALTERNATIVE	ADVANTAGES/ DISADVANTAGES
	• Tracers - salinity or conservative chemicals, combined with suction lysimeters EC probes and fiber optic optrodes	• Advantages - Same as for suction lysimeters - Can use redundant instrumentation • Disadvantages - Can only infer velocity of chemical, not moisture flux
	• Destructive tests	• Advantages - Provides detailed, one-time data, most accurate and precise • Disadvantages - Difficult to estimate temporal distribution - Can destroy the cover - Very laborious, very time consuming, stringent QA/QC - Errors during transport to lab
	• Instrument a replicate cover	• Advantages - No damage to cover, easy access - Can use all alternatives, supplies high quality data • Disadvantages - Does not actually measure the true cover performance

Suction lysimeters are an alternative to monitor moisture movement through the cover, although they are designed primarily to collect pore-liquid samples for chemical characterization. They do not produce quantitative estimates of moisture flux, only qualitative estimates.

The tracer alternative involves the use of soil-pore liquid samplers and/or probes designed to measure the presence of a tracer and thus infer the velocity of a chemical. These data can in turn be used to qualitatively estimate moisture flux. A tracer would conceivably be seeded into a portion of the cover. The tracer would be mobilized by the presence and passage of infiltering water.

In the destructive testing method, soil samples are collected at individual locations, at one point in time, and analyzed for moisture content and matric potential.

The final alternative involves replicating the cover in its final design within the cover and yet isolated from the functional portion of the cover. Into the replicate cover, a number of instrument stations may be installed that incorporate all or selected instruments. This is, in effect, a hybrid system or alternative. Figure 20.2 shows the layout of a feasible replicate cover monitoring networks. Several replicate covers could be instrumented to infer the distribution of the Darcy flux, estimate recharge through the cover via water budget calculations, estimate infiltration through the cover using tracers, and provide for destructive soil sampling. Alternatively, the flow in the drains and the flux through the replicate cover infiltration barrier may be measured directly by collecting and measuring flow into sumps and at the downgradient end of the replicate covers.

A variant of the replicate cover would be to make the whole cover a "replicate" cover. In effect, the detail of the replicate cover shown in Figure 20.2 would be placed over the whole cover. Seepage, if any, through the upper cover would be monitored in the lower drain, which would function as both a leak detection system and a second back up drain.

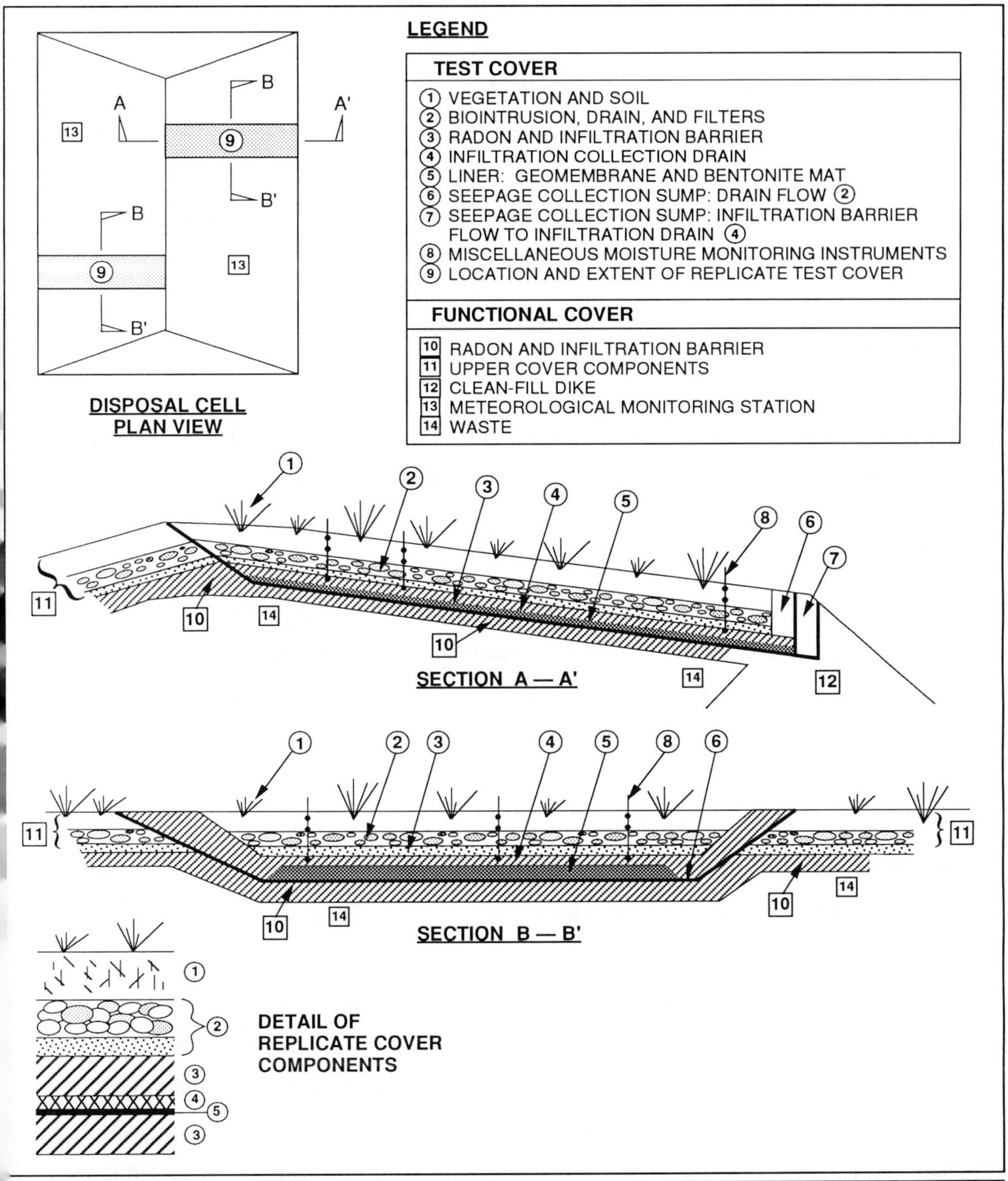

Construction on the disposal cell of an instrumented replicate cover facilitates accurate, nonintrusion cover performance monitoring.

FIGURE 20.2

COVER PERFORMANCE MONITORING - SCHEME I

Figure 20.3 shows a feasible layout for a replicate cover over the whole cell. Alternatively, this may be conceived of as a conventional multicomponent cover underlain by a leak detection system. The leak detection system would be a geogrid. This would be underlain by a layer of compacted soil or bentonite in a geomat; this would function both as a secondary infiltration barrier and the functional radon barrier.

Evaluation of Alternatives

Table 20.1 lists criteria that may be used to evaluate alternatives. Technical suitability is the most important criterion. Longevity and minimization of damage to the cover are of equal secondary importance.

Instruments in the cover are ranked lowest because they damage the cover. Although the water budget approach does not estimate moisture flux as accurately and precisely as the Darcy method, instruments are more easily replaced when they fail, and cover damage is reduced.

Destructive tests involve the most damage to the cover. Samples collected for this approach, however, provide the most accurate and precise data, including data which are used for calibration of the other monitoring instruments.

The best alternative is the replicate cover. Because its design replicates the actual cover, the monitor results are representative of the actual cover even though the replicate cover measures performance only at the precise location of the replicate and not globally over the whole cover. Several replicate covers can be instrumented to assess infiltration variability across the top cover of the cell. The instruments can be replaced if they fail, and damage to the functional portion of the cover is avoided.

DETAILS OF LEAK DETECTION SYSTEM BENEATH COVER

SCHEMATIC	COMPONENT	FUNCTION
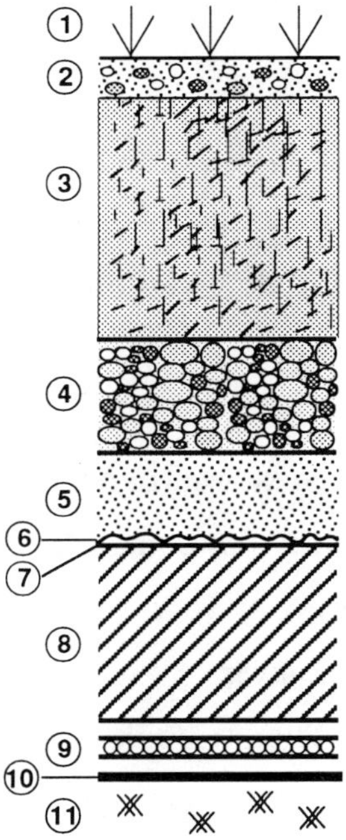	(1) VEGETATION	Erosion, Evapotranspiration
	(2) GRAVELLY SOIL	Erosion Control
	(3) RANDOM SOIL	Freeze/Thaw Protection, Rooting Zone
	(4) COBBLES	Biointrusion
	(5) DRAIN SAND	Drain, Bedding
	(6) GEOMEMBRANE	Infiltration (Short-term)
	(7) BENTONITE (IN GEOMAT)	Infiltration (Long-term)
	(8) CLAY/SILT	Infiltration, Radon Control
	(9) GEOGRID	Leak Detection, Water Conveyance
	(10) BENTONITE (IN GEOMAT)	Water Control, Radon Control
	(11) WASTE	

ANALYSIS OF LEAK DETECTION SYSTEM BENEATH COVER

ADVANTAGES	DISADVANTAGES
• Complete Monitoring of Cover Infiltration Control Performance. • Additional Components to Shed Water and Impede Infiltration.	• Bentonite Underliner Is Primary Radon Barrier, And Theoretically Sufficient. This May Be Difficult To Substantiate. • Difficult To Control Radon Flux From Cell Through Geogrid (Although Flux May Be Small). • Bentonite Underliner May Soak Up Water In Geogrid, Hence Distorting Monitoring Cover Water Flux.

If we accept that a layer of bentonite in a geomat will attenuate radon, a leak detection system beneath the cover is an advantageous system.

FIGURE 20.3

LEAK DETECTION BENEATH COVER: COMPARATIVE ANALYSIS

The criteria for selecting the number and locations of the replicate covers depend primarily on the construction of the cell, and also on trade-offs between numerous, smaller (by area) and fewer, larger replicate covers.

In addition, each replicate cover must be adequately instrumented to obtain representative estimates of moisture movement. Just as the cell can be subdivided into two or eight representative areas, each replicate cover can be subdivided such that representative sample/monitor locations are uniformly distributed across the cover.

Figure 20.3 lists the advantages and disadvantages of a leak detection system beneath the radon/infiltration barrier. The most significant advantage is that it provides complete, real time monitoring of the performance and integrity of the cover. A very significant disadvantage is that the functional radon barrier is no longer the thick layer of silt and clay, but the thin layer of bentonite in the geomat. Distress of the geomat could lead to increased radon flux from the cell.

To avoid the disadvantages associated with the geomat and geogrid, the geomat could be replaced with a second layer of compacted silt and clay. The geogrid could be replaced by a second layer of sand. In effect, the system normally used at the base of waste, a double liner and drain system with the lower drain acting as a leak detection system, is transferred to the cover. This is logical; the cover is unquestionably the first and most important cell component, limiting water inflow to the waste from the cell and to the environment. It may be appropriate to place the double liner and leak detection system above the waste and not below it. It may be appropriate to detect leaks before the water reaches the waste, not after it has passed through the waste and the primary liner.

Details of the Replicate Cover

A replicate cover can be constructed and instrumented to provide representative, accurate and precise data. If the cover is to be instrumented, then fiber optics, neutron probes, and time domain reflectometry (TDR) combined with a SEAMIST system (Keller 1991, Kramer et al. 1991) could be used to estimate infiltration. The results could be checked against water balance results and destructive testing, if necessary. This alternative offers the best demonstrated, technically suitable technology without damage to the functional portion of the cover. Instruments associated with this alternative facilitate adequate data collection and can be replaced if they fail.

Figure 20.1 shows how the preferred alternative, the replicate cover, can be incorporated into the final cover design. Two sections through the cover demonstrate that replicate covers can be constructed and integrated into the cell such that the replicate cover functions and responds as the actual cover, yet it is physically separated from the cover. Shown on Figure 20.4 are moisture content monitors (neutron access probes, TDR probes, moisture blocks, heat dissipation probes) and matric potential monitors (tensiometers, psychrometers, tension blocks, thermocouples) along vertical profiles in boreholes advanced across the cover to infer the Darcy flux. These monitors can be replaced when they fail.

Finally, soil samples can be collected at any point within the replicate cover by auguring through the cover and grouting the borehole to meet the density specifications of the compacted cover soils.

Separate sumps will be required to collect: (a) the flow in the drain above the replicate cover infiltration barrier; and (b) seepage through the replicate cover infiltration barrier to

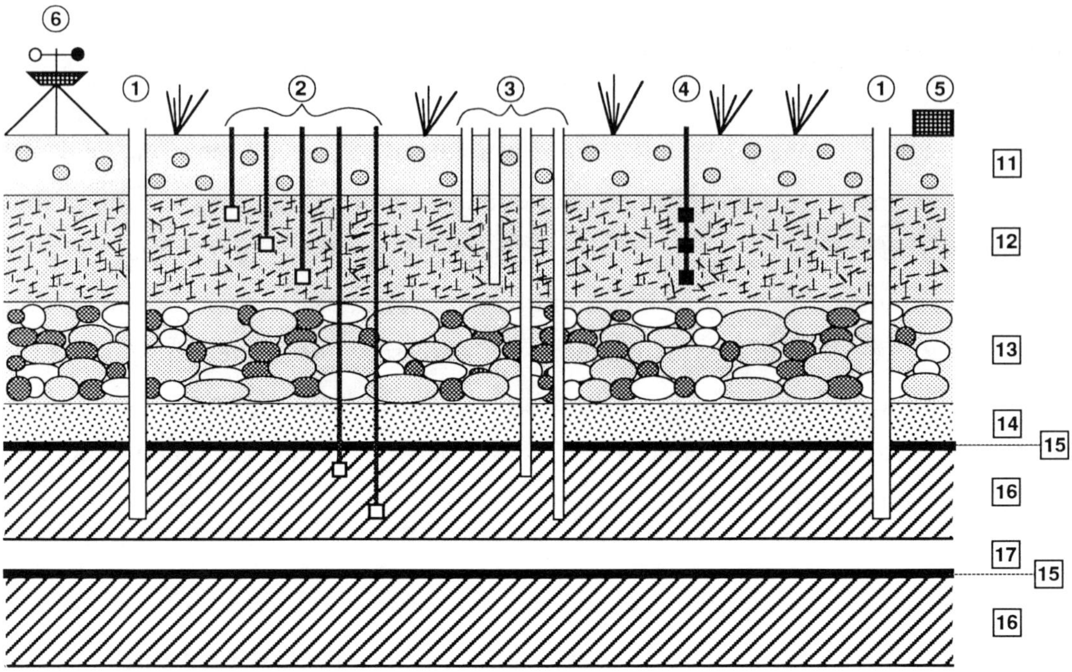

LEGEND

INSTRUMENTS

1. NEUTRON PROBE ACCESS TUBE
2. GYPSUM BLOCKS
3. TENSIOMETERS
4. THERMOCOUPLES
5. ELECTRONIC DATA LOGGER
6. WEATHER STATION

REPLICATE COVER COMPONENTS

11 SOIL WITH GRAVEL & VEGETATION
12 RANDOM SOIL
13 GRAVEL AND COBBLES – BIOINTRUSION
14 SAND DRAIN
15 BENTONITE IN GEOMAT
16 CLAY & SILT – INFILTRATION BARRIER
17 GEOGRID – INFILTRATION MONITORING

Various instruments may be installed in the replicate cover to measure soil moisture, soil suction and temperature.

FIGURE 20.
REPLICATE COVER INSTRUMENT STATION

the underlying collection layer. Some type of valved outlet from both the drain and the collection layer to their respective sumps may be required to preclude the need for continuous pumping from the sumps, or overflow at the surface. The replicate cover performance monitoring program should provide for regular sump emptying so that back up of water in the drains does not disturb replicate cover infiltration rates.

Figure 20.5 shows the monitoring and performance evaluation criteria for both the replicate cover and a functional cover underlain by a leak detection system. If the flow from the geogrid (or secondary drain) is less than an effective 10^{-7} $cm^3/cm^2/sec.$ of cover area, the cover may be said to be functioning satisfactorily as an infiltration barrier. If the effective flow exceeds this, increased cover inspection should be undertaken. If flow effective infiltration rates greater than 10^{-6} $cm^3/cm^2/sec.$ are measured, the source of the failure should be located and repaired.

The instruments installed in the replicate cover during construction will probably fail (or cease to function satisfactorily) within the first fifty, and certainly the first 200 years of the cell's design life.

As they fail, new instruments may be installed in the replicate cover. If new, technologically advanced instruments become available during the cover performance monitoring period, they may be installed in the replicate cover, either to replace or augment the instruments installed during construction. Failed instruments may also be left in place with proper abandonment procedures.

If the seepage collection drain clogs, or otherwise ceases to function, the complete replicate cover could be removed and replaced. Alternatively, additional instruments may be installed, or soil samples may be collected as part of a destructive testing program.

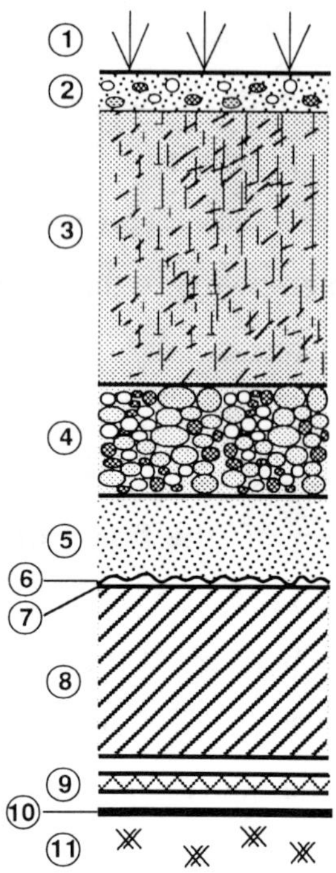

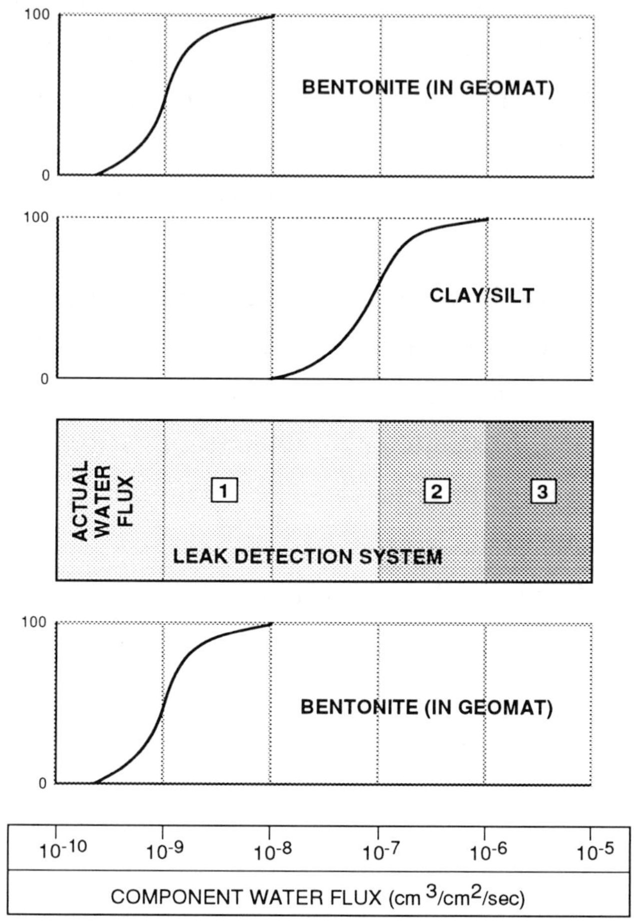

LEAK DETECTION SYSTEM MONITORED FLOW RATES

[1] Satisfactory Cover Infiltration Performance

[2] Marginal Cover Infiltration Performance/Inspect Cover

[3] Cover Failure - Locate And Repair Failure Zone

COVER COMPONENTS

(1) Vegetation
(2) Gravelly Soil
(3) Random Soil
(4) Biointrusion Cobbles
(5) Drain Sand
(6) Geomembrane
(7) Bentonite Layer (In Geomat)
(8) Clay Silt
(9) Geogrid Leak Detection System
(10) Bentonite Layer (In Geomat)
(11) Waste

The cover infiltration control system performance/failure may be established by monitoring the rate of flow from the geogrid.

FIGURE 20.5

COVER LEAK DETECTION SYSTEM: PERFORMANCE/FAILURE CRITERIA

If monitoring the replicate cover performance indicates a malfunction or failure of the replicate cover, then the actual cover itself will be visually inspected to determine if analogous conditions occur on the actual cover. The cause of malfunction or failure of the replicate cover (e.g., excess infiltration) shall be established. This may be done by examination of replicate cover performance records, by installing and monitoring new instruments, or by destructive testing of the replicate cover (e.g., digging holes into the various cover components to quantify their actual as-placed and successive characteristics).

WASTE SEEPAGE MONITORING

Functional and Performance Requirements

The function of a waste hydrologic performance monitoring system is to obtain data to establish that the seepage quantities and quality are as anticipated, and that excessive flows are not entering the waste from a malfunctioning cover. The waste monitoring system should facilitate location of the source of excess or unexpected seepage from the waste or through a malfunctioning cover. Figure 11.1 shows the predicted seepage rates from the waste due to transient drainage and infiltration through the cover. The waste monitoring system must be capable of measuring the variation of flow rates with time.

Access to the monitoring system should be simple so that regular measurements may be taken. The systems that may be used to monitor the hydrological performance of the waste are:

- Boreholes through the cover into the waste; instruments to measure waste moisture content, soil suction, etc., may be installed in the boreholes. The boreholes may be vertical, inclined, or horizontal, as shown in Figure 20.6.

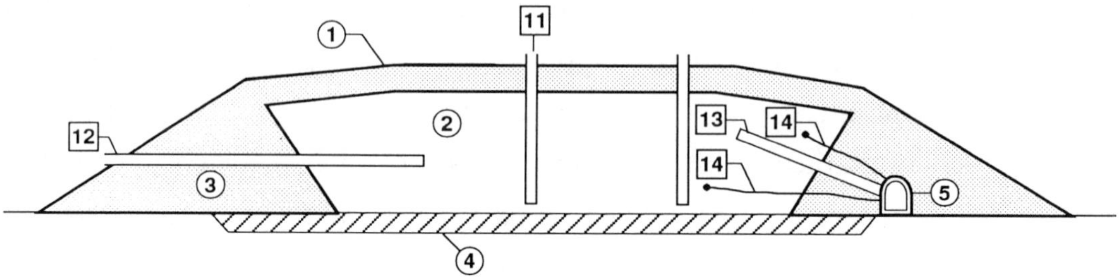

CELL SECTION SHOWING ALTERNATIVE MONITORING BOREHOLES AND INSTRUMENT INSTALLATIONS

LEGEND

CELL COMPONENTS

(1) COVER
(2) WASTE
(3) DIKE
(4) LINER & FOUNDATION
(5) ACCESS GALLERY

MONITORING INSTRUMENTS

[11] VERTICAL BOREHOLES
[12] HORIZONTAL BOREHOLES
[13] INCLINED BOREHOLES
[14] PNEUMATIC OR ELECTRIC LEADS

Boreholes and instrument leads may be used to provide access to the waste to monitor performance parameters. All breach the integrity of the encapsulation system.

FIGURE 20.6

WASTE PERFORMANCE MONITORING

- Instruments installed at placement in the waste; the leads to such instruments would be flexible pneumatic or electric tubes and wires, and they would be led to a convenient access port, see also Figure 20.6.

- The drain beneath the wastes; this acts both as a leachate collection and removal system and as a system at which waste seepage quantities and quality may be monitored. Figure 7.5 shows schematically alternative layouts for drains that may be used as seepage monitoring systems.

Evaluation of Alternatives

Boreholes into the waste are a distinct breach of the encapsulation system. As shown in Figure 10.20, they constitute routes by which excess water may enter the cell and from which excess radon gas may leave the cell. Inclined boreholes from perimeter access galleries in the clean-fill dikes are not as likely to be water ingress routes; they may, however, be radon egress routes.

Because the waste, in most cases, is heterogeneous, many instruments may be required to fully characterize flow in the waste. Instruments installed in the waste will not function for very long (5 to 20 years). Replacement is difficult, if not impossible. They provide only point readings at a limited number of discrete locations.

Drains beneath the waste provide complete, integrated coverage and monitoring of seepage from the waste. They are likely to function for a long time. Provided the underlying liner functions properly (i.e., remains intact with a very low permeability), the quantities monitored in the drains are an accurate measure of flow from the waste.

A drain system with a single outlet from the cell enables one to measure only the total flow from all the waste. By comparison, a drain layout with multiple outlets facilitates sensitive and accurate monitoring of the hydrologic performance of all parts of the cell. In the event of cover failure or unanticipated excess flows from the waste, the location of cover failure or other malfunction may be more readily identified with a drain system incorporating multiple outlets.

Figure 7.6 shows alternative monitoring drain layouts that account for site topographic constraints. The figure compares the schemes and notes the significant advantages associated with the system that has at least four individual drain outlets.

If different waste forms are placed at different locations in the cell, flow through the wastes may not be vertically downward. Considerable intermingling of flow from the different waste forms may occur, and significant lateral dispersion of flow within the wastes is possible. This will make it difficult to pinpoint the precise entry point of excess cover infiltration or the precise source of unanticipated excess flow from the waste.

Some of the water flowing in the drains may seep into the underlying liner. In particular, if the waste seepage rates are very low (less than 10^{-7} $cm^3/cm^2/sec.$), as they may well be, no lateral drain flow will occur, and it will not be possible to quantify waste seepage rates.

Blocking of the drain by physical and chemical processes may occur. If the drain system is designed to be totally interconnected, and to have more than one outlet through the dikes from the cell, such blocking will not render the whole system useless. It will, however, complicate interpretation of monitored drain flow rates.

Performance Evaluation Criteria

Provided the cumulative flow from the cell does not exceed rates predicted by performance assessment modeling, the waste may be said to be performing (hydrologically) satisfactorily.

Figure 10.16 shows the probable flow distributions through the cover, the waste, and the leak detection system for a feasible cell. Provided the flow from the various entities is less than the values indicated, the cover and the wastes may be considered to be performing adequately.

Potential Corrective Actions

If the drains clog or otherwise block as a result of physical or chemical action, no corrective action is possible. The outlet pipes may be cleaned if they are geometrically arranged to facilitate access (another advantage of multiple outlets).

If the cover fails, causing excess flows from the waste to be monitored, the location and cause of cover failure must be ascertained, and appropriate corrective action implemented.

In the event that excess flows from the waste (not due to cover infiltration) are monitored, no corrective action will be undertaken. In that the source of such flows is limited and excess flow is perforce of finite duration, the flow will be collected and treated as appropriate.

LINER AND FOUNDATION SEEPAGE MONITORING

Functional and Performance Requirements

Many regulations mandate a leak detection system beneath the liner that is placed beneath the leachate collection and removal system.

The hydrological performance monitoring systems must function to:

- Quantify leakage, if any, through the primary liner from the drain (leachate collection and removal system).

- Quantify and provide early warning of seepage from the cell to and through the lower liner and foundation to the vadose zone.

Figure 12.1 shows the conventional location of a basal leak detection system. Figure 20.7 shows alternative locations for leak detection systems. In particular, note that in addition to a leak detection system between a primary and secondary line, as shown in Figure 12.1, a leak detection system could be installed:

- Between the foundation soil and the lowermost liner of the cell, or

- Beneath the foundation layer.

If the disposal facility has clean-fill dikes, as shown in Figure 20.8, an access gallery may be constructed in the dike. The leachate collection and leak detection outlet systems may be led into the gallery.

Evaluation of Alternatives

A leak detection system between the primary and secondary liners is standard industry practice. The advantage is that it provides direct and early information about the performance of the primary liner. In addition, it acts as a backup or redundant leachate collection and removal system. The disadvantage is that it provides no information about the performance of the lower layers and hence the cell as a whole.

A leak detection system beneath the lowermost liner and the foundation has the advantage that the number of layers at the base of the cell is reduced. This location provides timely

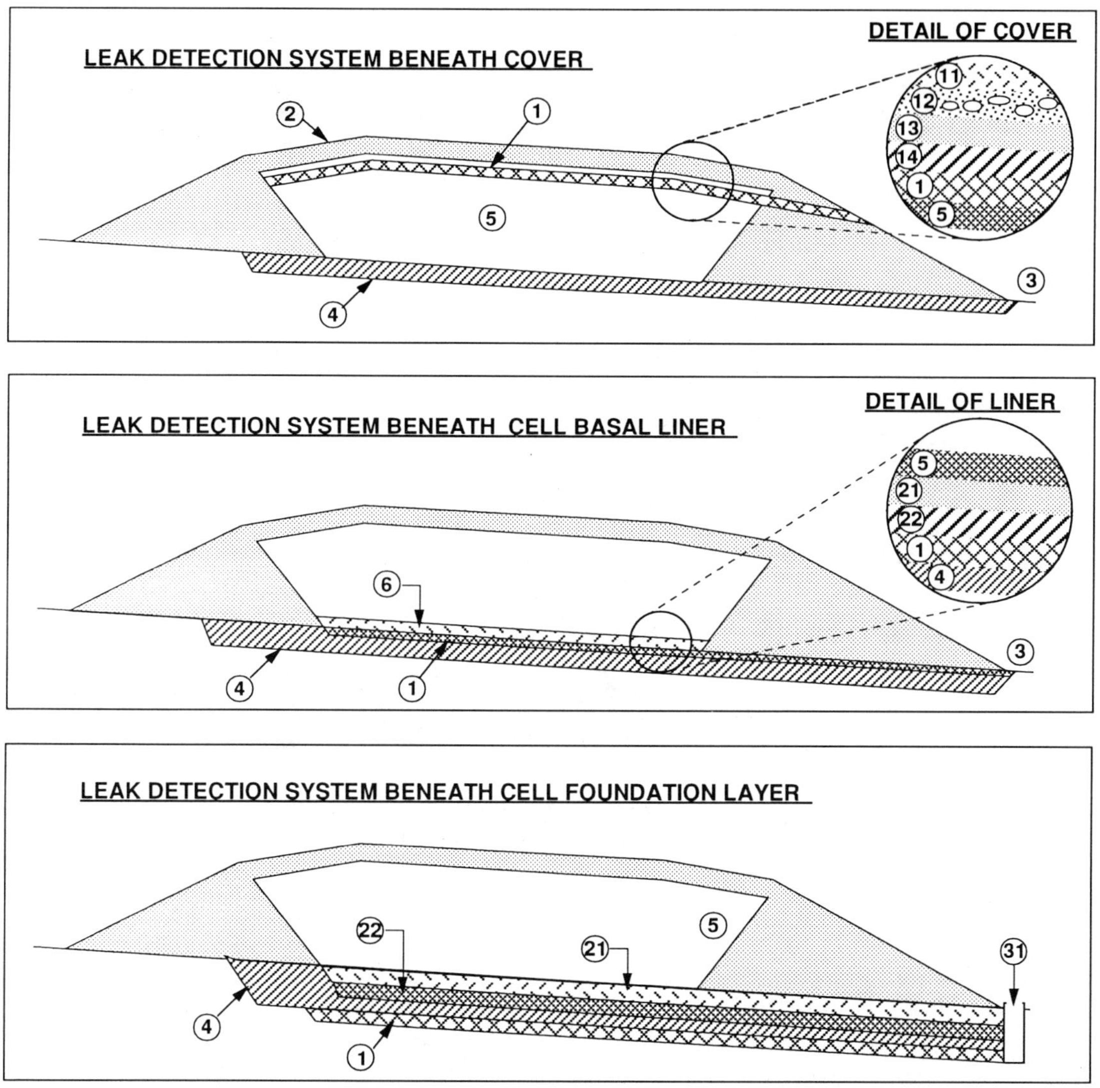

LEGEND

(1) LEAK DETECTION SYSTEM
(2) COVER-ALL COMPONENTS
(3) CLEAN-FILL DIKES
(4) FOUNDATION
(5) WASTE
(6) SOIL
(11) SOIL
(12) BIOINTRUSION
(13) DRAIN
(14) INFILTRATION/RADON BARRIER
(21) LEACHATE COLLECTION SYSTEM
(22) BASAL LINER
(31) SEEPAGE MONITORING WELL

To provide additional data about cell performance, additional leak detection systems may be installed.

FIGURE 20.7
LEAK DETECTION SYSTEMS: ADDITIONAL ALTERNATIVE LOCATIONS

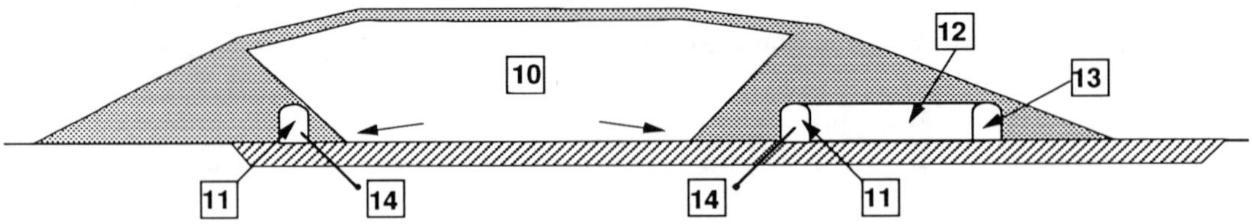

CROSS SECTION OF CELL SHOWING PERIMETER OBSERVATION GALLERY

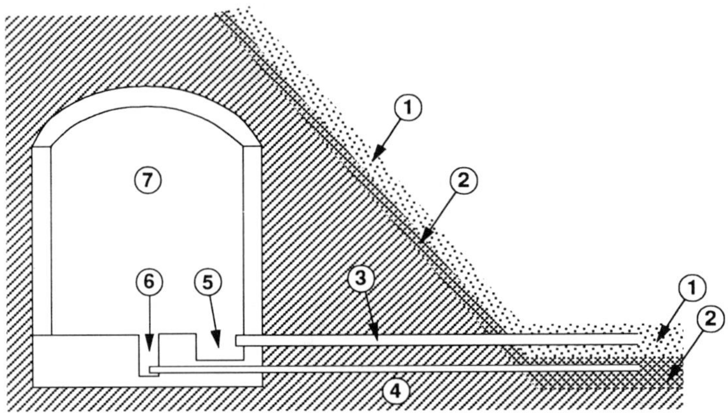

DETAIL OF PERIMETER OBSERVATION GALLERY AND DRAIN MONITORING SYSTEM

LEGEND

CELL LEACHATE CONTROL

1. LEACHATE COLLECTION SYSTEM
2. LEAK DETECTION SYSTEM
3. LEACHATE COLLECTION PIPE
4. LEAK DETECTION PIPE
5. LEACHATE MONITORING AND FLOW CHANNEL
6. LEAK DETECTION MONITORING WELL
7. PERIMETER OBSERVATION GALLERY

CELL & GALLERY COMPONENTS

10 WASTE
11 PERIMETER OBSERVATION GALLERY
12 ACCESS PORTAL
13 ACCESS RESTRICTION (AND RADON GAS CONTROL STATION)
14 VADOSE ZONE MONITORING INSTRUMENTS

A perimeter observation gallery may be used to collect leachate, monitor leakage and install and operate a vadose zone monitor system.

FIGURE 2

LEACHATE COLLECTION AND DRA MONITORING SYSTEM - SCHEM

information about the performance of the cell and its liners. It may also function as a backup seepage collection and removal system.

A leak detection system beneath the liner may be monitored and controlled, either as shown in Figure 20.7 or as shown in Figure 20.9. This figure notes the advantages and disadvantages of such a system. Figure 10.16 establishes the performance monitoring and evaluation criteria for this system.

VADOSE ZONE MONITORING

Functional and Performance Requirements

Vadose zone monitoring is not currently mandated by any laws or regulations. Vadose monitoring may, however, be undertaken if warranted by site- and project-specific circumstances and requirements.

The function of a vadose zone monitoring system, if used, is to obtain data to substantiate that the disposal cell and, in particular, its basal liners are performing as designed. If in fact the cell and basal liners are not performing as designed and anticipated, then the vadose zone monitoring system, if used, should function to identify a malfunction and, to the extent feasible, provide data on the extent and nature of the malfunction.

The vadose zone monitoring system, if used, must function to detect changes in moisture movement and contaminant migration from the cell and through the soils beneath the cell.

Also, as shown in Figure 10.7, seepage of clean water from the dikes and even the regraded area around the cell will seep into the vadose zone. A vadose zone monitoring system, if

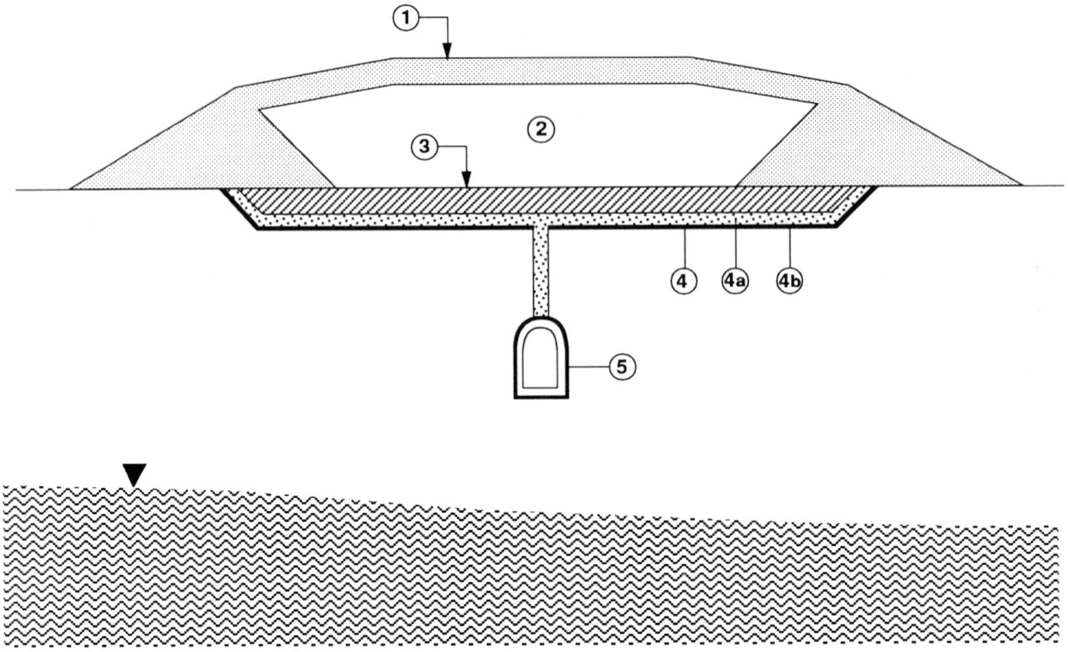

ANALYSIS OF LEAK DETECTION SYSTEM BENEATH CELL FOUNDATION

ADVANTAGES	DISADVANTAGES
• Complete, real time cell performance monitoring. • Additional drainage layer to remove cell seepage. • Additional seepage barrier and liner to protect vadose zone. • Gallery may be used also for vadose zone monitoring.	• Additional redundant system. • Gallery is a focus of potential failure to affect cell integrity. • Sensitive only to relatively large flows: bentonite layer may soak up cell seepage.

LEGEND

(1) COVER
(2) WASTE
(3) BASAL LINER & FOUNDATION
(4) LEAK DETECTION SYSTEM
(4a) SAND DRAIN
(4b) BENTONITE LAYER (IN GEOMAT)
(5) GALLERY FOR ACCESS AND MONITORING

A leak detection system beneath the cell may function also as a backup leachate control and vadose zone protection system - but could impair cell integrity.

FIGURE 20.

LEAK DETECTION SYSTEM BENEATH CEL
COMPARATIVE ANALYS

used, must be able to distinguish between the effects of seepage from the wastes, the dikes, and the surrounding area.

To the extent reasonable, the vadose zone monitoring system must provide for ready access to the base of the cell and the soils beneath the cell. Preferably, the vadose zone monitoring system should be nonintrusive, i.e, should not go through the disposal cell. In addition, the access system should be such that its design life is the same as that of the disposal cell; in the event of inadvertent or negligent abandonment of the cell, the access facilities will not be the focus of failure that could undermine or impair the unattended security of the cell as an encapsulation and isolation system.

There are three main aspects or issues associated with monitoring the vadose zone: ACCESS to install instruments; the INSTRUMENTS used to measure physical parameters; and the MODELS or analytical approaches used to interpret measured data.

Table 20.2 lists alternative procedures for gaining access to the vadose zone to install instruments or monitor potential seepage from the cell to the vadose zone. Figure 20.10 shows details of these. Physical access to the vadose zone may be obtained by boreholes through the cell or inclined from the perimeter of the cell. Horizontal access borings could be installed as shown in Figure 20.11. Alternatively, an access gallery may be constructed within the clean-fill dike or at the water table; from the gallery, boreholes may be advanced into the vadose zone.

Prior to instrumenting the vadose zone, soil samples need to be analyzed to determine the moisture characteristic curve and the unsaturated hydraulic conductivity curve. These relationships provide the necessary data for estimating moisture flux.

TABLE 20.2

ALTERNATIVE INFILTRATION: VADOSE ZONE

ALTERNATIVES	ADVANTAGES/DISADVANTAGES
• Access gallery in clean-fill dike	• Advantages - Can sample perimeter of cell - May result in less cell damage - Easily accessible and integratable • Disadvantages - Do not sample beneath cell - Mainly samples clean-fill dike zone
• Place in recompacted soil in excavated trenches	• Advantages - Simple, easy, and conventional - Soil properties well known • Disadvantages - One-time access only - Instrument leads may be very long - Instrument leads a built-in defect in cell foundation
• Access gallery at water table	• Advantages - Can be used to access entire vadose zone - Can be used for public acceptance - Can be integrated with groundwater monitor network • Disadvantages - May damage cell due to settlement - Very expensive - Requires abundant maintenance
• Directional boreholes	• Advantages - Can access areas beneath the cell from ground surface or underground gallery - Can be used for wide variety of instruments • Disadvantages - May damage cell - Difficult to install instruments and abandon - Does not provide effective access to sub-cell area
• Boreholes advanced through the disposal cell	• Advantages - Can monitor in cell and beneath cell - Easy access to borehole • Disadvantages - Damage to entire cell - May cause leakage into and out of cell - May cause cell failure

NOTE: No preferred alternative is designated. This is because the recommended approach is not to monitor the vadose zone. If mandated by law, then an access gallery construction in the contaminated soil removal trench at the south end of the cell is preferred.

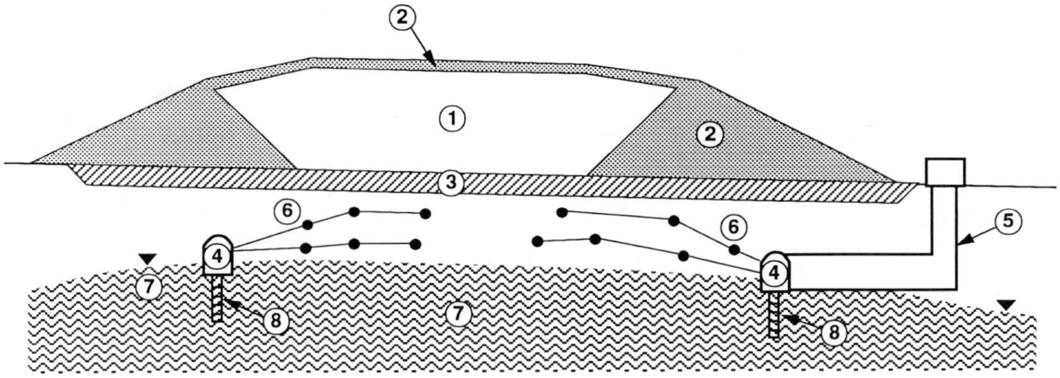

LEGEND

(1) WASTE
(2) CONTAINMENT SYSTEM
(3) LINERS AND FOUNDATION
(4) SUBSURFACE ACCESS GALLERY
(5) ACCESS SHAFT AND PORTAL
(6) VADOSE ZONE MONITORING INSTRUMENTS
(7) REGIONAL GROUNDWATER
(8) WATER QUALITY MONITORING WELLS

A subsurface access gallery beneath cell at groundwater table facilitates both vadose zone and groundwater monitoring.

FIGURE 20.10

VADOSE ZONE AND GROUNDWATER MONITORING ACCESS - SCHEME I

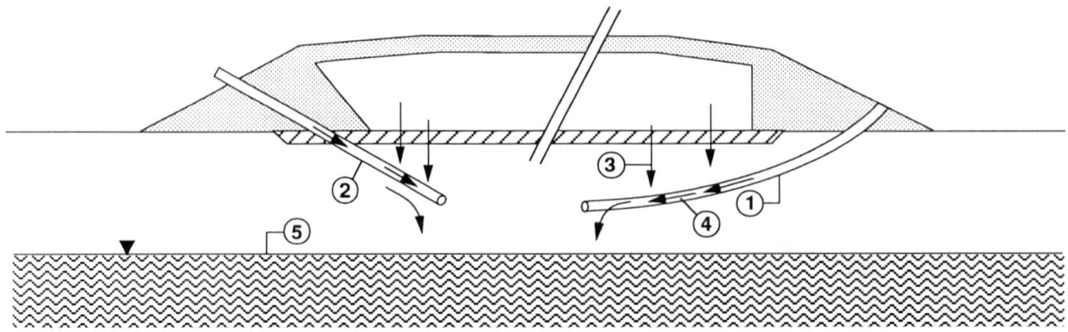

LEGEND

1. HORIZONTAL BOREHOLE
2. INCLINED BOREHOLE
3. SEEPAGE FROM THE CELL
4. PREFERENTIAL FLOW IN BOREHOLE VOID
5. GROUNDWATER

The void formed by an instrument borehole may become a preferential flow path for seepage from the cell, negating the impeding, protective role of the subsoils.

FIGURE 20.11

INSTRUMENT BOREHOLE FAILURE SCENARIO

Once access is obtained to the vadose zone and the moisture characteristic curve and unsaturated hydraulic conductivity curve are defined, instruments must be used to measure one or more of the following: soil moisture content, soil suction, and soil-pore liquid chemistry. Instruments that are currently available to do this are listed in Table 20.3.

Data measured with instruments must be interpreted to characterize the movement of moisture or contaminants from the cell through the vadose zone. This involves definition of a model of the system that constitutes the vadose zone. Alternative models and analytical approaches include: estimates of volumetric moisture flux using Darcy's law, unsaturated flow using the Richard's equation, infiltration estimates using the water balance approach, and qualitative estimates of moisture movement using tracers.

Evaluation of Alternatives

Criteria

Effects on cell integrity and interference on the natural system are considered the most important criteria to evaluate or select a vadose zone monitoring system. Consequently, the preferred approach is a nonintrusive technique that is practical, technically suitable, feasible, and has demonstrable longevity.

Alternative Access Approaches

The primary alternative approaches to accessing the vadose zone are installation of instruments in the vadose zone soils prior to cell construction, an access gallery in the clean-fill dike, an access gallery at the water table, directional boreholes, and boreholes advanced through the disposal cell. Any one or more of the instruments discussed above

TABLE 20.3

INSTRUMENTS TO MONITOR SOIL CONDITIONS FOR CELL HYDROLOGIC PERFORMANCE

MOISTURE CONTENT	MATRIC POTENTIAL	SOIL-PORE LIQUID CHEMISTRY
• Neutron sources and detectors • Gamma-ray sources and detectors • Gypsum and fiberglass blocks • Heat dissipation probes • Time-domain reflectometry (TDR) instruments • Hybrid neutron source/TDR probes • Fiber-optic optrodes	• Tensiometers • Transiometers • Thermocouple psychrometers • Heat dissipation probes • Gypsum and fiberglass blocks • Vibrating wire-line piezometers • Fiber-optic optrodes	• Suction samplers - Vacuum lysimeters - Pressure-vacuum lysimeters - High-pressure-vacuum lysimeters - Filter tip samplers • Experimental suction samplers - Cellulose-acetate hollow-fiber samplers - Membrane filter samplers - Vacuum plate samplers - Barrel lysimeters • Experimental absorption samplers - Sponge samplers - Ceramic rod samplers • Fiber-optic optrodes • Electrical conductivity (EC) probes • pH

may be placed in these, or from these, access portals or facilities. These alternatives are compared in Table 20.2.

An access gallery in the clean-fill dikes may function both as potential location from which to install instruments into the vadose zone and as a place to conveniently collect seepage from the cell drainage system. In addition, the leak detection system may be lead directlyinto such an access gallery. In reality, monitoring the performance of the leak detection system in a discrete, proximate manner is a practical and appropriate alternative to monitoring the more remote, inaccessible vadose zone.

Alternatively, an access gallery could be installed beneath any part of the disposed cell at any convenient location in the vadose zone. In particular, an access gallery could be constructed in the trenches excavated to remove contaminated soils and other facilities; the gallery could be constructed prior to trench backfilling. An access gallery at the water table could function both as a groundwater and a vadose zone monitoring system.

Access galleries at any location in the clean-fill dikes, beneath the cell, or at the water table would be very expensive and difficult to maintain. Problems may include build up of radon in the gallery and hence a health risk to those operating the monitoring system. Inadvertent or negligent abandonment of the galleries or inappropriate closure of the galleries might cause a situation that would lead to significant adverse impact on the integrity of the disposal cell and impairment of its ability to function unattended as an encapsulation and isolation facility. As shown in Figure 20.12, failure of an improperly closed gallery could severely impact the integrity and performance of the disposal cell.

Directional boreholes have a very limited range. Their range is restricted by the geometry of the cell and the clean-fill dikes; they are not able to reach all parts of the vadose zone. It may be difficult to install instruments in inclined boreholes.

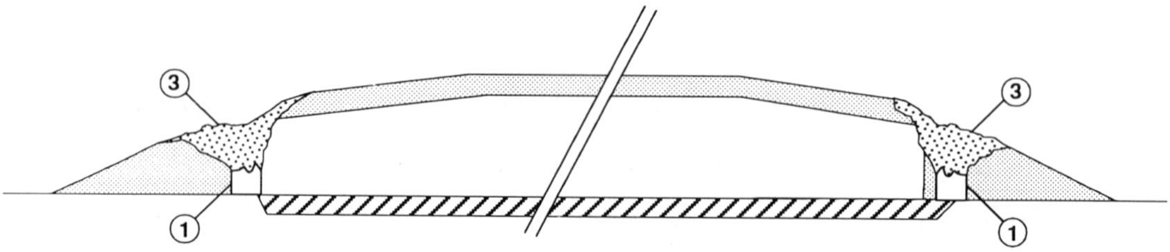

ALTERNATIVE GALLERY LOCATION
COLLAPSE SCENARIO CONSEQUENCES

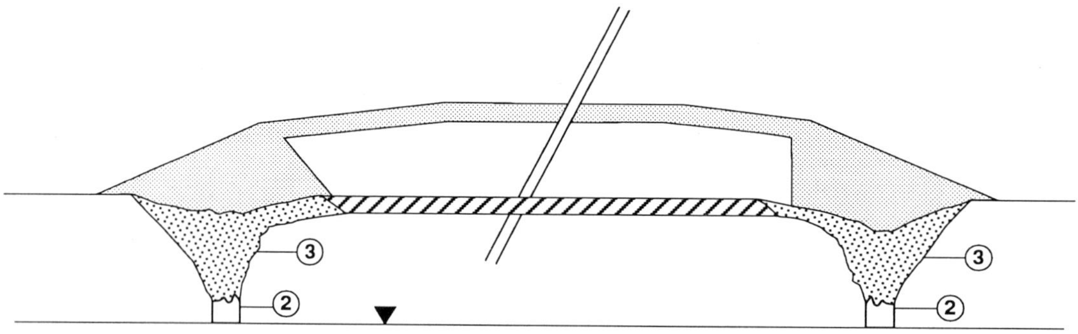

ALTERNATIVE DEEP GALLERY LOCATION
COLLAPSE SCENARIO CONSEQUENCES

LEGEND

(1) DIKE GALLERY – COLLAPSED
(2) DEEP GALLERY – COLLAPSED
(3) AFFECTED ZONE OF CELL AND WASTE

Collapse of the roof of the gallery could cause deformation of the cell and disruption of the waste.

FIGURE 20.12
GALLERY FAILURE SCENARIOS

Boreholes through the disposal cell negate the objectives of the cell: encapsulate wastes, isolate waste, reduce infiltration, and limit radon flux. Instrument access holes through the cover into or through the wastes are potential paths for increased infiltration, for radon emanation, and inevitably a focus of cell and cover distress. It is difficult to conceive how they can be sealed for closure or cell abandonment. Conventional sealing with bentonite grout at closure cannot assure a positive seal. In the event of inadvertent or negligent abandonment, the penetrating holes may be left open to continue as paths for free migration of air and water to and from the wastes.

Figure 20.10 shows that a complete leak detection system may be placed beneath the zone of compacted fill before it is placed. The leak detection system could be arranged to flow to an access gallery beneath the cell. As noted on the figure, the advantages of this approach include potential real-time monitoring of the cell and an additional redundant layer to preclude the passage of seepage to the vadose zone. The disadvantages are, however, significant. In particular, the potential impact of improper closure must be considered.

Alternative Instrument Approaches

Instruments to monitor the soil moisture content or the soil matrix potential may be installed from boreholes into the vadose zone. None of the alternatives listed in Table 20.3 to obtain such measurements have significant longevity. They all involve difficulties of placement, sensitivity, and operation control. If the vadose zone, as expected, is very close to or at saturation, these instruments may not be sufficiently sensitive for practical use.

If the soil is near or at saturation, it may be feasible to monitor or obtain samples of the pore fluid. The chemistry of the pore fluid may be monitored in place, or a sample may be collected and tested in a laboratory. In-situ sampling of soil-pore fluid may be done with experimental suction samplers, experimental absorption samplers, and free-drainage samplers (Dorrance et al. 1991). Chemical analyses of the soil-pore liquids may provide an early warning of potential groundwater pollution. Disadvantages include alteration of the liquid samples by the sampling process.

Tracers may be placed in the soils in the bottom of the cell to demonstrate the movement of moisture and chemically-similar constituents through the liner and into the vadose zone. If temperatures are high, most tracer dyes may be destroyed. Tracers provide only a one-time measure of moisture movement.

Destructive tests, i.e., physically taking a sample of the soil from a borehole, provide samples for chemistry, moisture content, and matric potentials measurement. This process involves the disadvantages of interfering with the natural system, affecting the cell integrity, and preventing repetitive sampling from the same location.

GROUNDWATER MONITORING TECHNICAL APPROACH

Functional and Performance Requirements

It is standard practice at most waste disposal facilities to undertake groundwater monitoring. This is done because groundwater monitoring is a technically practicable way to monitor the performance of a disposal facility. Moreover, most regulations mandate or imply a need for groundwater monitoring. Accordingly, the approach recommended in this section is to undertake groundwater monitoring consistent with site characteristics and surveillance of cell performance.

The function of a disposal cell performance groundwater monitoring system is to obtain data about the physical and chemical state of the groundwater system at, and in the vicinity of, the disposal cell. The groundwater monitoring system must function to detect and quantify the impact, if any, of the disposal cell on the groundwater. The groundwater monitoring system must, in addition, function to provide information about potential migration of contaminants (if any) from the site, and to assess the need for corrective or remedial action (see Figure 20.13).

The disposal cell performance groundwater monitoring system must perform for as long as is practical. To the extent reasonable, the monitoring system should perform for the short-term, i.e., the first fifty years of the cell. During that period, and thereafter, the components of the system should be replaceable. Alternatively, as technology advances, it should be possible to replace or augment the cell performance groundwater monitoring system.

The system should be flexible, i.e., able to be installed at discrete or diverse points to respond to the specifics of the site, the cell, and the groundwater layout.

It should be possible with the system to obtain data about both the geometry of the groundwater system and the quality of the groundwater. The system should be designed to obtain data of sufficient sensitivity to establish deviations from background conditions that indicate changes from the designed performance of the cell.

Table 20.4 lists alternative groundwater monitoring systems.

The first alternative involves the use of monitoring wells. Monitoring wells may be installed upgradient and downgradient of the disposal cell. The elevation (level) of the water in the monitoring wells is used to establish the groundwater surface topography, and hence flow patterns. If pore-water pressure measurement devices are installed in the monitoring wells, the data from these may be used to establish the hydraulic potentials and gradients in the groundwater flow system.

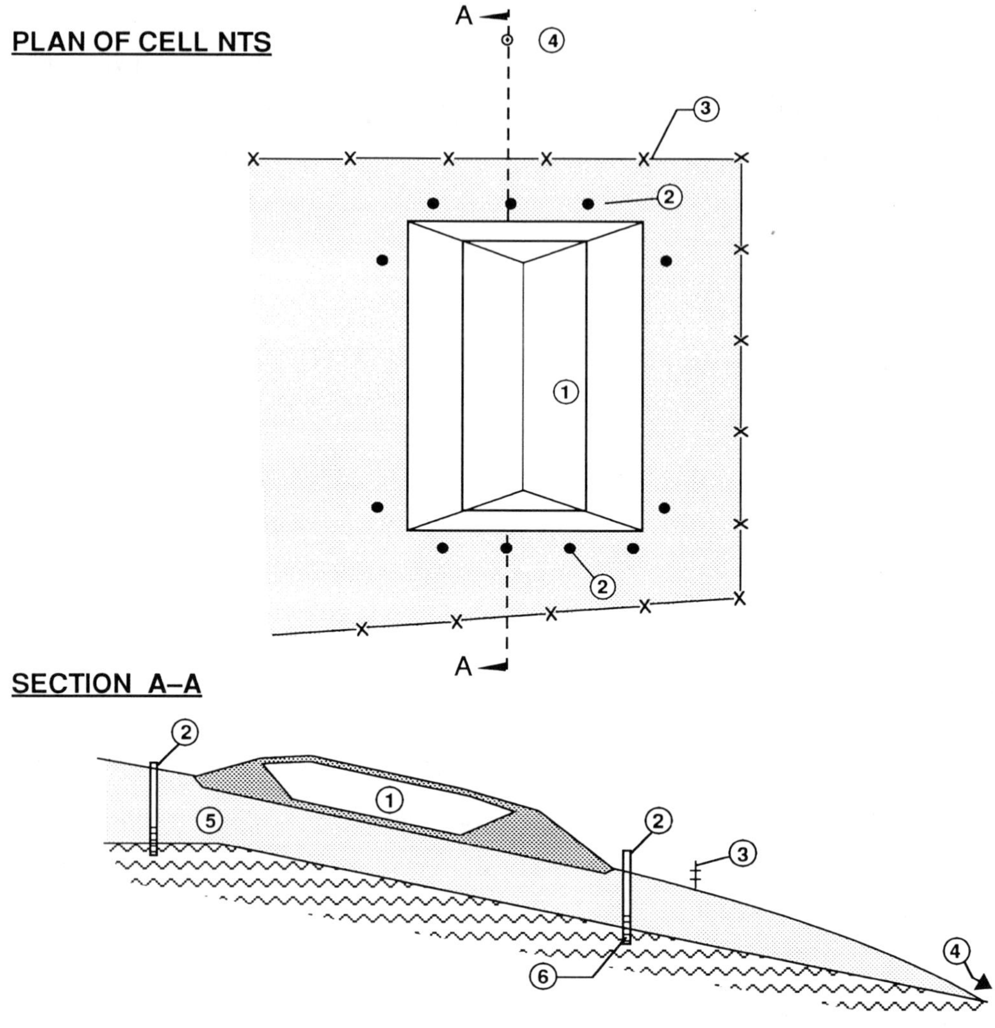

LEGEND

① DISPOSAL CELL

② POINT OF COMPLIANCE (GROUNDWATER MONITORING WELLS IN MONITORING PHASE)

③ PROPERTY BOUNDARY (FENCE)

④ SPRING

⑤ GROUNDWATER DIVIDE

⑥ GROUNDWATER QUALITY BELOW MAXIMUM CONCENTRATION LIMITS (MCLs)

In both the monitoring and post-monitoring period, groundwater quality due to disposal cell will not exceed MCLs at boundary of disposal cell complex.

FIGURE 20.13

LONG-TERM GROUNDWATER COMPLIANCE STRATEGY - CONCEPTUAL MODEL

TABLE 20.4 ALTERNATIVE GROUNDWATER MONITORING SYSTEMS

EVALUATION CRITERIA	ALTERNATIVES[a]	ADVANTAGES/DISADVANTAGES
• Practical and technically suitable - Demonstrated technology, must be accurate and precise • Simplicity - must be easy to construct using standard installation techniques • Longevity - system must operate for the monitoring and maintenance phases • Effect on cell integrity • Interference with the natural system	• Monitor Wells	• Advantages - Meets regulatory requirements - Can be used spatially and temporally - Can detect contaminant changes prior to groundwater reaching point of compliance repetitively - Can monitor beneath the cell at numerous locations • Disadvantages - Will require abundant measurements and maintenance - Expensive - Point measurements - will need several wells to assess changes with depth - Requires special techniques to collect groundwater samples
	• Springs/Surface Water	• Advantages - Can sample groundwater discharging of ground surface or springs or along gaining segments of streams - Does not require special sampling techniques - Inexpensive, repetitive sampling - Can sample high and low flows • Disadvantages - Sample locations are predefined - Cannot sample directly beneath the cell - May not sample at point of compliance - May not detect contaminant changes in aquifer prior to groundwater reaching point of compliance
	• Structural access: galleries, portals	• Advantages - Can sample directly beneath the cell at multiple locations - Can be used to advance directional boreholes and complete monitor wells • Disadvantages - Very expensive - Interferes with natural system - May affect cell integrity
	• Nonintrusive techniques	• Advantages - No effect on cell integrity, can conduct surveys from air or ground surface - Subsurface instrumentation not required, therefore no effect on cell integrity and does not interfere with natural system - Longevity is not a problem • Disadvantages - Mainly used for characterization, subtle changes may be difficult to identify - Cannot quantify contamination, or locate leaks in cells - May not be adequate as an early warning system

[a]Preferred alternative includes monitor wells and sampling springs and surface water

Water samples may be taken from the monitoring wells, the chemical quality of the water may be characterized in a laboratory, and the data used to assess performance of the cell. A third way to monitor groundwater to characterize disposal cell performance is to construct observation galleries extending to the water table beneath the cell. Groundwater samples may be collected from the galleries, or from boreholes advanced from the gallery faces, and completed as monitoring wells.

The fourth alternative includes nonintrusive, innovative technologies currently being evaluated as part of the DOE-EM program (DOE, 1991). These technologies include airborne infrared, integrated remote sensing and computer technology, Ultrasonic Ranging and Data Systems, and EM-31 Terrain Conductivity. These methods have been used for monitoring groundwater movement during site characterizations and assessment.

Evaluation of Alternatives

Table 20.4 summarizes the advantages and disadvantages of the alternative disposal cell performance groundwater monitoring systems. Monitoring the groundwater using wells is generally the favored approach because such systems meet regulatory requirements if designed properly.

Groundwater monitoring involves determining the quality of the groundwater for selected species. The constituents for which the water is tested are selected on the basis of current groundwater quality, the quality of the water as it may be affected by remedial actions (if any), and the species present in the waste. In addition, it may be feasible to spike the lowermost liner, or the recompacted foundation soils beneath a cell with a chemical constituent that is not normally found at the site, and which is not detrimental to the environment. If seepage through the liner were to exceed specified levels, the spiked constituent would be mobilized, transported to the groundwater, and monitored in groundwater monitoring wells.

Access galleries at the water table beneath the cell would be expensive to construct and maintain. They may damage the cell through settlement. An access gallery would, however, provide relatively convenient access to the aquifer. Boreholes could be advanced into both the saturated and the unsaturated zones, and experiments designed and conducted to demonstrate cell performance requirements and compliance in portions of the aquifer system that would otherwise be inaccessible with the other alternatives.

Details of a Groundwater Monitoring System

The final groundwater monitoring design for a disposal cell should include monitoring wells and spring monitoring, both upgradient and downgradient of the site. They are used to monitor changes in groundwater quality as well as the rate and the direction of the groundwater flux. This system should allow for some redundancy in techniques and locations.

A plan for a cell performance surveillance groundwater monitor should include details about:

- Monitoring well locations: upgradient, downgradient, and at the point of compliance
- Monitoring well design: i.e., open standpipe or pressure monitoring
- Depth and strata into which the wells will be placed and screened interval
- Monitoring well sampling frequency
- Spring sampling locations and sampling frequency

- Water quality parameter measurements
- Water level measurements
- The groundwater model that will be used to analyze data and assess disposal cell performance.

Specifics depend on existing conditions, the layout of the final disposal cell, the proposed groundwater operable unit remedial action, and the anticipated performance of the disposal cell and its interaction with the environment and, in particular, its impact on the groundwater.

Additional details about the disposal cell groundwater performance monitoring systems are provided in the DOE Technical Approach Document (DOE, 1989).

CHAPTER 21

GEOTECHNICAL MONITORING

Technical Approach

In Section III we divided the cell into hydrological, geotechnical, and ecological systems. For each system we defined models that may be used to quantify the performance of the cell. As described in previous chapters, monitoring of the performance of the cell involves measuring those parameters that are quantified in a performance assessment calculation, and comparing predicted to actual values.

To monitor the geotechnical and ecological performance of an encapsulation system we, in reality, need to do no more than walk the site at regular intervals and note how well (or badly) the vegetation is growing, and how much deformation or erosion has occurred. And if we notice failure, we repair it. Such an approach is the heart of any surveillance and maintenance program.

Preparation of a long-term surveillance plan for a disposal and encapsulation facility may be done during or after design, or during or after construction. Preferably, the long-term surveillance plan is compiled concurrent with design. This approach forces the designer to ponder long-term changes that the cover may undergo, how such changes may detrimentally affect cell performance, and what alternative details may better achieve the goals of long-term cover and dike stability.

Design objectives or performance criteria for waste disposal facility covers frequently include the requirement for low, minimal, or even zero maintenance. Achievement of this design objective does not, however, obviate the need for a long-term surveillance plan or an active long-term maintenance program. The state-of-the-art of cover and dike design and construction and the accumulated wisdom of practical experience has not yet progressed to a point where the engineer can construct the cover and walk away, leaving the structure to unmonitored satisfactory environmental stress response in perpetuity.

The standard contents of a long-term surveillance plan are best described by the DOE (1990). In regard to cover and dike design and performance, the basic elements of a long-term surveillance program include, but are not restricted to:

- Site inspections at specified intervals
- Site inspection procedures
- Inspector qualifications (degrees in engineering and/or geology)
- Care of pertinent design and construction documents
- Care of as-built drawings
- Ground and aerial photographs (as-built and at regular intervals thereafter)
- Local and regional topography maps (to monitor geomorphologic damage in the vicinity of the disposal cell)
- Survey beacons
- Settlement markers

- Boundary or perimeter monuments or markers
- Custodial maintenance or repair procedures
- Corrective action procedures
- Groundwater monitoring plans (an alteration in groundwater quality may indicate inappropriate cover performance)
- Quality assurance procedures.

Surveillance and maintenance is perforce restricted to observation and repair of the cover on the top of the cell and the dikes, and to geomorphological change of the surrounding environment. For this reason, the following section discusses the surveillance and maintenance implications of alternative cover and dike outer component designs.

EVALUATION OF ALTERNATIVE COVERS

General Guidelines

What is important for comparison of alternative covers is the interaction or balancing of the basic philosophies of long-term cover surveillance, maintenance, acceptance of a given probability of corrective action, present and future cost commitment, and fundamental cover design features.

To date, there is no generally agreed procedure for comparing the long-term surveillance needs of cover designs. At best, some simple rules of thumb may be enumerated:

- Covers for which the top component is durable, erosion-resistant rock will probably require less surveillance or maintenance than will vegetative covers to deal with run-off-induced erosion.

- In a humid environment, vegetation growth in a cover with an upper component of rock will probably necessitate greater surveillance and regular maintenance to control, remove, or kill unwanted plants than will a cover with a vegetated soil layer.

- Covers that rely on significant vegetation for erosion control may require surveillance and custodial care to ensure proper growth of planned plants. For example, a prairie grass community may require mowing or burning every few years. At Title II (active uranium mill tailings) piles in Texas, monthly grass cutting is undertaken during the maximum growth season.

- Inclusion of a reasonable amount of gravel in a soil that nevertheless has a satisfactory vegetation supporting capacity, may be the optimum approach to reducing the need for frequent surveillance, regular maintenance, and corrective action. Thus, such a cover incorporates, yet seeks to balance, erosion control with both rock and roots, and provides for the controlled incorporation of the inevitability of vegetation growth.

As noted in DOE (1988), vegetative covers may promote achievement of the objective of minimal long-term maintenance for these reasons:

- <u>Climax Vegetation Equilibrium</u>: If a locally-adapted species is a favorable medium, the vegetation will reproduce and the cover will be self-perpetuating. The plant community will renew itself after disturbances and will adapt to climatic change. If the cover is properly designed, the community will become part of the natural "climax" vegetation.

- Maintenance Minimization: A vegetative cover with a biointrusion layer will require less long-term maintenance than a rock cover that is perpetually invaded by deep-rooted species that may have to be removed annually. Some repair of rills may be necessary after severe rainfall that may occur before the plants are well established. Irrigation or reseeding may be needed after extraordinary disturbances by fire or drought. Mowing may be required to maintain shallow-rooted species.

- Prevention of Organism Intrusion: A thick biobarrier will resist intrusion by plants and animals, and will be an impediment to the progress of humans digging with anything other than heavy equipment.

It is reasonable to conclude that the basic alternative covers considered in this book 1) will require some degree of regular surveillance, 2) are likely to require some degree of regular maintenance, 3) have varying but unquantifiable (or rankable) probabilities of needing corrective action during the design life, and 4) nevertheless represent, in their own way, reasonable compromises among the competing demands for minimum long-term care.

Evaluation of Specific Covers

Because the erosion protection layer used on the cover surface has a significant effect on the surveillance and maintenance requirements of the cover, the evaluation of specific covers can be performed in only a general way. The following is a general discussion of the maintenance needs of the five basic covers if they were to be constructed at a site in the Midwest.

Cover 1 - The Simple Cover: If suitable soils are used to establish a prairie community, deep-rooted species should be excluded by occasional burning or perhaps mowing. The lack of a biobarrier will allow deep-rooted species to cross into the radon/infiltration barrier, therefore close surveillance will be needed. The soil would have to be inspected annually for rills or gullies: these would have to be filled in if they develop.

Cover 2 - The RCRA Cover: This would require essentially the same maintenance approach as Cover 1, but may be more resistant to deep-rooted species because of the drain and the geomembrane.

Cover 3 - The Double-Drain Cover: The rock would have to be inspected each decade or so to monitor degradation of the rock, or filling of the rock voids by windblown sands. The rock cover would need to be inspected annually to monitor and, as necessary, remove unplanned vegetation. Only vegetation that would, if not removed, lead to cover degradation or damage need be removed. Such removal might involve cutting down, pulling out, or killing with herbicide.

Cover 4 - The Multicomponent Cover: Depending on the actual soil gradation of the upper component (i.e., the percentage of gravel), different vegetation would establish. The cover would be able to accommodate the climax vegetation without significant probability of cover damage. Accordingly, annual inspections and occasional burning or mowing may be all that is necessary. If rills or gullies develop, they would have to be repaired. If unanticipated, excessively deep-rooting species establish, they would have to be removed.

Cover 5 - The Erosion Resister: Proper functioning of this cover depends on the presence of a very limited thickness of grass. To maintain this type of unnatural vegetation, annual or more frequent mowing of the grass may be necessary. In addition, rills and gullies would have to be repaired immediately.

CHAPTER 22

ECOLOGICAL MONITORING

GENERAL

As mentioned in Chapter 8, short-term monitoring of vegetation is a necessary step in the establishment of a vegetated cover. Patterns of germination and early plant growth on the cover need to be documented, and bare spots must be examined to identify factors that impede establishment of vegetation. The bare spots must be interseeded, and the entire cover treated with an additional seed application, as necessary, to assure that an uninterrupted coverage of perennial vegetation is achieved. Rills may be repaired and straw mulch crimped in along the contour as needed.

Long-term monitoring will be little more than a formality for a well-constructed vegetated cover on a well-conceived cell. The natural processes affecting the cell will be beneficial -- carbon capture, particle deposition, soil development -- and these can be reported with satisfaction by the design team and their descendants. Phenomena such as invasion by deep-rooters and disruption by large burrowers may temporarily affect the cell's upper layers, but the natural repair mechanisms inherent in a vegetative cover will, in almost every case, prevent these factors from breaching the containment system. Truly, the monitoring effort for a good cover should consist of visiting the site after a major storm event to determine if that storm happened to impinge upon the cell during a period of unusual vulnerability; for instance, after a fire or during a protracted drought. In such a case, some gully repair and interseeding may be appropriate. Alternatively, the maintainers may choose a prophylactic approach to their mission, taking note of reduced vegetation abundance -- for instance, during drought -- and effecting the appropriate corrective action. The cover could be seeded and irrigated within the guidelines of Chapter 8, or could be mulched with straw crimped along the contour.

The plant community on a cell is intended to be in a climax condition, meaning that it will not depart under ordinary circumstance, from its current state of development. Prairies are somewhat unusual as climax communities in that they require regular, though not particularly frequent, disturbances to maintain themselves. Accordingly, natural fires, if mild, should be allowed to burn on a prairie-covered cell. If this is unacceptable for social reasons, then controlled scorching should be practiced to selectively remove occasional deep rooters, to stimulate grasses, and to mobilize tied-up nutrients. Mowing can be substituted for scorching, but then fertilization will be needed to compensate for the fact that nutrients are not liberated.

Mowing, fertilization, and supplemental irrigation may be a required practice for vegetation on a soil cover that lacks sufficient rooting media to qualify as a vegetated cover. The mowing will serve to control woody species and will enforce a sod-like growth habit among the grasses. Fertilization and irrigation may be needed to replenish the water and nutrient reservoirs that are inadequately supported by the thin rooting medium. The thinner the rooting medium, or the more rock in a rock soil matrix, the more mowing, manicuring, and supplementation will be needed.

Where rock covers have been constructed, the maintainers are sentenced to perpetual programs of plucking and poisoning. Burning or cutting are not recommended because these will, for many woody species, stimulate root sprouting and stump sprouting. If the client desires immediate plant removal to save face, as was recently prescribed at the Burrel cell, then the freshly cut stumps must be treated with a systemic herbicide. Too much plucking obviously harms the cell, and herbicides are not highly desirable from a chemical avoidance perspective. So retroactive modification of covers from rock to vegetated may someday be practiced. If the shape of the cell allows this to be done, then the upper layer of rock must be choked with gravel prior to the placement of a properly sized layer of rooting medium. Where rock is, for erosion purposes, the only viable alternative, then a choke layer may be established with no overlying soil. The choke layer will do much to discourage further plant invasion for a generation or two.

The best approach to maintenance is to design a cover that requires none. Cover maintenance crews, like Maytag repairmen, will have plenty of time to read this book ... over and over again.

FACTORS TO MONITOR

The following elements of the cover may be monitored on a regular basis where institutional factors so dictate. The reason for monitoring each of the parameters and the use to which the information will be put is briefly described in the following section.

- Vegetation

  - The number of plants
  - Root penetration depth
  - Plant height, spread, vigor
  - Percent groundcover
  - Seed germination and plant growth
  - Buffer zone encroachment

- Soil

  - Burrow depths
  - Soil loss
  - Deformation - surface roughness change

- Soil chemistry (pH, nutrients)
- Physical soil parameters (moisture content, cation exchange, organic content)
- Soil/rock/vegetation mixing and movement

• Moisture

- Infiltration: lysimeters/underdrains
- Run-off quality

• Maintenance

- Frequency
- Cost
- Man hours required
- Water quantity
- Biomass when mowing
- Fertilizer needs

MONITORING PROCEDURES

Full-blown vegetation and soil monitoring may be appropriate during the first few years after a cell has been constructed. This may be done for experimental purposes, or for validating the design relative to the intended performance objectives. Following are suggested monitoring procedures.

Vegetation

Number of Plants

The number of plants per unit area should be measured quarterly by setting up at least four square quadrants on each plot. Each quadrant should be approximately one square meter in size and should contain a representative sample of the types of vegetation growing on each plot. An alternative method might be to establish a transect on each plot and determine the plant count using a random numbers table.

Pure biomass may be measured as follows: prior to mowing/cutting/eradicating/etc., establish a quadrant (enclosing at least one square meter area); cut all aerial portions of the vegetation in the quadrant; segregate the cuttings on a per-species basis; place the vegetation in a dehumidifying environment; after several days, weigh the dried vegetation. Repeat this process at regular intervals during the growing season. Biomass production rates can then be corrected with seasonal and climatic influences, soil and rock types, and fertilizer and water applications.

Root Penetration Depth

Root penetration measurements should consider spurts in growth, i.e., first year growth is generally more intense than subsequent years. Root penetration depths should be evaluated for each plant species. If excavation is the method of evaluation selected, care should be exercised when measurements are taken so as not to disturb adjacent vegetation.

One method recommended for monitoring general root penetration is to use transparent tubes or ports. This may be as simple as a plexiglass tube that penetrates the cover components. The tube should have flanges or other flow impeders to prevent the plane of migration. The diameter of the plexiglass tube should be sufficient to accommodate a camera that may be used down the tube to photograph the soils and any roots penetrating successive layers.

Plant Height and Vigor

When gathering plant count data, quarterly monitoring activities should include measurement of the plant height of each species growing in the quadrant being monitored. Minimum and maximum heights should be noted as well as an average or typical height.

Vigor is a somewhat subjective quality, but one that gives a general description of the status of the vegetal growth. Vigorous growth would be defined as active, healthy, and well-balanced growth. Descriptive notes should be kept quarterly. Indications of stress shall be included in such descriptions.

Percent Groundcover

Percent groundcover should be estimated in marked quadrants. Measurements may be extrapolated to estimate the percent cover for the entire plot. These estimates should be made concurrently with the other quarterly measurements.

Seed Germination and Plant Growth

During the period immediately following the establishment of the vegetation, seed fetch and germination success shall be evaluated daily. Measure the number and height of shoots, in a quadrant as small as 300 mm by 300 mm.

Soil

Burrow Depths

Covers should be monitored for the presence of mammal and reptile burrows. In the event that any burrows are located, they may be filled with a quick-setting foam or similar substance and then excavated to determine the extent of the burrow. Invertebrate burrows such as ant tunnels should be noted, with records indicating approximate quantity of displaced soil.

Soil Loss

Several methods of monitoring or estimating soil loss are available. If the cover is equipped with a run-off collection system, soil transported off the test plots may accumulate as sediment in this collection system. The quantity of sediment can then be measured (preferably on a dry-weight basis). All rills or gullies that develop should be measured with respect to width, depth, and length.

Deformation - Surface Roughness Change

Stakes may be set out over specified distances (5 to 10 m) and base measurements taken in order to determine the extent of deformation taking place over specified time intervals.

Soil Chemistry

Soil core samples should be collected periodically and submitted to a laboratory for analysis to quantify soil chemistry, such as pH and nutrients.

Physical Soil Parameters

Measurements should be taken on site and recorded, and/or soil core samples shall be collected periodically and submitted to a laboratory for analysis. Parameters to be measured include moisture content, cation exchange capacity, and organic matter content.

Soil/Rock/Vegetation Mixing and Movement

Visual observations will be made and photographs taken of any soil or rock movement or mixing that may result from freezing and thawing or other physical or biological processes.

Run-off and Seepage Quality

At least once per season, samples of the precipitation run-off shall be collected, and the quality of the water characterized. In those test pads provided with an infiltration underdrainage system and sump, samples of underdrain flow shall be collected and submitted to a laboratory for analysis to quantify the chemistry of the water.

REPORTING FORMAT

Table 22.1 provides the form to be used to monitor and record plant growth. All other data based on visual observation of the cover shall be recorded in narrative, sketches, and with photographs. An annual report on the cover performance should be prepared.

TABLE 22.1

COVER VEGETATION PERFORMANCE MONITOR

COVER OPERATIONS AND MONITORING PLAN			
DATE	**DATA COLLECTOR**	**ITEM (Method of measurement used)**	**OBSERVATION**
		Number of plants	
		Root penetration depth	
		Plant height and vigor	
		Percent groundcover	
		Seed germination and weed growth	
		Buffer zone encroachment	

GUIDELINES FOR MAINTENANCE AND REPAIR TO COVERS

If seeds fail to germinate, new or additional seeds should be spread. If seed germination and initial growth is sparse or unsatisfactory, water or fertilizer may be added. If vegetation death is greater than may reasonably be expected, appropriate specialists shall be consulted.

If and when erosion occurs, any rills or gullies that have formed shall be filled with soil and revegetated. Hay bales may be strategically placed to limit future erosion. Any materials that degrade shall be removed and replaced. A record shall be kept documenting the nature and extent of material degradation and actions taken to correct the problem. If animal intrusion becomes a problem, it is recommended that the intruder be trapped and relocated, or otherwise eliminated.

SECTION V

CASE HISTORY

CHAPTER

Chapter 23

CASE HISTORY: ANDUJAR

INTRODUCTION

In the south of Spain on the outskirts of the town of Andujar an inactive uranium mill tailings pile was stabilized in place. A cover system was placed over the pile. This chapter describes the technical procedures used in Spain to design the Andujar pile cover and, in particular, discusses the approaches used to ev251
aluate cover infiltration and cracking.

SITE AND PILE CHARACTERISTICS

The Andujar Uranium Milling Plant is in the province of Jaen (Andalucía) on the southern flood plain of the Guadalquivir River at 1.5 kilometers south from the urban center of Andujar. The site is trapezoidal in shape, covers an area of approximately 17.5 hectares, and is about 150 m from the river.

The site is on alluvial clays and gravel terraces, underlain by a very low permeability shale. Most groundwater flow is in the gravels beneath the pile toward the river.

The climate of the Andujar area is characterized by low to medium precipitation, abundant sunshine, medium relative humidity, and moderate to high temperatures with large diurnal and annual ranges. The regional climate may be classified as mediterranean subtropical with a dry and warm summer.

The Andujar area exhibits a large diurnal range in temperatures with a mean daily minimum of 25°C and a mean daily maximum of 12°C. The mean daily average is 17.6°C with a maximum of 42°C and a minimum of 0°C.

Most of the precipitation in the Andujar area occurs during the fall and winter seasons. The annual average precipitation is 550 mm, with a maximum monthly precipitation in December of 93.4 mm. The mean annual potential evapotranspiration in the area reaches 823 mm, and about 50 percent of that figure occurs from May through September.

The plant was designed for processing low grade uranium ore (0.15% of U_3O_8) and produced 80% concentrate of U_3O_8 in the form of sodium and ammonium uranate at a rate of 60 to 80 tons per year. The plant began operation in 1959 and continued until 1981. During this period, 1.2 million tons of uranium ore were processed to produce 1,350 tons of U_3O_8 with a fineness of 80-85%. Recovery of the uranium involved sulphuric acid leaching followed by ion exchange or by tertiary amine/kerosene extraction. Solid wastes were stored in the tailings piles and liquid wastes were treated before disposal to the Guadalquivir River.

The configuration of the Andujar mill site is shown in Figure 23.1 and includes the following areas: the tailings pile, the processing plant, the wastewater treatment area, the auxiliary and administrative buildings, and the housing area.

All solid waste generated during the plant's operation (1.2 million tons) are contained in the tailings pile, which covers an area of 9.4 hectares and has a total volume of 980,000 cubic meters. The pile was constructed in five cells by upstream construction to a height of 20 meters in the central and eastern parts and to a height of 10 meters in the western part. The sideslopes vary from 25 to 35 degrees and are currently covered with asphalt.

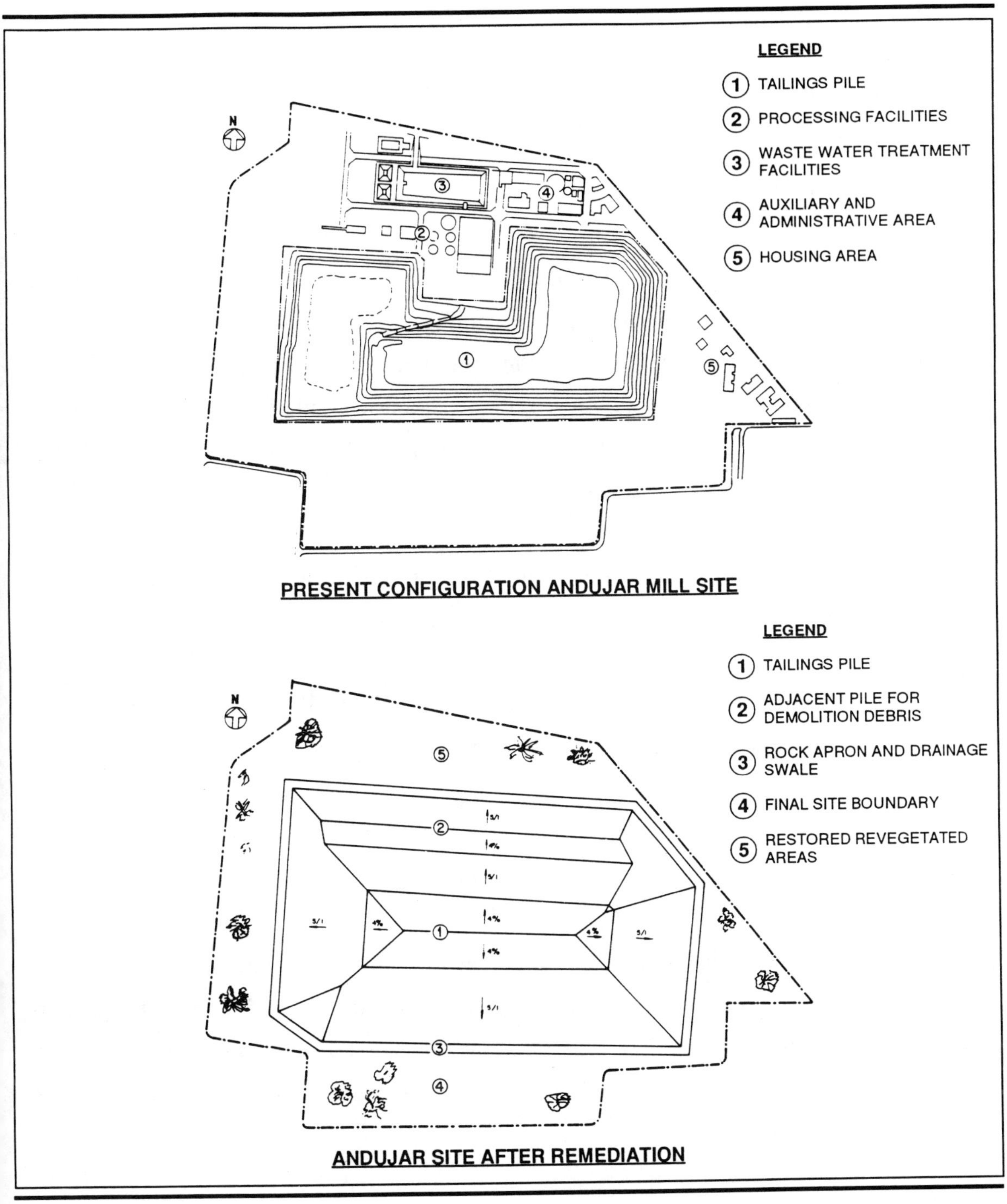

FIGURE 23.1

ANDUJAR MILL SITE CONFIGURATION AND REMEDIATED SITE SHOWING FINAL PILE CONFIGURATION

The pile is surrounded by olive fields, and an olive oil processing factory is nearby.

REGULATORY OBJECTIVES

The regulations and standards that govern the remediation activities at Andujar have been established by the Spanish Nuclear Safety Council (CSN), taking into account the recommendations of international organizations, the standards promulgated by the EPA for the remediation of uranium mill tailings, and the Spanish regulations, specifically those related to groundwater protection and the long-term disposal of radioactive wastes. These regulations may be summarized as follows:

- Dispersion Control: prevent inadvertent human intrusion and dispersion of contaminated materials by wind and water erosion.

- Long-term Radiation Protection: achieve an effective equivalent dose to the individual in the critical group below 0.1 mSv/year.

- Design Life: Remain stable for 1,000 years to the extent reasonably achievable, and in any case for at least 200 years.

- Soil Clean-up: reduce the residual concentration of radium-226 in land, averaged over an area of 100 m^2, so that the background level is not exceeded by more than 5 pCi/g (averaged over the first 15 cm soil) and is less than 15 pCi/g (averaged over 15 cm-thick layers of soil more than 15 cm below the surface).

- Radon Control: reduce radon flux over the surface of the final pile to an average release rate of less than 20 pCi/m^2 s.

- Groundwater Quality Protection: control groundwater contamination so that background water quality or maximum concentration levels (in accordance with Spanish regulations and CSN guidelines for radioactive constituents) are achieved in the long-term. These maximum levels are: combined radium-226 and radium-228 0.18 Bq/l (4.86 pCi/l), combined uranium-234 and uranium-238 1.2 Bq/1 (32.4 pCi/l) and gross alpha activity, excluding radon and uranium, 0.5 Bq/1 (13.5 pCi/l).

- Long-term Maintenance: minimize the need for long-term maintenance.

- Construction Works: minimize hazards to the workers and the environment.

- Regulations: comply with other applicable and relevant Spanish regulations governing air and water quality.

With regard to groundwater quality protection, it is also required that for short-term conditions the cover system be designed to limit infiltration to ensure that, at the end of the compliance period (minimum 10 years), the combined uranium-234 and uranium-238 concentration in groundwater complies with the two following conditions:

- Be less than 6.15 Bq/l (166 pCi/l) at the point of compliance, at the downgradient boundary of the disposal site

- Be less than 3.5 B1/l (94.5 pCi/l) at the wells in the vicinity of the site.

In addition to the above design standards, a performance standard has been established for Andujar: groundwater quality must be monitored during the compliance period (minimum 10 years) to confirm adequate performance of the cover and compliance with the maximum concentration limits established for short-term conditions.

REMEDIAL ACTION DESIGN

The remedial action plan proposed for the Andujar mill site involved stabilizing and consolidating the uranium mill tailings and contaminated materials in place. The tailings pile was reshaped by flattening the sideslopes to improve stability. Tailings from sideslope flattening were relocated around the existing pile and on the top of the lower pile. Mill equipment, buildings, and process facilities were dismantled and demolished and placed in a pile adjoining the main tailings pile. Off-pile contaminated soils were excavated and placed on top of the tailings pile in order to reduce the radon flux.

The final pile configuration, as shown in Figure 23.1, was designed to minimize the movement of tailings and the size of the restricted disposal area. The pile was constructed with four percent topslopes and 20 percent sideslopes which provide static and dynamic slope stability without requiring excessively large rock to resist erosion. Protection against upland watershed run-off is provided by channeling run-off around and away from the pile via drainage diversions wales along the perimeter of the pile. Protection against floods associated with the Guadalquivir River is provided by a rock apron around the perimeter of the pile and riprap layers on the sideslopes.

The pile is covered with a multilayer system to meet the three simultaneous demands of erosion control, infiltration and radon control. In order to comply with the standards, the following design elements are incorporated into the cover system:

- Stabilization control for up to 1,000 years: only natural materials are used and the cover is designed to resist extreme events such as probable maximum precipitation, probable maximum flood, maximum credible earthquake.

- Dispersion and intrusion control: erosion protection layers and biointrusion barriers within the cover.

- Soil clean-up: remove contaminated soils and incorporate them within the tailings underneath the cover system. This reduces the thickness of the radon barrier.

- Radon control: a radon barrier of natural soils.

- Protection of groundwater quality: a multiple redundant cover system to limit infiltration. This system includes soil/vegetation layers, drain layers and a low permeability infiltration barrier.

- Long-term maintenance: a rooting medium for the establishment of climax vegetation.

DEFINITION OF ALTERNATIVE COVERS

Six covers were selected for evaluation as alternatives for the Andujar remedial action project. The three primary topslope covers are shown in Figure 23.2. Following is a brief description of each and the reason why it was selected for further evaluation.

Topslope Covers

Cover 1: Vegetated Multicomponent Cover

This cover incorporates all the elements of the Checklist Cover, but may not provide adequate long-term erosion protection. From top down, this cover includes the following components:

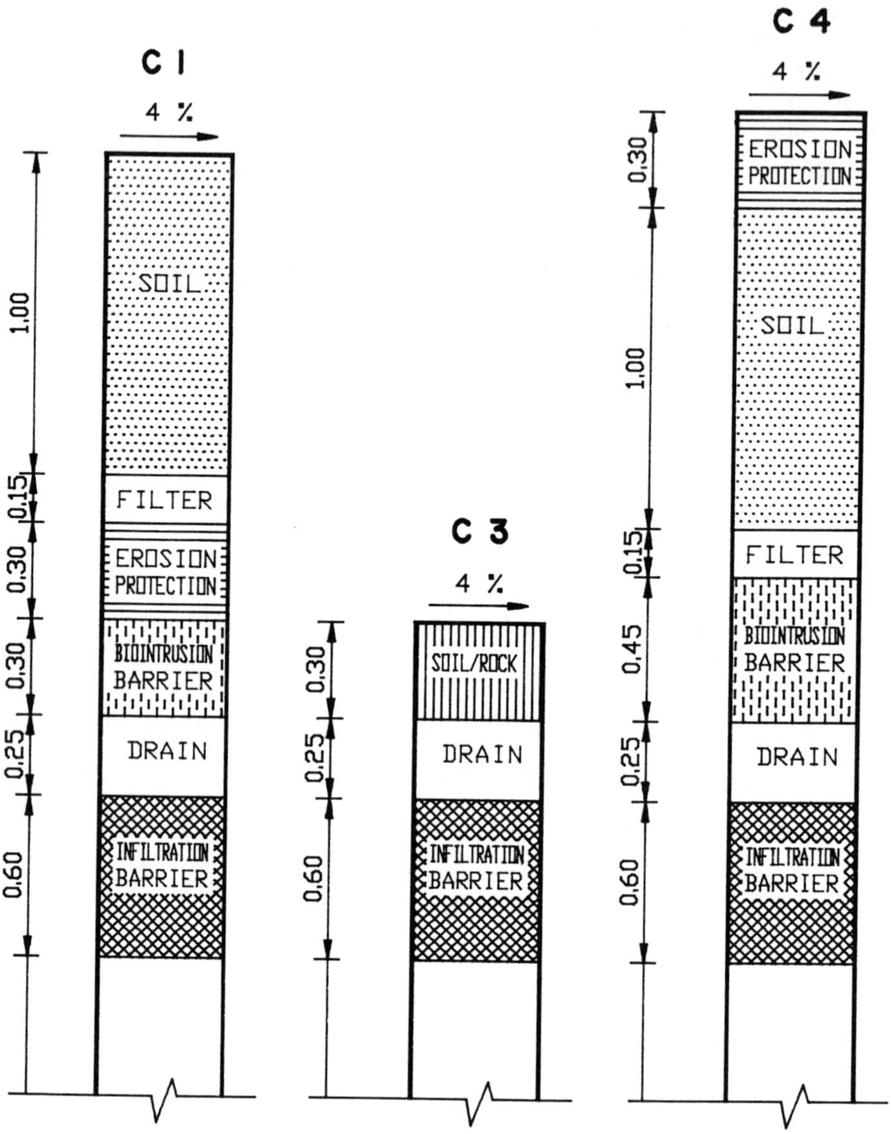

FIGURE 23.2

TOPSLOPE COVER ALTERNATIVES

- A vegetation growth and rooting medium consisting of a layer of random soil, which may also double as protection of the radon barrier from potential desiccation during the hot and dry summer of Andujar.

- A filter of clean sand to prevent migration of the fine particles of the random soil into the underlying erosion barrier.

- A biointrusion barrier of large gravel and coarse rock to deter root penetration and impede burrowing animals. Another function of this layer is to provide erosion protection in the event of gully formation in the overlying soil.

- A drain of clean sand to shed water laterally off the pile.

- A radon and infiltration barrier consisting of silty clay amended with bentonite or a bentonite mat.

This cover has been selected because of its ability to deal effectively with vegetation and to reduce infiltration to the pile because of effective evapotranspiration.

Cover 3: The Simple Cover

From the waste up, the three components of this cover are the radon/infiltration barrier of compacted silty clay, the drain and bedding layer of clean sand and the erosion resistant rock with soil filling the voids. This soil/rock matrix will allow the establishment of vegetation and improve the aesthetics of the remediated pile. No biointrusion barrier is incorporated in this cover. This function would be provided by the bedding layer which would be as permeable as possible to preclude an environment conducive to plant growth.

This cover provides a basis for comparison with the other covers and was evaluated because of its simplicity and low cost.

Cover 4: The Rock Multicomponent Cover

In this cover, maximum achievable erosion protection is provided by the uppermost layer of durable rock. From the waste up, the components of this cover are the radon/infiltration barrier, the drain of clean sand, the biointrusion barrier, the filter or bedding layer, the zone of random soil, the purpose of which is to increase the depth of the cover and to act as a growth medium for vegetation, and the erosion resistant rock.

This cover was evaluated because of its effective erosion protection properties, its ability to provide protection against drying of the radon barrier and the existence of a controlled zone -the random soil- for vegetation that might establish through the riprap. A disadvantage of this cover is that the rock may reduce evaporation/evapotranspiration from the cover, therefore potentially increasing infiltration.

Sideslope Covers

The sideslope covers are essentially the same as the topslope covers with some minor differences in the erosion protection layers due to the change in slope. The greater flow velocities in the sideslope covers T1 and T5 shown in Figure 23.3 require an increased rock size to resist water erosion, therefore increasing the thickness of the erosion resistant layers.

In T5, the coarse rock erosion barrier is changed to a soil/matrix to help the establishment of vegetation through the riprap and reduce the visual impact of the remediated facility.

Bentonite mats have not been considered as infiltration barriers for the sideslopes due to the difficulty of providing adequate long-term stability on 20 percent slopes.

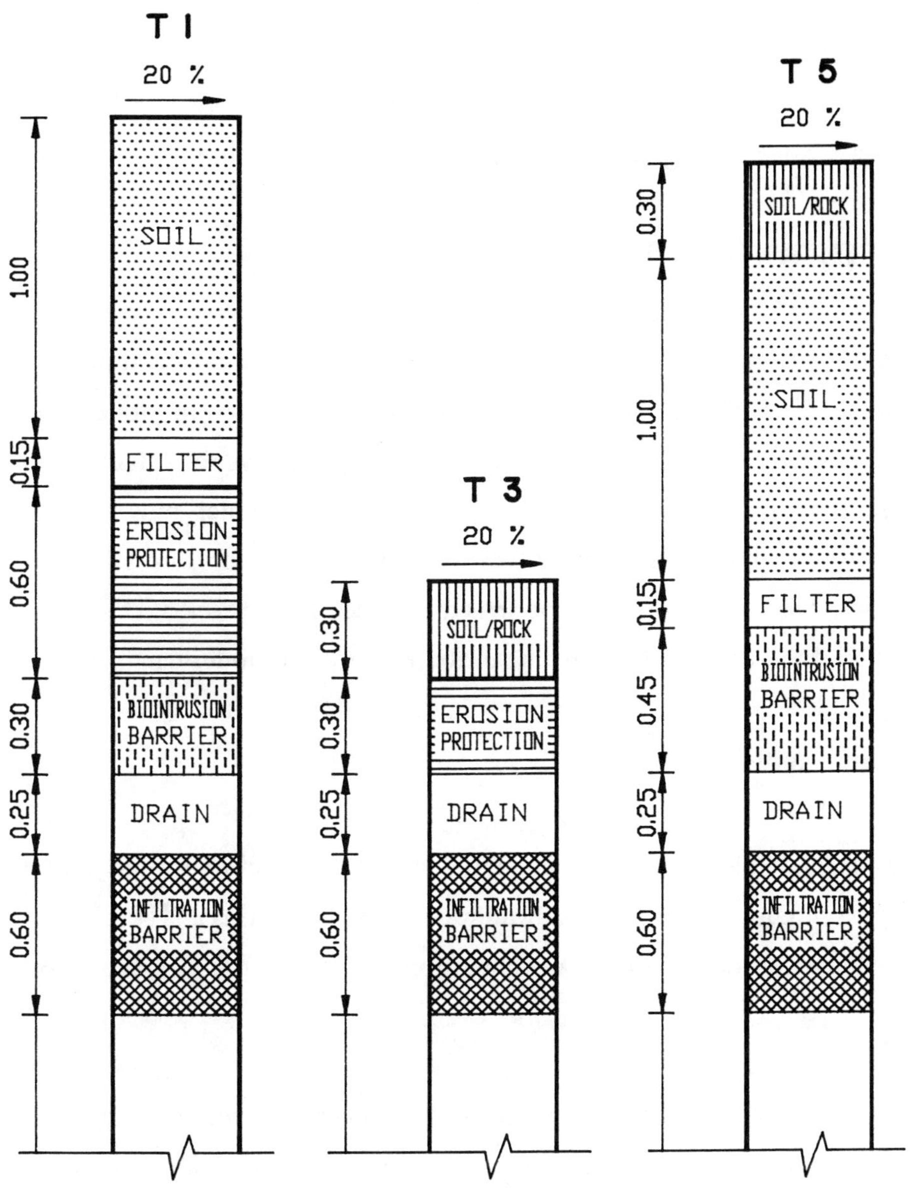

FIGURE 23.3

SIDESLOPE COVER ALTERNATIVES

Each of the six covers selected was evaluated to examine its performance with respect to the following factors: infiltration, erosion control, desiccation, and cover cracking, biointrusion control and aesthetics. Detailed studies were performed to analyze cover infiltration and cover cracking, and they are presented in the following sections.

COVER INFILTRATION

Cover infiltration evaluations are generally performed using standard computer codes such as HELP, CREAMS, UNSAT which use a water balance approach to model the water flux through a layered cover system. These estimates are very sensitive to surface water balances and do not account for the effects of the upward vapor flux which may occur during dry periods. Experimental measurements of evaporation and vapor flux taken at Shiprock tailings pile, New Mexico, USA, have confirmed the need for a proper consideration of the vapor phase and, therefore, heat transport in the infiltration modeling.

Mathematical Model

Seepage rates through the covers were calculated using the one-dimensional finite difference model Transport of Unsaturated Groundwater and Heat (TOUGH). This code uses a mass and energy balance approach and allows simulation of water, vapor and heat fluxes through the cover.

A complete analysis of all the cover infiltration mechanisms requires the solution of a complex system of seven coupled differential equations in partial derivatives. In order to simplify the numerical calculations, the following assumptions are considered in the code: equilibrium between liquid and gaseous phases, equal temperature in all phases, equilibrium between the air dissolved in water and the gaseous phase (Henry's law).

Improvements and modifications were made to this code to simulate precipitation, plant uptake and evapotranspiration, and lateral drainage. Key features of the model are given below:

- Precipitation is calculated using an artificial weather generator which uses the past available rainfall data to provide an artificial weather history which maintains the actual monthly precipitation and the duration of the storms.

- Run-off is evaluated by the Soil Conservation Service (SCS) Curve method.

- Evaporation is calculated on the basis of aerodynamic diffusion using Milly's equation.

- Evapotranspiration is obtained from the modified Penman's equation affected by a reduction factor which takes into account the actual availability of water to evapotranspirate and is a function of soil moisture and plant interception.

- Lateral drainage is evaluated by Darcy's law corrected by a factor which accounts for the two-dimensional effects in the sideslopes.

The material properties which are required for the analysis are:

- Soil-water retention curves (soil tension vs. moisture content) where the Van Genuchten equation can be used.

- Saturated and unsaturated permeabilities. Irmay's equation may be used for the evaluation of unsaturated permeabilities of granular materials and Van Genuchten's approach for nongranular materials.

- Thermal and mechanical properties such as a density, thermal conductivity, gas and solid phase compressibility and specific heat.

Analysis of Infiltration

The TOUGH code was used to model the performance of the vegetated multicomponent cover type C1, which consists of, from top down, vegetation, random soil, filter sands, biointrusion gravels, drain sands, and clay infiltration barrier. In order to investigate the cover infiltration mechanism, the evaluation of parameters, such as soil degree of saturation, percolation through the soil, drain degree of saturation, lateral drainage, infiltration, and clay degree of saturation was analyzed. On the basis of these results, the following conclusions are applicable to a multilayer cover system:

- The degree of saturation of the soil increases when the soil/vegetation evapotranspiration capacity is exceeded and the soil begins accumulating water. Soil saturation is maximum in winter when precipitation is important and evapotranspiration is minimum (see Figure 23.4).

- The percolation through the soil is negligible except in those periods when the evaopotranspirative capacity of the soil is exceeded and the soil degree of saturation reaches a maximum. This coincides with the periods of minimum evapotranspiration.

- Infiltration through the clay vs. time follows a trend similar to the variation of degree of saturation with time (see Figure 23.5). Infiltration will always occur if there is water in the drain. Infiltration is very sensitive to the degree of saturation of the clay, and it was greater for the first peak when the clay was dry and suction pressures were higher.

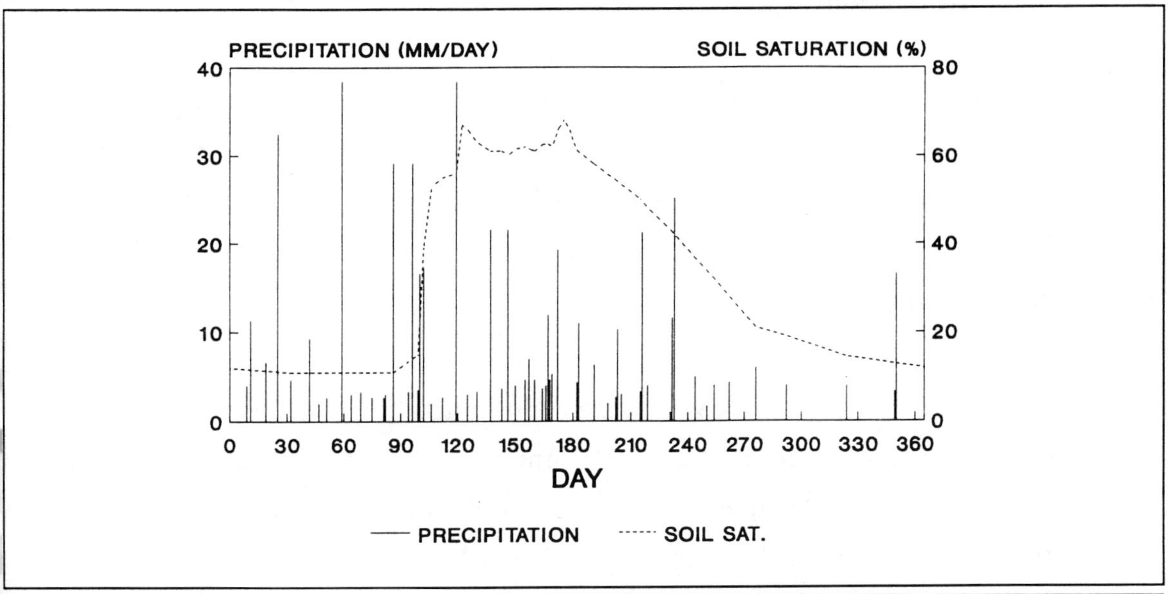

FIGURE 23.4
SOIL SATURATION

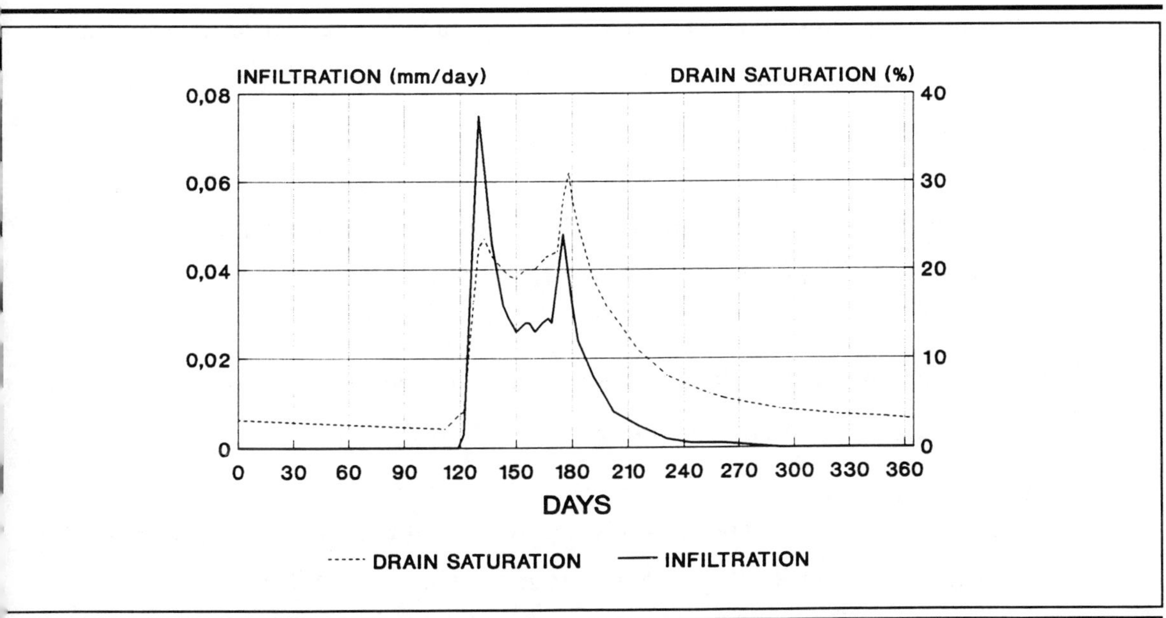

FIGURE 23.5
INFILTRATION

- Variations in the degree of saturation of the clay layer are relatively small. The degree of saturation of the clay is constant at the base of the layer and responds to the variation of infiltration at the top.

Sensitivity Analyses

Sensitivity analyses were performed to study the contribution of the various cover elements to the overall cover performance. Cover performance sensitivity was analyzed for the following parameters: soil hydraulic conductivity, soil thickness, root length, potential evapotranspiration, drain hydraulic conductivity, clay hydraulic conductivity, clay thickness, and precipitation.

The results of the analyses may be summarized as follows:

- An increase in soil hydraulic conductivity leads to a reduction in run-off, and an increase in later flow but has no effect on infiltration (see Figure 23.6).

- An increase in soil thickness has no effect on the overall cover performance. However, percolation is more uniformly distributed over time and leads to a small increase in infiltration (see Figure 23.7).

- An increase in root length leads to an increase in evapotranspiration and reduces drainage and infiltration (see Figure 23.8).

- An increase in evapotranspiration reduces percolation and, accordingly, drainage and infiltration are also reduced (see Figure 23.9).

- An increase in drain hydraulic conductivity leads to a reduction in the degree of saturation of the drain and a reduction in infiltration (see Figure 23.10).

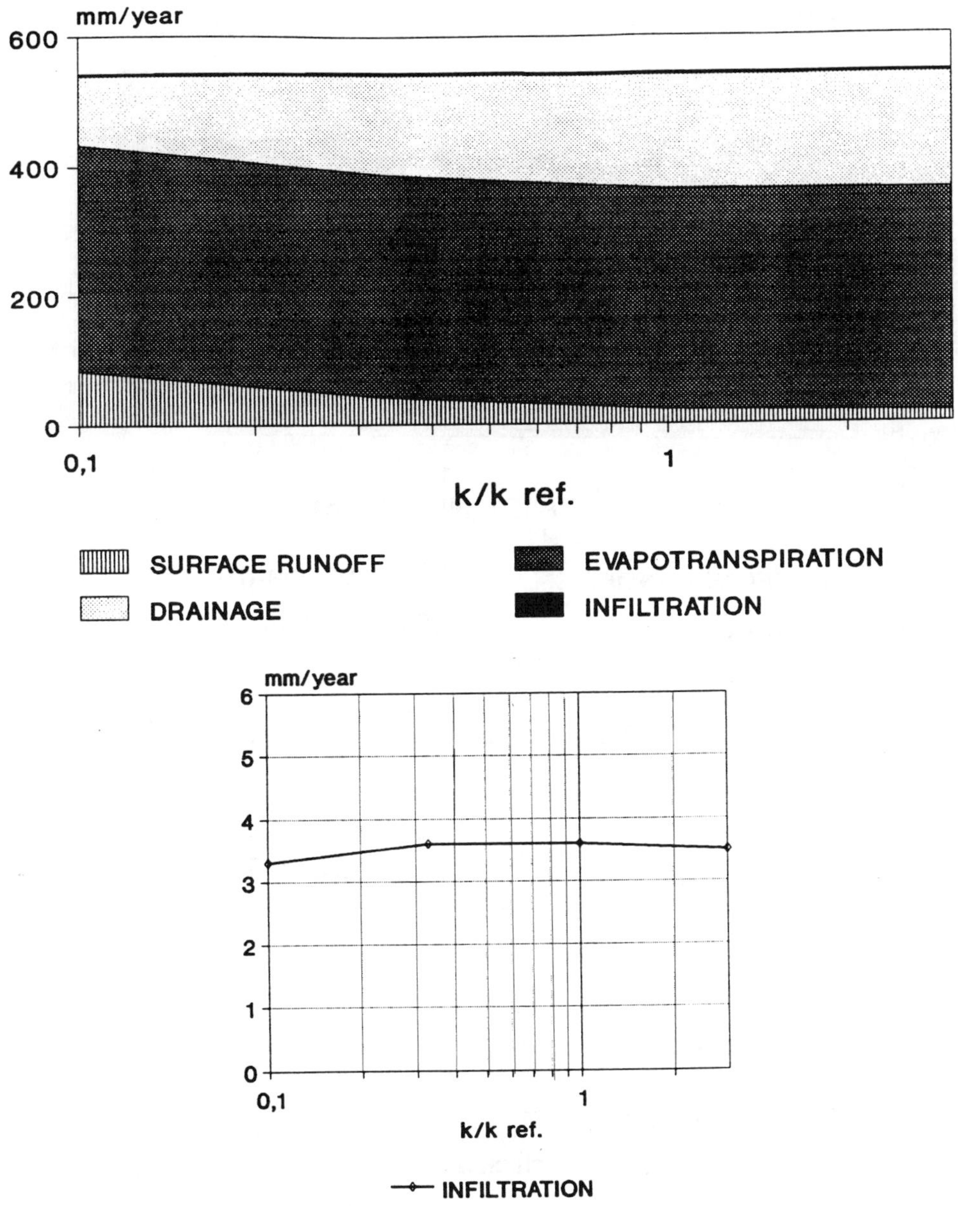

WITH VEGETATION
k ref. 10E-4 cm/sec.

FIGURE 23.6

SOIL HYDRAULIC CONDUCTIVITY

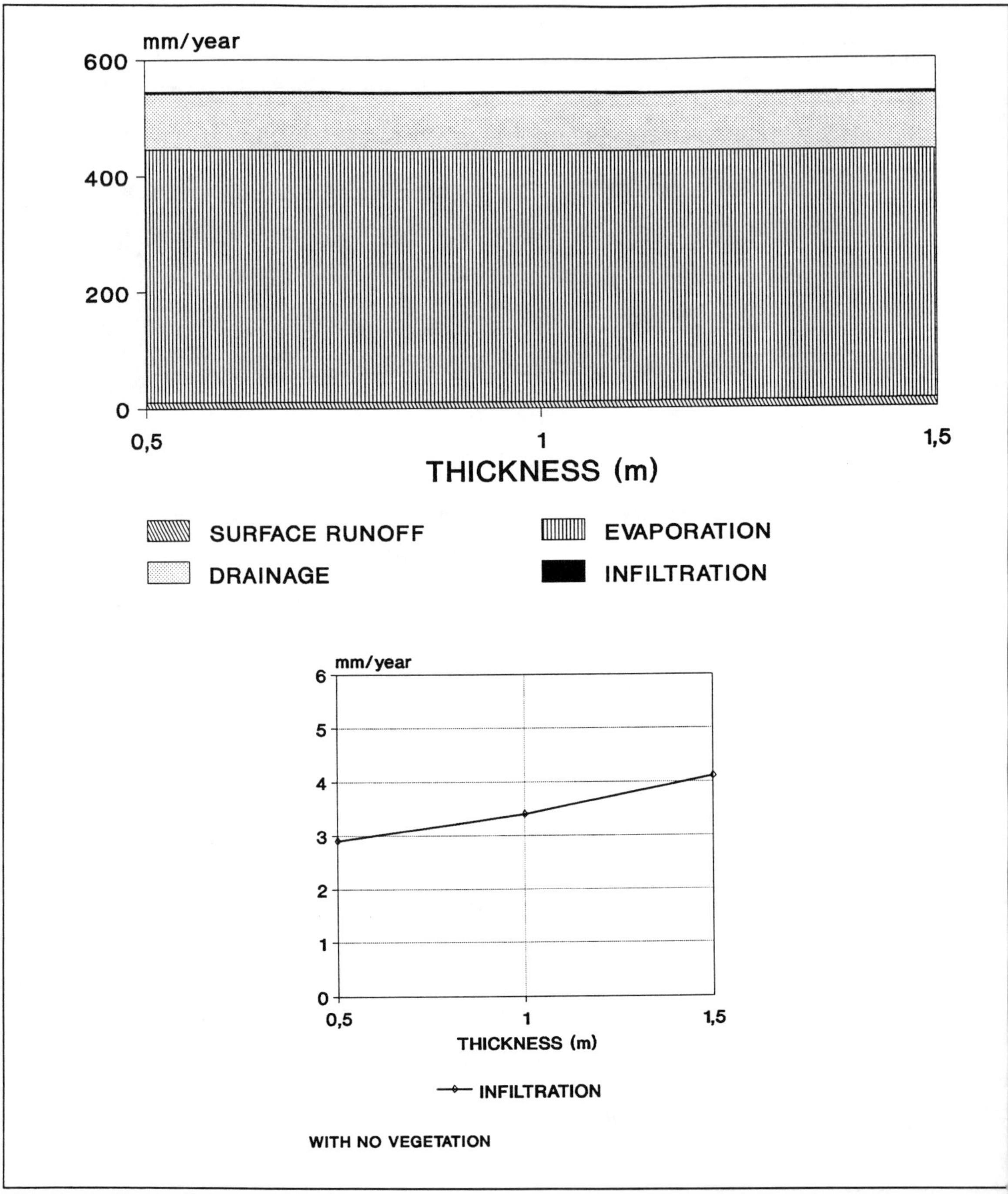

FIGURE 23.7
SOIL THICKNESS

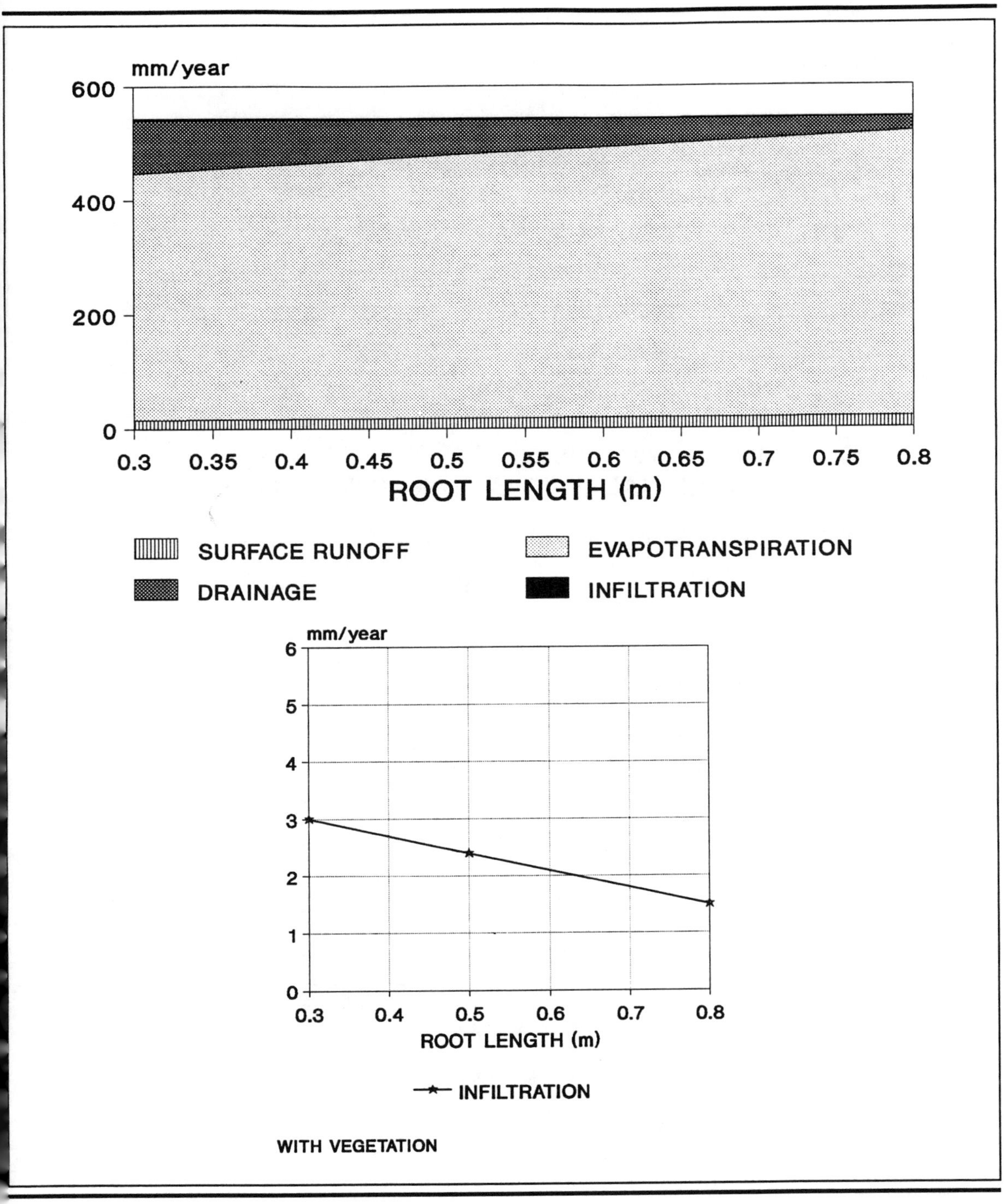

FIGURE 23.8
ROOT LENGTH

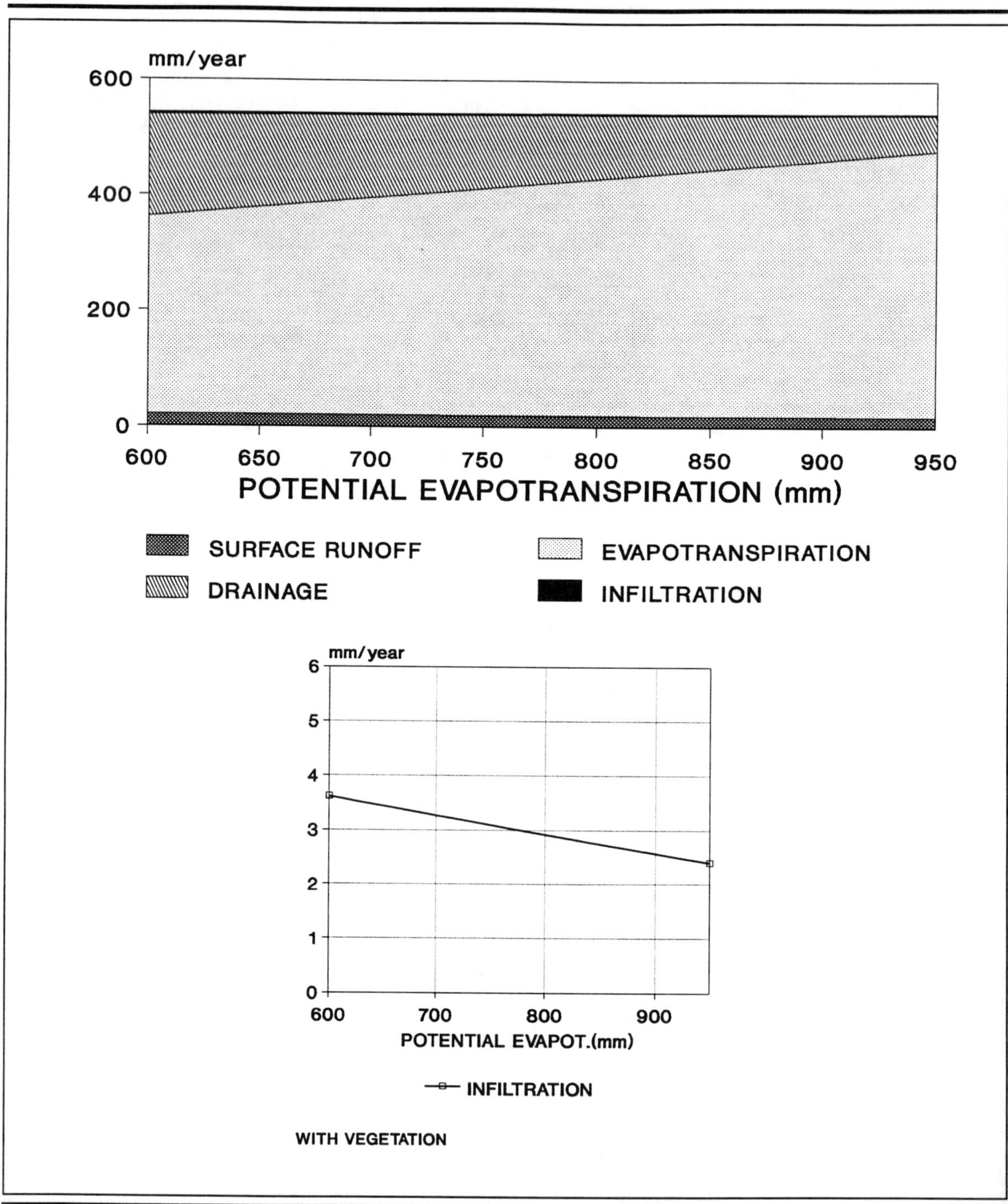

FIGURE 23.9

POTENTIAL EVAPOTRANSPIRATION

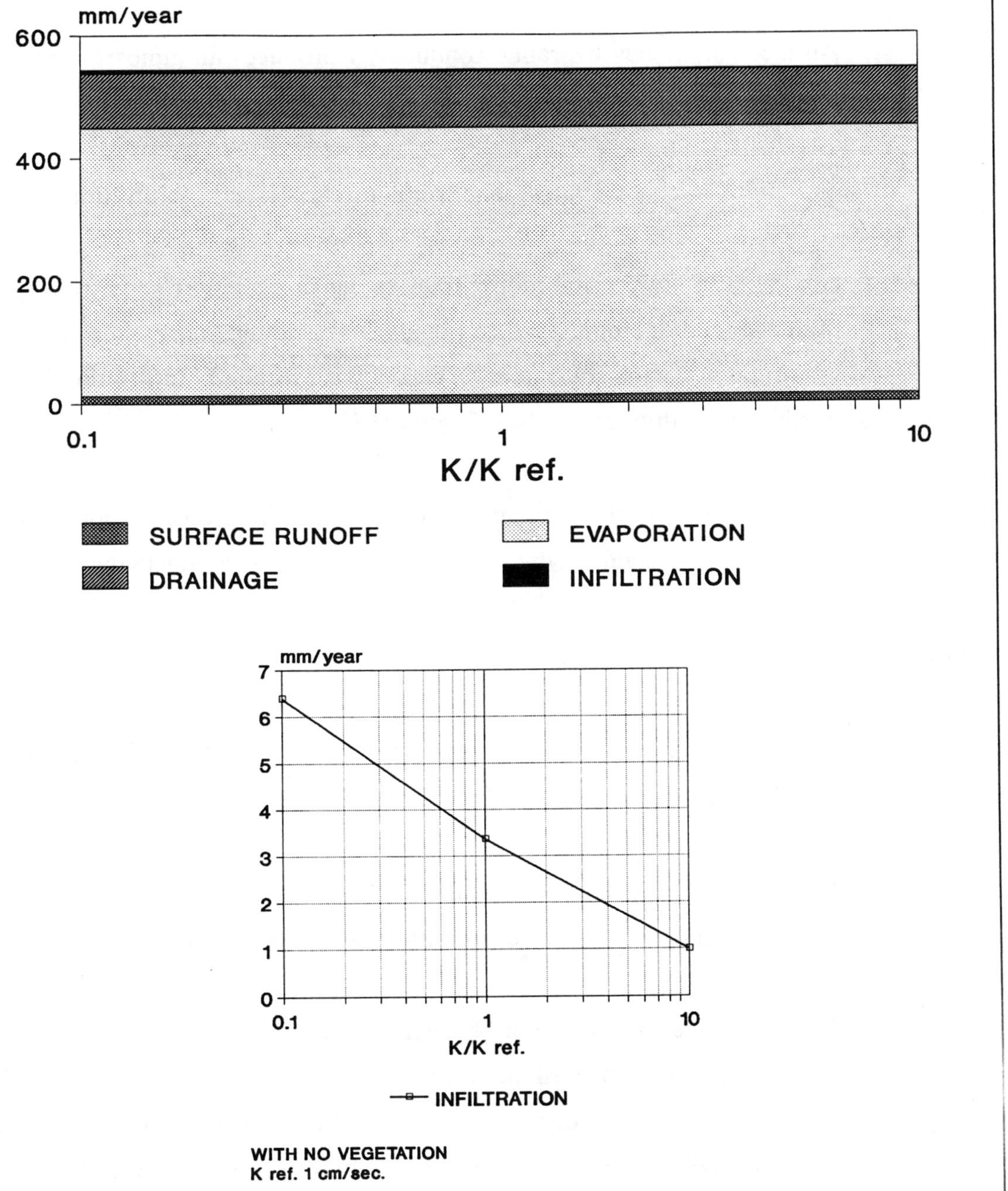

FIGURE 23.10

DRAIN HYDRAULIC CONDUCTIVITY

- An increase in clay hydraulic conductivity produces an almost proportional increase in infiltration (see Figure 23.11).

- Variations in clay thickness lead to minor variations in infiltration.

- An increase in precipitation produces maximum effects on run-off and drainage and increases infiltration. Drainage efficiency becomes more important when the soil retention capacity is exceeded. Infiltration becomes negligible for dry periods (see Figure 23.12).

- Key parameters for infiltration are the evapotranspiration and root length, the drain hydraulic conductivity, and the clay hydraulic conductivity.

Evaluation of Alternative Covers

The infiltration through the six alternative covers was evaluated using the TOUGH model. The following are the general conclusions of this study:

- The covers with a layer of random soil are the most efficient with a low infiltration due to effective evapotranspiration (see Figure 23.13).

- The difference in infiltration between the various covers is relatively small even though their overall performance is quite different in terms of the contribution of the various layers. Covers with soil vegetation layers make an effective use of evapotranspiration and require less drainage from the drain elements to limit infiltration.

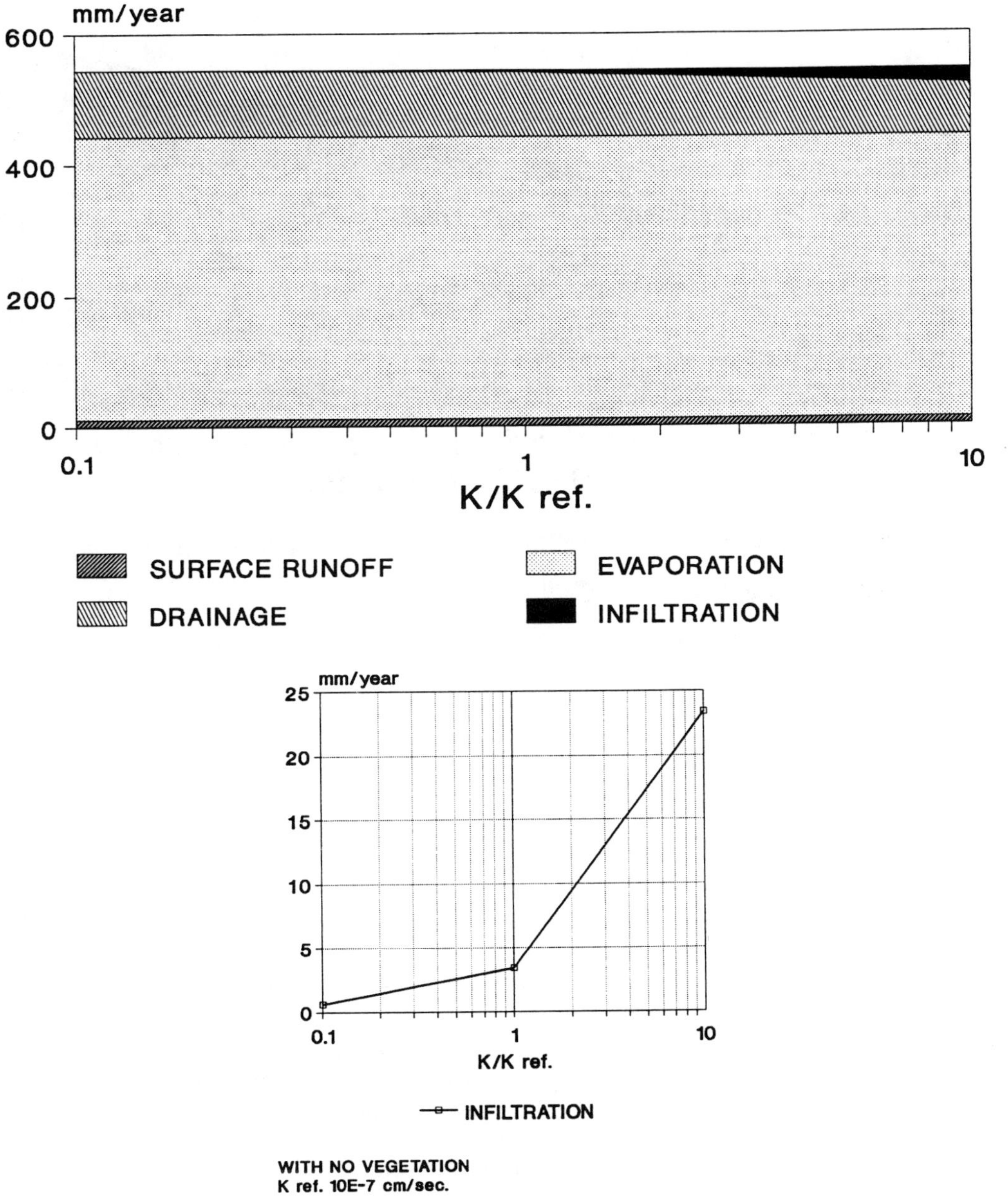

FIGURE 23.11

CLAY HYDRAULIC CONDUCTIVITY

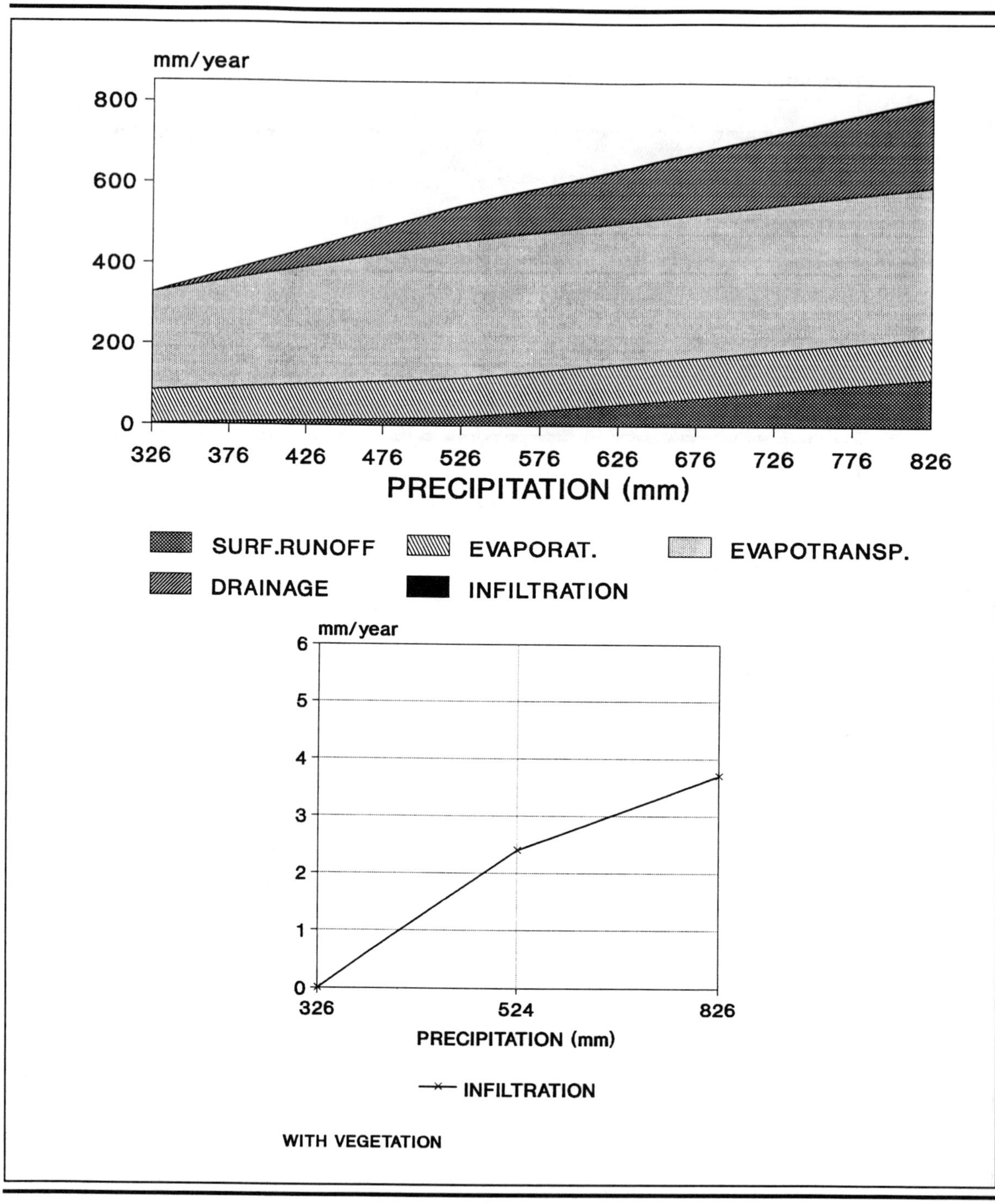

FIGURE 23.12
PRECIPITATION

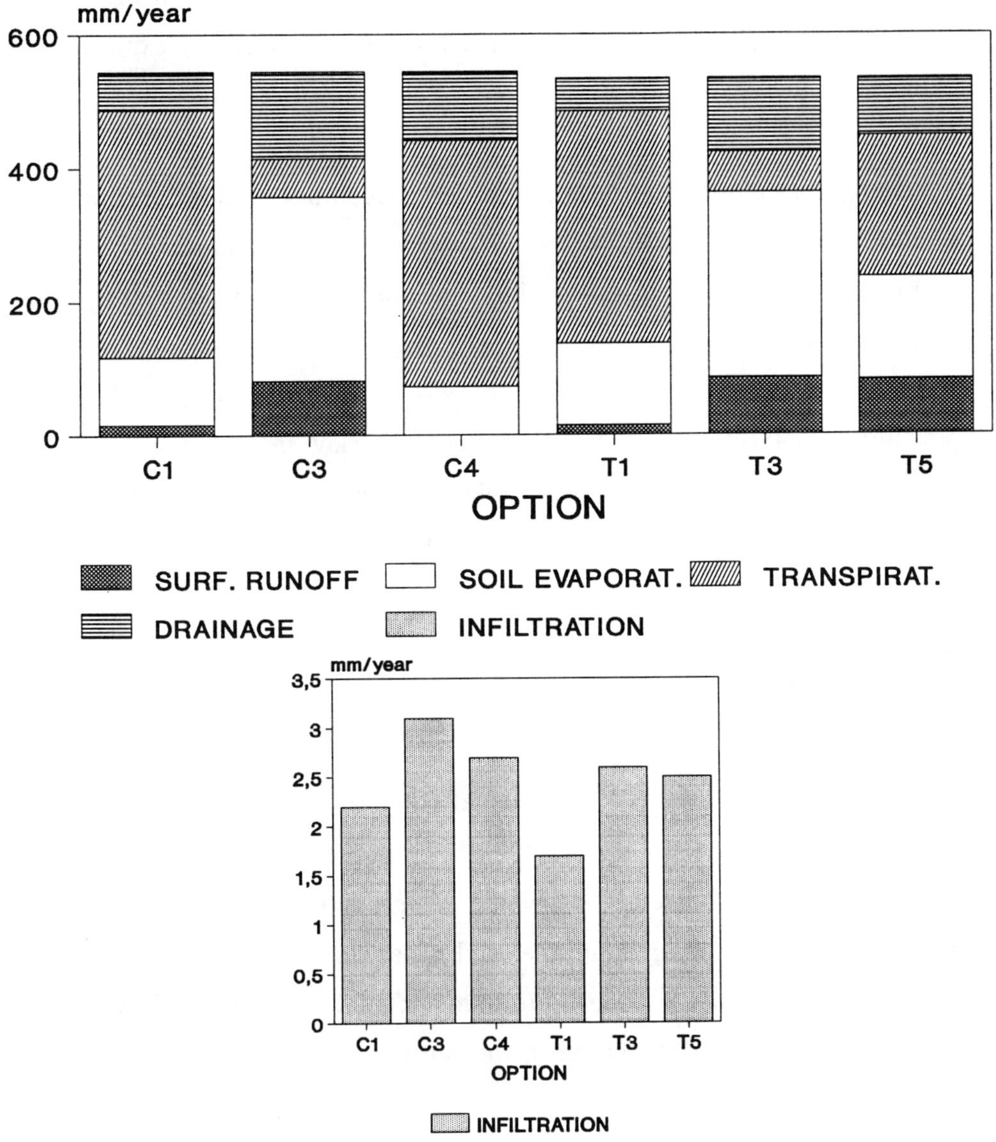

FIGURE 23.13

GENERAL COVER PERFORMANCE

- Covers without soil layer are more sensitive to changes in precipitation and climatic conditions. Covers with a layer of random soil experience more continuous, less sensitive variations of clay moisture with time.

- There is a unique relationship between percolation and infiltration as shown in Figure 23.14. Soil/vegetation and drain/clay elements are therefore uncoupled in terms of infiltration and may then be analyzed separately, if necessary.

- There is a unique relationship between clay degree of saturation and infiltration (see Figure 23.15). A high degree of saturation of the clay barrier is associated with a high infiltration. On the other hand, reduced infiltration may cause desiccation of the clay barrier.

COVER CRACKING

Cover cracking may result from either soil desiccation or excess cover deformation.

In this section, desiccation of the selected covers is investigated using the TOUGH model which allows simulation of the evaporative and solar radiation effects on the soils. Deformation cracking is evaluated using finite element analyses procedures which permit the determination of the strain distribution in the radon/infiltration barrier.

Desiccation Cracking

Desiccation can occur as a result of either exposure to evaporation or because of evapotranspiration resulting from root penetration.

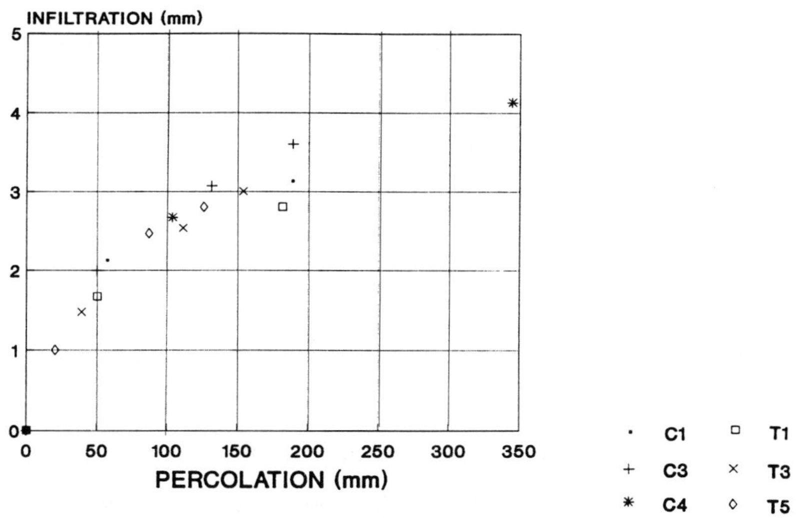

FIGURE 23.14

RELATIONSHIP BETWEEN INFILTRATION AND PERCOLATION

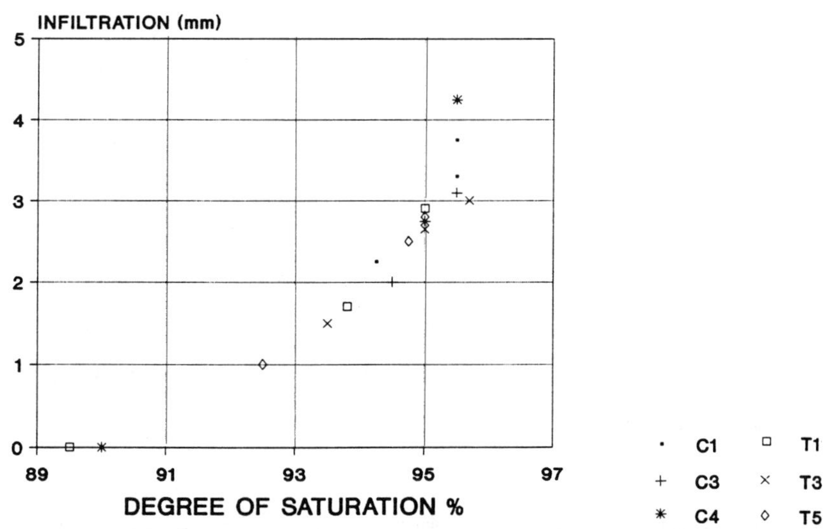

FIGURE 23.15

RELATIONSHIP BETWEEN INFILTRATION AND CLAY SATURATION

Desiccation on the selected covers was analyzed using the TOUGH computer code which accounts for the vapor and heat fluxes through the cover system. The performance of the selected covers was also compared with that of a cover similar to C3 without the soil in the upper layer of soil/rock. This was done in order to study the effect of soil on the desiccation of the underlying soil layers.

Results of these analyses are:

- Covers without soil are very sensitive to changes in climatic conditions. Severe oscillations in clay moisture and an upward vapor flux from the clay are found for dry periods.

- Soil is needed to avoid clay desiccation and cracking. However, an excessive depth of soil may also lead to desiccation in dry periods, as for covers C1 and C4, due to the absence of infiltration.

- Air and vapor flux are not significant for soil covers. However, they play an important role in covers without soil.

Deformation Cracking

Differential settlements of the disposal cell contents, and hence of the cover, may cause tensile or shear strains in the radon/infiltration barrier and potential cracking if the soil strength is exceeded.

The strain distribution in the radon barrier was determined using the ISBILD finite element computer code which incorporates hyperbolic stress-strain relationships for the soil and allows simulation of the consolidation/deformation process.

A finite element grid was compiled incorporating tailings and cover elements (see Figure 23.16) and element properties were evaluated from the results of laboratory consolidation testing and in-situ load testing. The computer code simulates the placement and build-up of compacted tailings over the existing lower pile and calculates settlements under the cover as well as stresses and strains within the cover elements.

Tensile strains were found to occur in areas close to the intersection between the top and sideslopes and are defined by changes in the curvature of the deformed shape of the surface of the compacted tailings. Tensile strains are also controlled by the presence of an internal sand dike which separated two different settling ponds during the formation of the tailings pile.

Tensile strains were compared with the soil yielding strains, but no cracking was predicted for the radon barrier as a result of differential settlement.

Evaluation of Alternative Covers

In terms of desiccation cracking, the covers with soil elements, i.e., covers C1, C4, T1 and T5 are found to be adequate to protect the radon barrier with a layer of random soil which attenuates seasonal moisture fluctuations and prevents long-term exposure to evaporation.

Deformation-induced cover cracking is primarily dependent upon the deformability of the underlying tailings and the stress-strain properties of the radon barrier. Since these two parameters are the same for all alternative covers, the consideration of deformation cracking is not a significant factor in selecting one cover over another.

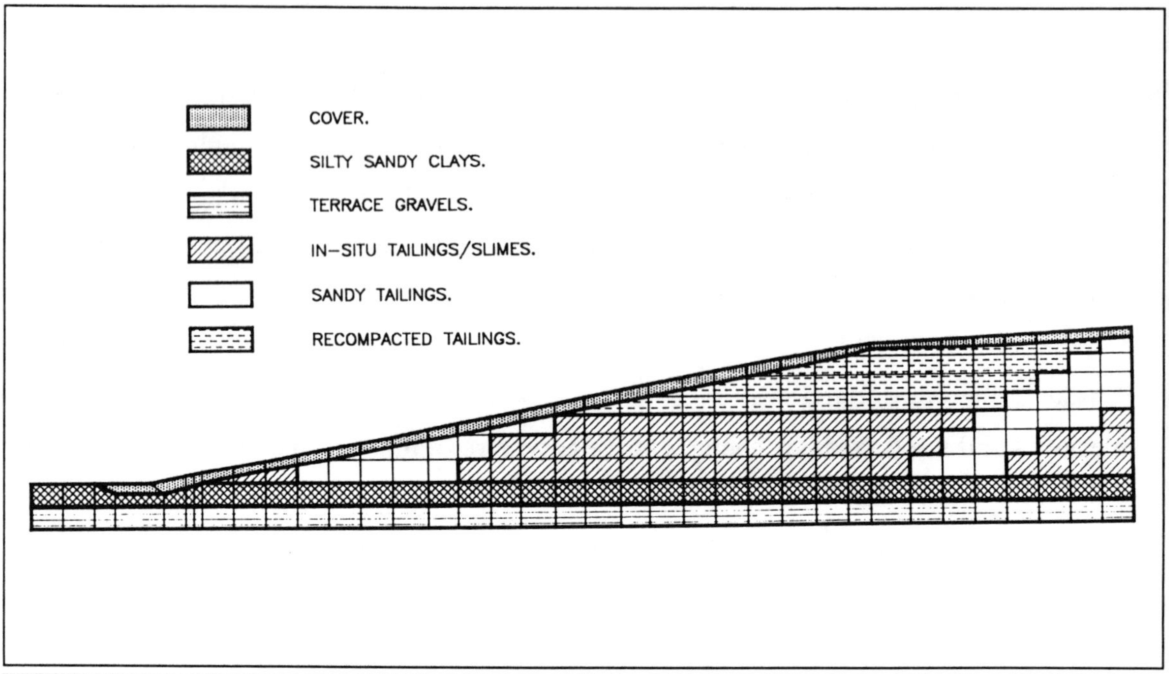

FIGURE 23.16

FINITE ELEMENT MODEL FOR COVER CRACKING ANALYSES

SELECTION OF THE PREFERRED COVER

In selecting the preferred cover, factors such as infiltration, cover cracking, erosion protection, biointrusion and aesthetics, were considered.

Cover C1 was selected for the topslope for these reasons:

- It provides the lowest infiltration for the type C covers.
- It includes a layer of random soil which protects the radon barrier from desiccation. The depth of soil is reduced to 0.5 m to preclude desiccation in dry periods due to the absence of infiltration.
- It incorporates a biointrusion barrier, which also acts as an erosion protection barrier. In addition, a gravel pavement is placed over the soil to limit erosion by rainfall splashing.
- It permits establishment of vegetation in the soil growth layer which will improve aesthetics and blending in with the surrounding area.

Cover T5 was selected for the sideslope for these reasons:

- Its uppermost rock layer provides the required erosion protection.
- Infiltration is reduced to levels similar to cover C1 selected for the topslope.
- The layer of random soil provides effective desiccation protection to the radon barrier. Similarly to the topslope, the depth of soil is reduced to 0.5 m to avoid desiccation in dry periods.

- It incorporates a biointrusion barrier, to limit plant and animal intrusion into the radon barrier.

- Soil is added to the uppermost layer of rock to help the establishment of vegetation through the riprap and reduce the visual impact of the remediated pile.

Preferred covers for topslope and sideslope are shown in Figure 23.17.

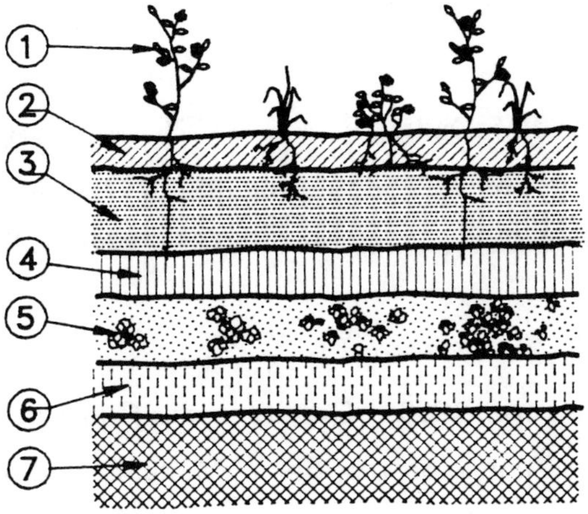

LEGEND

1. VEGETATION
2. PAVEMENT: SOIL/GRAVEL
3. VEGETATION GROWTH LAYER: RANDOM SOIL
4. FILTER: CLEAN SAND
5. BIOINTRUSION BARRIER: COBBLES–TOP CHOCKED
6. DRAIN: CLEAN SAND
7. RADON/INFILTRATION: SILTY CLAY

TOPSLOPE COVER

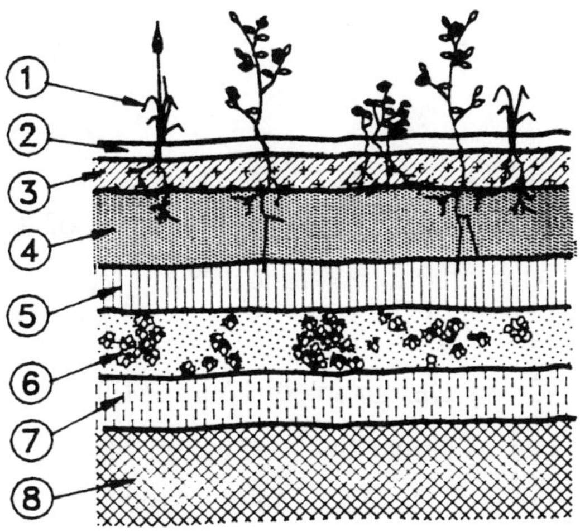

LEGEND

1. VEGETATION
2. SOIL TO MIGRATE INTO ROCK
3. EROSION PROTECTION ROCK
4. VEGETATION GROWTH LAYER: RANDOM SOIL
5. FILTER: CLEAN SAND
6. BIOINTRUSION BARRIER: COBBLES–TOP CHOCKED
7. DRAIN: CLEAN SAND
8. RADON/INFILTRATION: SILTY CLAY

SIDESLOPE COVER

FIGURE 23.17

COVER DESIGN FOR ANDUJAR TAILINGS PILE

REFERENCES

Abt, S.R., High, K.M., Gongloff, D.M., and McDaniel, R.D., 1989. "Radon Barrier Erosion and Bedding Drainage Criteria for UMTRA Sites." prepared by Department of Civil Engineering, Engineering Research Center, Colorado State University. Fort Collins, Colorado. September 1989.

ASCE (American Society of Civil Engineers), 1990. "Quality in the Constructed Project: a Guide for Owners, Designers, and Constructors." Manuals and Reports on Engineering Practice No 73. New York, New York.-2398

Bedinger, M.S. and Stevens, P.R. 1987. Safe Disposal of Radionuclides in Low-Level Radioactive Waste Repository Sites: Low-Level Radioactive Waste Disposal Workshop, U.S.G.S. July 11-16, 1987 Big Bear, California, Proceedings. U.S.G.S. Circular 1036.

Caldwell, J.A., and Dery, V., 1989. "Report on Performance Assessment of Select Covers and Compliance with Standards Memorandum." June 1989.

Caldwell, J.A., and Reith, C.C., 1990. "Cover Design for Uranium Mill Tailings Piles: A Standard for the Future." Western Regional Symposium on Mining and Mineral Processing Wastes. Berkeley, California. May 10 - June 1, 1990.

Caldwell, J.A., and Reith, C.C., 1990. "Vegetative Covers for Uranium Mill Tailings."

Caldwell, J.A. 1991. "Pile Stabilization and Groundwater Protection at Title I and Title II Uranium Mill Tailings Sites." Remedial Action Program Conference. Department of Energy. U.S. DOE Office of Environmental Restoration. Albuquerque, New Mexico, April 1990.

Chow, V.T. 1959. "Open Channel Hydraulics." McGraw Hill. New York.

DOE (U.S. Department of Energy), 1988. "Final Report on the Sodium-Amended Radon Barrier, A Special Study Finding," UMTRA-DOE/AL. Albuquerque, New Mexico.

DOE (U.S. Department of Energy), 1988. "Final Report on the Sodium-Amended Radon Barrier: A Special Study Finding." UMTRA-DOE/AL. October 1988.

DOE (U.S. Department of Energy), 1988. "Geomembrane Special Study, Final Report." UMTRA-DOE/AL-400642.0000. Albuquerque, New Mexico.

DOE (U.S. Department of Energy), 1988. "Alternate Cover Materials." UMTRA-DOE/AL-Albuquerque, New Mexico.

DOE (U.S. Department of Energy), 1988. "Effect of Freezing and Thawing on UMTRA Covers, A Special Study Finding." UMTRA-DOE/AL- Albuquerque, New Mexico.

DOE (U.S. Department of Energy), 1989. "Moisture Contents and Unsaturated Conditions in UMTRA Project Radon Barriers." UMTRA-DOE/AL. Albuquerque, New Mexico.

DOE (U.S. Department of Energy), 1989. "Design Method for Erosion Resistant Vegetated Topslopes for UMTRA Project Disposal Cells." Albuquerque, New Mexico. April 1989.

DOE (U.S. Department of Energy), 1989. "Technical Approach Document, Revision II." UMTRA-DOE/AL 050425.0002. Albuquerque, New Mexico.

DOE (U.S. Department of Energy), 1989. "Bentonite-Amended Soil Special Study, Phase I Evaluation Report." UMTRA-DOE/AL-400660.0000. Albuquerque, New Mexico.

DOE (U.S. Department of Energy), 1989. "Technical Approach Document." UMTRA-DOE/AL-050425.0002. Albuquerque, New Mexico.

DOE (U.S. Department of Energy), 1989. "Remedial Action Planning and Disposal Cell Design, To Comply with the Proposed EPA Standards (40 CFR Part 192)." UMTRA-DOE/AL-400503.0000. Albuquerque, New Mexico.

DOE (U.S. Department of Energy), 1989. "Bentonite-Amended Soils Special Study - Phase I Evaluation Report." UMTRA-DOE/AL-400660.0000. Albuquerque, New Mexico.

DOE (U.S. Department of Energy), 1989. Uranium Mill Tailings Remedial Action Report. "Technical Approach Document - Revision II." UMTRA-DOE/AL-050425.0002. December 1989.

DOE (U.S. Department of Energy), 1990. "Guidance for Surveillance and Monitoring for the UMTRA Project Long-term Care Program, Draft for Review." UMTRA-DOE/AL-350124.0001. Albuquerque, New Mexico.

DOE (U.S. Department of Energy), 1990. "Bentonite-Amended Soils Special Study, Phase II Evaluation Report." UMTRA DOE/AL-400665.0000. Albuquerque, New Mexico.

DOE (U.S. Department of Energy), 1990a. "An Assessment of Plant Biointrusion at the Uranium Mill Tailings Remedial Action Project Rock-Covered Disposal Cells" UMTRA-DOE/AL-400662.0000. Albuquerque, New Mexico.

DOE (U.S. Department of Energy), 1990b. "Infiltration Uncertainty Assessment for UMTRA Project Disposal Cell Covers." UMTRA-DOE/AL-400664.0000. Albuquerque, New Mexico.

DOE (U.S. Department of Energy), 1990c. "Bentonite-Amended Soil Special Study - Phase II Evaluation Report." UMTRA-DOE/AL-400665.0000. Albuquerque, New Mexico.

DOE (U.S. Department of Energy), 1991. Environmental Restoration and Waste Management (EM) Program, An Introduction, U.S. Department of Energy Office of Environmental Restoration and Waste Management, June 1991, DOE/EM-0013P.

DOE (U.S. Department of Energy), 1991b. Uranium Mill Tailings Remedial Action Project. "Technical Framework for Groundwater Restoration." UMTRA-DOE/AL-400671.0000. April 1991.

DOE (U.S. Department of Energy), 1991c. Uranium Mill Tailings Remedial Action Project. "Preplanning Guidance Document for Groundwater Restoration". UMTRA-DOE/AL-400670.0000. June 1991.

Dorrance, D.W., Wilson, L.B., Everett, L.G., and Cullen, S.J., 1991. Compendium of In-Situ Pore-Liquid Sampler for Vadose Zone, Chapter 19 in Groundwater Residue Sampling Design, Eds. R.G. Nash, and A.R. Leslie, ARCS Symposium 465, American Chemical Society, Washington, D.C., April 22-27, 1990.

EPA (U.S. Environmental Protection Agency), 1987a. "Joint NRC-EPA Guidance on a Conceptual Design Approach for Commercial Mixed Low-Level Radioactive and Hazardous Waste Disposal Facilities -- Action Memorandum." Prepared by Marcia E. Williams, Director, Office of Solid Waste. Washington, D.C.

EPA (U.S. Environmental Protection Agency), 1987b. "Minimum Technology Guidance on Final Covers for Landfills and Surface Impoundments." Prepared for Hazardous Waste Engineering Research Laboratory, Office of Research and Development. Cincinnati, Ohio.

EPA (U.S. Environmental Protection Agency), 1987c. "Project Summary, Design, Construction, and Maintenance of Cover Systems for Hazardous Waste: An Engineering Guidance Document." Prepared by Hazardous Waste Engineering Research Laboratory. Cincinnati, Ohio.

EPA (U.S. Environmental Protection Agency), 1989. "Considerations in Groundwater Remediation at Superfund Sites." U.S. EPA Directive No 9355.4-00.

Ferguson, MK Company, 1991. "Vegetative Covers and the NRC Staff Technical Position Paper on Erosion Protection Covers - UMTRA General." January 1991.

Hirshfield, T., Deaton, T., Mitarovich, F., and Klainer, S., 1983. Feasibility of Using Fiber Optics for Monitoring Groundwater Contaminants. Optical Engineering, V.22 #5, pp. 527-531 (September/October 1983).

International Atomic Energy Agency (1990). "Draft Report on Uranium Mine and Mill Decommissioning." IAEA. Vienna.

Kane, J.D., and Widmayer, D.A. "Considerations for the Long-Term Performance of Geosynthetics at Radioactive Waste Disposal Facilities." prepared by Technical Branch, Division of Low-Level Waste Management and Decommissioning, Office of Nuclear Material Safety and Safeguards, U.S. Nuclear Regulatory Commission. Washington, D.C.

Kane, J.D. and Widmayer, D.A. (1988) "Considerations for the Long-Term Performance of Geosynthetics at Radioactive Waste Disposal Facilities." Technical Branch, Division of Low-Level Waste Management and Decommissioning, Office of Nuclear Materials Safety and Safeguards, U.S. Nuclear Regulatory Commission. Washington D.C.

Keller, C., 1991. "So What is the Practical Value at SEAMIST?", Proceedings of the Fifth National Outdoor Action Conference on Aquifer Restoration, Ground Water Monitoring and Geophysical Methods, pp. 59-73, May 1991, Las Vegas, Nevada, presented by the Association of Ground Water Scientists and Engineers (division of NWWA).

Kierner, R.M., Halse, Y.H., and Lord, A.E., (1990). "Long-Term Durability and Aging of Geomembranes" in "Waste Containment Systems: Construction, Regulation, and Performance." Edited by Rudolph Bonapaite. Geotechnical Special Publication No. 26. American Society of Civil Engineers. New York, N.Y.

Kramer, J.H., Everett, L.G., and Cullen, S. 1991. Innovative Vadose Zone Monitoring at a Landfill Using the Neutron Probe, Proceedings of the Fifth National Outdoor Action Conference on Aquifer Restoration, Ground Water Monitoring and Geophysical Methods, pp. 196-214, 135-149, May 1991, Las Vegas, Nevada, presented by the Association of Ground Water Scientists and Engineers (division of NWWA).

Morris, R.C. and Fraley, J., 1989. "Effects of Vegetation, a Clay Cap and Environmental Variables on Rn-222 Fluence Rate from Reclaimed Mill Tailings." Health Physics, Volume 56, pp. 431-440.

Morrison-Knudsen Engineers, Inc., 1986. "MKE UMTRA Design Procedures, Chapter 5, Erosion Protection." UMTRA Project.

Morrison-Knudsen Engineers, Inc., 1986. "MKE UMTRA Design Procedures, Chapter 14, Radon Barrier Cracking, Contents"

Morrison-Knudsen Engineers, Inc., 1986. "MKE UMTRA Design Procedures, Chapter 10, Stability Analysis, Contents."

Morrison-Knudsen Engineers, Inc., 1991. "Weldon Spring Site Remedial Action Project - Operational Scenario for Disposal Cell Construction - Task 824." Draft Design Document Rev. A. May 1991.

NRC (U.S. Nuclear Regulatory Commission), 1990. "Final Staff Technical Position Design of Erosion Protection Covers for Stabilization of Uranium Mill Tailings Sites." Washington, D.C. August 1990.

NRC (U.S. Nuclear Regulatory Commission), 1990. "Final Staff Technical Position: Design of Erosion Protection Covers for Stabilization of Uranium Mill Tailings Sites." Washington D.C.

Peck, R.B., 1969. "The Advantages and Limitations of the Observational Method." Geotechnique 1969. The Institution of Civil Engineers. London.

Reith, C., (1990) "Cost Optimization on the Jackpile Paguate Reclamation Project." Western Regional Symposium on Mining and Mineral Processing Wastes. Berkeley, California.

Schulz, R.K., Ridky, R.W., and O'Donnell, E., 1987. "Control of Water Infiltration into Surface LLW Disposal Units: Annual Report, October 1985 - September 1986." Prepared for Division of Engineering, Office of Nuclear Regulatory Research, U.S. Nuclear Regulatory Commission. NRC FIN B8958. Washington, D.C.

Schulz, R.K., Ridky, R.W., and O'Donnell, E., 1988. "Control of Water Infiltration Into Near Surface LLW Disposal Units, Task Report - A Discussion." Prepared for Division of Engineering, Office of Nuclear Regulatory Research, U.S. Nuclear Regulatory Commission. NRC FIN B8958. NUREG/CR-4918, Vol. 2. Washington, D.C.

Schulz, R.K., Ridky, R.W., and O'Donnell, E., 1989. "Control of Water Infiltration Into Near Surface LLW Disposal Units, Progress Report." Prepared for Division of Engineering, Office of Nuclear Regulatory Research, U.S. Nuclear Regulatory Commission. NRC FIN 8958. NUREG/CR-4918, Vol. 3. Washington, D.C.

Schulz, R.K., Ridky, R.W., and O'Donnell, E., 1989. "Control of Water Infiltration Into Near Surface LLW Disposal Units, Progress Report on Field Experiments at a Humid Region Site, Beltsville, Maryland." Prepared for Division of Engineering, Office of Nuclear Regulatory Research, U.S. Nuclear Regulatory Commission. NRC FIN B8958. NUREG/CR-4918. Vol. 4. Washington D.C.

Temple, D.M., 1980. "Tractive Force Design of Vegetated Channels." In Transactions, American Society of Agricultural Engineers, Vol. 23, No. 4, pp. 884-890.

Temple, D.M. et. al., 1987. "Stability Design of Grass-Lined Open Channels." U.S. Department of Agriculture, Agriculture Research Service, Agriculture Handbook Number 667.

"The Effect of Desiccation on UMTRA Project Radon Barrier Materials." November 1990.

White, G.J., Ferns, T.W., (Part I) Marts, S.T., DeHaan, M.S., Schwaller, R.G., and White, G.J., 1990. "(Part II), Low-Level Radioactive Waste Disposal Facility Closure, Part I: Long-Term Environmental Conditions Affecting Low-Level Waste Disposal Site Performance, Part II: Performance Monitoring to Support Regulatory Decisions." Prepared for U.S. Nuclear Regulatory Commission. NRC FIN A6853. NUREG/CR-5615, EGG-2604. Washington, D.C.

Index